Brain Storm

operationalizing by putting values and pretheoretical assumptions into practice

Brain Storm

The Flaws in the Science of

Sex Differences

REBECCA M. JORDAN-YOUNG

HARVARD UNIVERSITY PRESS
Cambridge, Massachusetts
London, England

First Harvard University Press paperback edition, 2011

Library of Congress Cataloging-in-Publication Data

Jordan-Young, Rebecca M., 1963–
Brain storm : the flaws in the science of sex differences / Rebecca M.
Jordan-Young.
p.   cm.
Includes bibliographical references and index.
ISBN 978-0-674-05730-2 (cloth : alk. paper)
ISBN 978-0-674-06351-8 (pbk.)
1. Brain—Sex differences.   I. Title.
QP81.5.J67 2010
612.8'2—dc22   2010010165

For Sal

Contents

Preface

Most books set out to answer questions. This book sets out to question answers. The answers I question have to do with the nature and causes of differences between men and women, and between straight people and gay people. Specifically, I question what we "know" about male and female brains, or gay and straight brains.

When Simon LeVay reported in 1991 that he had found a difference in brain structure between gay and heterosexual men, which was trumpeted as the discovery of "The Gay Brain," I found it interesting but also puzzling. How could gayness take a single identifiable form in the brain when it takes such varied forms in people's lives?

At the time, I had already been engaged for several years in large-scale sexuality research related to the AIDS epidemic. In an outreach storefront in Washington, DC, I ran a project that focused on injection drug users. It was there that I first met a lot of gay men. These men were not the poster children of the gay rights movement, but were poor, struggling with addiction and recovery, and trying to avoid or outlive AIDS. And so were their heterosexual brothers, with whom I also worked. In fact, these men were so similar in demeanor, dress, and daily struggles that without our detailed interviews, it was impossible to tell the difference between the gay and straight men. In LeVay's study, the homosexual men were a singular "type" unlike the (also homogeneous) heterosexual men, and they were also somehow similar to (presumably straight) women. Those are common enough ideas both inside and outside of science, but our research challenged these notions.

The reason for this was our research methods. Since our main questions didn't have to do with sexual orientation, our preconceptions about sexual orientation didn't shape the way we gathered our information. We asked people to join the study based on their drug use or their connection with drug users, not their sexuality—not their sexual practices or identities, not the gay or straight reputation of the bars or neighborhoods they hung out in, nor even how we perceived them as straight or gay. We asked everyone extremely detailed questions about their sexual practices with both same-sex and other-sex partners. Scientifically, this approach is a much more reliable way to get information about the nature of sexuality in a population than to go out looking for gay and straight people to compare. The number and variety of people who talked about same-sex relationships surprised all of us—including the gay and lesbian staff members who thought we had finely tuned "gaydar." The bias in epidemiology at the time was to see same-sex behavior among people outside of well-defined gay communities as being "instrumental"—meaning that it was due to drug use, incarceration, or sex work. But as we spent hours and weeks and eventually years with our participants, growing to know and love many of them, it was clear that they had same-sex relationships for the same reasons they had heterosexual relationships: desire, affection, and love. The thousand or so sexual histories we gathered fed into an enormous pool of research that eventually included information on the sexual and drug-using behavior of tens of thousands of people from more than fifty cities. And on the basis of this research, it would be very hard to suggest that either men who desire and have sex with men, or women who desire and have sex with women, are a distinct type of person, or are somehow "like" heterosexuals of the other sex (Young, Weissman, and Cohen 1992; Young et al. 2000; Young and Meyer 2005).

So how could I make sense of what LeVay had found? What functions might relate to the brain structures that he suggested were somehow connected to both sex and sexual orientation? Where did the brain differences come from? And are they the *cause* of differences in behavior, personality, or desires, or the result of them?

Some years later, when I was in graduate school, one particular semester had my brain stretching in almost too many directions to bear. I was studying psychometrics, observational epidemiology, and biostatistics, and I decided to add a class on "the gay brain." LeVay's 1991 study was on the syllabus, and by then, I knew enough to spot quite a few problems with it. But other studies seemed to point in the same direction as his: Dick Swaab and Michel Hofman (1990) had reported another structural difference between the brains of heterosexual and gay men, and Laura Allen and Roger Gorski (1992) reported yet a third. Each of these research teams was also looking at male versus female differences in the brain. It turned out that none of the studies could answer my earlier questions about function, and none of them even entertained the idea that the structural differences might come from behavior and experience, rather than the other way around. Instead, through these studies, I learned about the largely unquestioned theory that was guiding the work of these research teams and many others. According to this theory, prenatal hormone exposures cause sexual differentiation of the brain—that is, early hormones create permanent masculine or feminine patterns of desire, personality, temperament, and cognition. Further, hormones later in life could "activate" behavioral predispositions, but the predispositions themselves result from the initial "organizing" effect of hormones very early in development, before birth. Intrigued, I began to look for other research related to this theory, which some scientists call the "organization-activation hypothesis," and some call the "neurohormonal theory." I was particularly interested in studies that explore the earlier, organizing role of hormones, the time when hormones presumably cause sex-typed predispositions. I think the clearest way to refer to that work is by the term *brain organization research,* so that is the term I use in this book.

Once I began looking into this theory, I couldn't look away, and I have now spent thirteen years exploring brain organization research. I was not initially interested in using my analysis to reflect back on how well or how poorly brain organization theory was supported by the evidence from these studies. I was more interested in methods, particularly in how scientists resolved the problem of measuring something as complex as sexuality or gender in such a way that these in turn could be associated with brain structure or hormone exposures. I was also curious to see how scientists differed in their approaches, and what kinds of methods they used to translate research findings across different study designs. Focusing on original, peer-reviewed research in English-language scientific journals, I used a combination of strategies to identify studies, beginning with a small set of high-profile studies and searching for the research those works cited, as well as subsequent studies that cited my index cases. I identified the "founding paper" in the field (Phoenix et al. 1959), in which the theory of brain organization was first proposed, and systematically searched for research reports on humans that cited that paper. Using the ISI Web of Science, Medline, and PsychInfo databases, I combined keywords about hormonal "inputs" (such as prenatal, in utero, organizing effects, hormones, testosterone, estrogen, progesterone) with psychosexual "outputs" (such as masculinity, femininity, eroticism, sexual behavior, psychosexuality, sexual orientation). Early on, I limited myself to analyzing studies that explored the connection between prenatal hormone exposures and human sexuality. I ultimately analyzed virtually every study on the ostensible prenatal hormone-sexuality connection published from 1967, when the theory was first applied to humans, up to the year 2000, when the increased flow of research in this area made it no longer possible to examine every published study in depth. I continued to examine all major studies (those published in the most important journals, those that garnered a lot of scientific attention, and those by well-established scientists) through 2008. Further, because brain organization research had always addressed broader questions of masculinity and femininity, I expanded my ongoing search strategy to identify studies that focus on those variables, too. I have now done a close analysis of over three hundred studies that span all of the many research designs used to explore the hypothetical connection between prenatal hormone exposures, on the one hand, and human sexuality or gender, on the other.

I also interviewed scientists. I did an influence analysis to identify the twenty-five most influential people doing brain organization research, based on how their studies were cited by other scientists. Twenty-one of them generously agreed to be interviewed; five spoke with me at least twice. They talked with me about the theory and about the nitty-gritty details of their studies: subject recruitment, questionnaires, statistical analysis, and so on. I enjoyed those conversations, and many of the scientists told me that they were happy for the chance to step back from their work and think about the field as a whole, or about conceptual issues that sometimes take a backseat to daily research practice. I am grateful for their time and their openness with me. The interviews were of immeasurable value to me in making sure I understood scientists' varied research approaches, as well as their disagreements with one another. I use quotations from the interviews somewhat sparingly, though, because I am less interested, in this particular book, in what scientists *say* than in what they *do* in their research practice. Several scientists were wary about controversies that had erupted shortly before I began my interviews, so I follow standard ethnographic practice, and treat the interviews as confidential. In a few places where I have used material that bears directly on a particular scientist's work, I have used their names with permission. Elsewhere, they are identified by simple pseudonyms (Dr. A, Dr. B, etc.).

The closer I looked at brain organization research, the less it made sense. My initial focus on methods gradually gave way to the realization that the evidence simply does not support the theory. In this book, I ask readers to follow the same journey I have taken, looking closely at the measures and methods in order to understand gaps and fundamental contradictions in the data. Although I find much fault with research done in connection with brain organization theory, I hope that the researchers who generously shared their time with me, as well as others who read this book, will take it in the spirit of constructive criticism in which it is offered.

Brain organization research is important because it engages deep and enduring questions that practically everyone shares: how do we come to be the sort of people we are? How do our bodies matter to our personalities, skills, interests, and desires? These are great questions, but sometimes we treat them like rhetorical questions, and plug in answers that seem right just because they are familiar.

For instance, when I talk to people about the idea that people are born with a male brain or a female brain, I almost always end up hearing about their experiences with children. A few people relate tales of unexpected flexibility or gender-bending, but most relay their experiences with boys who practically radiate "boyness" from some deep space within (interestingly, I hear many fewer stories of girls who are sugar and spice and everything nice). Recently, my mother happened to be present for one of these conversations, and she very matter-of-factly presented what is considered these days to be a radical idea. Before relaying her idea, let me explain that my mother is a lovely southern lady who has raised an impressive number of children: four boys and four girls of her own, in addition to playing a major role in raising some half dozen of her more than two score grandchildren and great-grandchildren (I think she's drawing the line at the great-great-grandchildren). When my friend talked about her girl being so different from her younger brother, who is "all boy," my mother literally snorted. "That's because you only have two," she said. Mama went on to explain that with just a couple of children, gender looms large—it's the most obvious explanation for every difference you see between them, and unless your children are really unusual, it's going to be easiest to see their personalities as "boy" versus "girl." But when you have a lot of children, you begin to notice that they all come with personalities of their own, and they are all quite different from one another. Gender recedes in importance.

Of course, that doesn't mean that gender isn't real. One particular question about the "realness" of gender is at the heart of this book: how is gender connected to bodies, specifically, to the brain? And how does sexuality figure in that connection?

Brain organization theory offers a pretty tidy answer: it's all down to "sex hormones" that shape the brain before birth. Get enough "male hormones" and you'll be masculine in desires, interests, and personality. Maybe it's the Missouri in me, but I am never happy with an answer if it's too easy. I really do demand evidence: "show me." And I expect you want the same, so that's what I do in this book.

CHAPTER ONE

Sexual Brains and Body Politics

A LITTLE GIRL, given trucks to play with, calls them "daddy truck" and "baby truck." A child, raised as a girl after a tragic accident that destroyed his penis, rejects a female identity and insists that he is a boy. A woman wisecracks that the difference between how females and males learn is that you can teach her how the dishwasher works, but you can't make her interested in it.

Stories help us understand the world. These modern parables about sex differences underscore what it means to be male or female, and suggest answers to questions that intrigue us, such as how changeable or resistant to "tampering" sexual natures are. For many people today, the most authoritative stories come from science, and the dominant scientific tale about sex differences goes like this: Because of early exposure to different sex hormones, males and females have different brains. Moreover, the same process that makes men and women fundamentally different is responsible for within-sex differences in sexual orientation, so "gay brains" are also different from "straight brains," courtesy of early hormone exposures.

Increasingly, scientists in the public eye have characterized skeptics of this idea as politically correct ideologues who are, in Steven Pinker's words, "on a collision course with the findings of science and the spirit of free inquiry" (2005, 15). Writing for *Scientific American* a few years ago, neuroscientist Doreen Kimura (2002, 32) put it this way:

> For the past few decades, it has been ideologically fashionable to insist that [male–female] behavioral differences are minimal and are the consequence of variations in experience during development before and after adolescence. Evidence accumulated more recently, however, suggests that the effects of sex hormones on brain organization occur so early in life that from the start the environment is acting on differently wired brains in boys and girls.

Of course, science itself isn't just one thing, and there isn't just one story that "science" tells about sex in the brain. But this particular story—brain organization theory—is especially prominent. Unlike tales about childhood play or a woman's level of interest in how her dishwasher works, it's hard for the average person to critically engage with a story that comes from the rarified domain of cognitive neuroscience. (It's not that easy for scientists to critically engage with the story, either: there's nothing more chilling to a scientist than being called "ideologically fashionable.")

In daily life, people understand that it is important not only to listen to talk about how things work but also to directly observe how things work. Likewise, it could be very informative to go behind scientists' theories about sex in the brain and look instead at the research they do. But how? The volume of studies is enormous, and the technology involved seems ever more sophisticated, making it very difficult for laypeople—or even for scientists outside of particular subspecialties—to do anything but stand back and be impressed with the sheer quantity of this work. But the first premise of this book is that ordinary people can take a hard look at how research supporting brain organization theory is done. In the chapters that follow, I present a synthetic, critical analysis of the evidence that human brains are "hardwired" for sex-typed preferences and skills by early hormone exposures, and I build the analysis on simple rules of "symmetry" that are easily understood by nonspecialists.

Despite the high profile of this theory, this is the first synthetic analysis of organization theory research on humans, including more than three hundred studies conducted from the late 1960s through 2008. There have been other summaries of brain organization research (Berenbaum 1998; Hines 2004; Cohen-Bendahan, van de Beek, and Berenbaum 2005), and past critics have pointed out methodological weaknesses that call the results of particular studies into question (Longino and Doell 1983; Schmidt and Clement 1990; Byne and Parsons 1993; Fausto-Sterling 1985 and 2000). But these summaries haven't been systematic in the same sense as my analysis here. What has been missing is a rigorous and comprehensive *synthesis* of the studies, considering all the different findings from studies across the many designs that scientists use to test the theory. If the studies on brain organization were more similar, it would be ideal to do a meta-analysis—a statistical approach in which data from many studies are pooled and analyzed together. Unfortunately, the network of research on brain organization theory is too diverse to do a conventional meta-analysis.

But there is something special about these studies that offers another way of doing synthetic analysis. Each study of how prenatal hormones sexually "organize" the human brain is a *quasi experiment,* rather than a true experiment. In true experiments, subjects would be randomly assigned to receive particular hormone exposures, and their development would be observed over the life span, keeping rearing experiences and environments constant across experimental and control groups. Of course, this isn't possible with humans, so scientists must piece together evidence from animal studies, and from individual human quasi experiments that are by definition partial and uncontrolled. The interpretation of every quasi experiment depends on carefully placing that study within the overall body of evidence. So, a synthetic analysis of quasi experiments can actually be done by mapping the *structure* of studies, to see how well the studies fit together.

The second premise of this book is that it is not just possible, but urgently necessary, to reopen the questions that have been closed by accepting brain organization as a done deal. To broadly anticipate my findings, evidence that human brains are hormonally organized to be either masculine or feminine turns out to be surprisingly disjointed, and even contradictory—and the stakes involved in prematurely promoting this theory to a "fact" of human development are high, both for the advancement of science and for social debates that draw on science. In the last chapter, I begin to sketch some ways in which breaking free of brain organization theory as it is currently conceived may stimulate better science, specifically more dynamic research on human development. As for the importance of this theory to social debates, I don't devote much space in the rest of this book to how such research matters in the broad world outside of science. So I want to point here to just a few of the reasons why a hard-nosed understanding of brain organization research is so important.

The idea that men and women "naturally" think and act differently because they have distinctly male or female brains is not new, but a number of high-profile controversies over the past few years have given the idea new prominence, especially as an explanation for why there aren't more women in science, engineering, and mathematics. Shortly after a report from Massachusetts Institute of Technology cited widespread discrimination against women at MIT (1999, 15), the debate over women in math, science, and engineering moved from its usual simmer into one of its periodic fierce eruptions. While the report drew widespread praise and attention (Loder 2000; Goldberg 1999), it also provoked skepticism and counterarguments suggesting that the dearth of senior women at MIT was less a reflection of discrimination than an inevitable reflection of the difference in male and female brains. In a presentation to the National Academy of Engineering that garnered a good deal of press, Patricia Hausman flatly asserted that barriers for women in science and engineering "don't exist."[1] Instead, she argued, occupational gender differences can be explained by "exposure to testosterone during a key phase of fetal development [that] appears to influence spatial ability and some aspects of personality" (P. Hausman 2000, 4).

A few years later, in *The Essential Difference: The Truth about the Male and Female Brain,* psychologist Simon Baron-Cohen (2003a) put some new clothes on the male brain/female brain idea in the way of updated terminology for these sex-differentiated organs: "systemizers" and "empathizers." Baron-Cohen notes that "professions such as maths, physics, and engineering, which require high systemising, are also largely male-chosen disciplines," and points to high fetal testosterone as explaining the connection (Baron-Cohen 2003b, 4). Although Baron-Cohen's work has mostly been warmly reviewed, some such pronouncements have provoked fierce controversy. Most famously, Larry Summers, former president of Harvard University, suggested that innate "taste-differences" between the sexes and a male predominance in high-end aptitude are probably the most important reasons women are underrepresented among tenured scientists and engineers at top research institutions (Summers 2005). Summers didn't directly invoke brain organization theory to support this idea, but almost all the scientists and pundits who came to his defense in the ensuing firestorm did (for instance, Pinker 2005; Klienfeld 2005).[2] The notion that there are male brains and female brains has continued to crop up in controversies over women in math and science, but it has also recently shown up in discussions of single-sex education, sex disparities in wages, "abstinence-only" sexuality education curricula, the protocol for medical treatment of children born with ambiguous genitalia, and yes, even the out-of-control, risky trading that recently brought the economy to its knees (see, for instance, Klienfeld 2005; Byne 2006; Stein 2009; Weil 2008; Tyre 2005; Sax 2005; U.S. House of Representatives 2004; Kay and Jackson 2008; Ceci and Williams 2007; Wilson and Reiner 1998; National Academy of Sciences 2006).

The same theory connects sex differences in the brain to sexual orientation, and it's common for scientists to explain the theory by connecting "sex-typical" patterns of interests, cognitive skills, and sexuality. For example, in explaining evidence that many scientists find to be some of the strongest for demonstrating that the human brain is hormonally organized to be masculine or feminine, neuroscientist Sandra Witelson cites "girls who have a hormone disorder that causes them to have higher testosterone levels in utero. 'In these girls, their play patterns, their spatial ability and even their sexual orientation are much closer to the male pattern'" (Weise 2006, quoting Witelson).

In an era where diversity is celebrated, the idea of "sex in the brain" no longer equals an endorsement of male superiority, and critics of the idea are increasingly cast as not only antiscience, but antidiversity. The most straightforward claims of this sort are made in connection with the rights of sexual minorities. In a popular media campaign, the Gil Foundation uses a cartoon dog named Norman who moos instead of barks: Norman was "born different." The gist of the ad is that gay men, lesbians, and transgender people of both sexes have brains that are "wired" differently from most people, and that accepting this is an important component of combating anti-gay prejudice. And while the notion of innately different preferences in men and women was once politically suspect, it is now often suggested that accepting these innate differences will encourage a more rational approach to equality. Should boys and girls be taught differently, because the sexes have innately different patterns of learning? Perhaps we should stop striving for parity in the professions, for example, or for an equal division of parenting labor, because women and men want different things out of life and are temperamentally suited both for different work and for a different balance between career and family (Hewlett 2002; Story 2005; Young 2006). Sociologist Linda Gottfredson warns, "If you insist on using parity as your measure of social justice, it means you will have to keep many men and women out of the work they like best and push them into work they don't like" (quoted in Holden 2000, 380).

No area of social policy has been so charged by brain organization theory as the protocol for the medical treatment of intersex infants, children born with genitalia that either are ambiguous or don't "agree" with genetic sex. Research on intersex conditions is a cornerstone of studies related to brain organization, and the various conditions are discussed throughout this book. It is also important to consider one high-profile case that concerns someone who was *not* born intersex, but on which ideas about both brain organization in general and intersex management in particular have disproportionately hinged. David Reimer, also known in the research literature as "John/Joan" (Money and Ehrhardt 1972) and in the popular media as "the boy who was raised as a girl" (Colapinto 2001), was born a normal male infant, and an identical twin. Reimer, whose life tragically ended in suicide a few years ago, was raised as a girl after his penis was irreparably damaged in infancy. Reimer's doctors, especially the famous sexologist John Money, had believed that Reimer could not develop a normal male identity without a penis. Money's prior research suggested that gender (one's sense of self as male or female) was "flexible" until about age 2 and was largely determined by socialization rather than by biology. Money's advice was to rear the child as a girl, and Money cited "Joan's" successful adjustment as especially compelling evidence for the importance of rearing in gender development (for example, Money and Ehrhardt 1972).

In the late 1990s, however, news began to emerge that Reimer had rejected a female identity, insisting that he was really male. Another group of scientists, especially Money's archrival Milton Diamond, interpreted this to indicate that Reimer's male identity was "fixed" in his brain because of the testosterone he had produced as a normal male fetus. After a revision of the case in the medical literature, and a BBC documentary, Reimer's story was broken to a broader public with an article in *Rolling Stone* (Colapinto 1997). The article was followed by a popular book (Colapinto 2001), which in turn was followed by almost countless popular articles, television and radio stories, and even a special science documentary for the PBS science series *NOVA*. It is safe to say that nothing has done more to dramatize the theory of brain organization. So it is particularly fascinating that the psychologist John Money, who is presented in the popular version of that story as the absolute embodiment of "blank-slate thinking" about gender, was in fact the very first scientist to extend brain organization theory to humans. In an especially odd coincidence, Money first tentatively applied brain organization theory to humans in 1965—the year David Reimer was born (Money 1965a).

The riveting story of "the boy who was raised as a girl" is perhaps an allegory for the story of brain organization theory. Things aren't always what they seem. In spite of the much-heralded early reports, it is not "easy" to turn a child who is born a normal boy into a girl. And in spite of later reports that painted him as an extreme proponent of socialization theory, John Money turns out to have been one of the most important and prolific researchers producing studies of hormone effects on human behavior. There are more complications to tell and corrections to be made, but these must wait until later chapters. Suffice it to say that while we should honor David Reimer's fundamental sense of masculinity, and take sober account of the ways in which his treatment was mismanaged, we should be cautious about accepting the conclusion that so many have drawn from his difficult life: that testosterone in the womb was responsible for his unshakable sense of masculinity.[3]

Brain Organization Research and the Biosocial Model

Given the popular image of John Money as a radical proponent of social factors in sexual development, it may be particularly confusing that feminist biologists and philosophers of science usually describe Money and his longtime collaborator Anke Ehrhardt as primarily "biologically oriented" researchers (Bleier 1984; van den Wijngaard 1997). Money and Ehrhardt, as well as *her* collaborator of more than thirty years, Heino Meyer-Bahlburg, have thought of themselves all along as "interactionists" who believe in the importance of both biology and the rearing environment. When I interviewed Anke Ehrhardt in 1998, she indicated that the idea that hormones would create inborn, behavioral predispositions for masculinity or femininity was unexpected and went against the grain of the working hypotheses among Money's team:

> [It was] really in some ways, against what we all believed in Money's unit. . . . as you know, at the same time, Money had broken the taboo about gender identity, that gender identity could develop in contrast to biological variables and in concordance with the sex of rearing. But Money was really an early interactionist and thought always that you can't split the body from the mind. But he thought, well . . . here are these interesting animal experimental data . . . so I think we were truly open, because of these different kinds of mental sets. (Ehrhardt interview, December 4, 1998)

Among scientists conducting brain organization research, Money, Ehrhardt, and their respective collaborators have indeed devoted more ink than others to emphasizing social aspects of development. In the interviews I conducted, some scientists praised their approach as especially thoughtful (e.g., Drs. H and J), while others ridiculed them for using "disclaimers galore" about the effects of hormones (Drs. A and F). The range of opinions about Money, especially, mirrors a number of other differences among brain organization researchers and helps counter any simplistic notion that these scientists pursue biological research on sex differences out of some uniform agenda. This is not to say that scientists conducting this work do not have agendas, some more explicit than others. Rather, the point is that their ideas diverge in significant ways from one another, so the story of how brain organization research unfolded should not be read as the story of a tight scientific club. It's especially important to keep that in mind when I describe the "network of brain organization research," because the term *network* is often employed to describe groups of connected people, and in science studies especially to describe how personal and professional connections among scientists shape the scientific knowledge they produce. Certainly there are quite deep connections among many scientists who study brain organization, and such connections undoubtedly influence their work, but that is simply not the subject of this book. Instead, I use the idea of a *network of studies* to signify that all such studies are connected by a single theory about how hormones affect the developing brain. This is not a sociological claim, but a scientific one: because these studies all relate to the same theory, they must also all relate to one another. That point is important for understanding how strong the theory itself is, as will become increasingly clear throughout this book.

The fact that scientists doing brain organization research disagree about the relative importance of social experiences versus hormone exposures can obscure a fundamental issue on which they do agree, namely, the *way* in which nature and nurture contribute to development. Money and Ehrhardt's analytic model, which set the stage for decades of research on hormones and sexuality, is not truly interactionist, because it does not attempt to account for how physical and social variables actually *work in tandem*. Biologist Ruth Doell and philosopher Helen Longino put their finger on this problem years ago when they dubbed Money and Ehrhardt's approach as a "linear model" of development that treats the brain as a "black box with prenatal hormonal input and later behavioral output" (Doell and Longino 1988, 59).

These days, opposing "interactionism" is about as popular as being against "freedom," so brain organization researchers all call themselves interactionists, though they actually continue to employ the same model. This additive "biosocial" model can allow for both biological and social variables, but each is seen making a separate contribution that can be added or subtracted without changing the *way* the other affects development. An interactionist model, on the other hand, suggests that the *character* (not just the amount) of biological influence is affected by specific aspects of the environment, and vice versa. I explore the idea of interaction in much more detail in Chapter 10. I also show, especially in Chapter 9, that most brain organization studies do not include social variables in their models at all, let alone explore potential interactions between these and the hormones that they see as the main actors.

Some Notes on the Aims of This Book

Because the research I've described is so politically important, and so politicized, it's worth pausing briefly for some comments on what this book is, and is not. First, this book is *not* about the political commitments of scientists who do the research. Scientists who conduct brain organization research often participate in policy debates, sometimes quite directly. Richard Udry, for example, who has reported correlations between prenatal testosterone levels and women's "gendered" behavior, concludes that "males and females have different and biologically influenced behavioral predispositions." That sounds neutral enough, but the lessons he suggests for society are rather more ominous: "If [societies] depart too far from the underlying sex-dimorphism of biological predispositions, they will generate social malaise and social pressures to drift back toward closer alignment with biology" (Udry 2000, 454). Other scientists are usually somewhat more cautious in this regard. Psychologists Anke Ehrhardt and Susan Baker (1974, 50) suggested more than three decades ago that prenatal hormone effects "in human beings are subtle and can in no way be taken as a basis for prescribing social roles." More recently, Sheri Berenbaum and Melissa Hines have both cautioned against abandoning efforts "to increase the participation of girls and women in science and engineering, because it is likely that other factors are also involved in the underrepresentation of girls and women in these fields" (Berenbaum 1999, 108; Hines 2007).

Nonetheless, while comments like this indicate that scientists are interested in—and sometimes perhaps a bit worried about—the social implications of their work, I think it would be a mistake to view research related to brain organization theory as motivated primarily by the political commitments of scientists. In fact, after analyzing hundreds of studies, listening to talks that scientists have given on the topic, and interviewing nearly two dozen of the most important scientists in this research area, I've come to the conclusion that there is a huge diversity of opinion among scientists themselves about what the idea of brain organization means for social questions like programs to address gender disparities, or policies regarding the rights of sexual minorities. Thus, although scientists' political and social ideas are sometimes quite interesting, and no doubt important for the way they conduct their own research, I don't think it's the really big story for understanding the research overall. Ditto for who does the research: though the research is often deployed in debates about sexual difference and equality, the important scientific actors in this field have always included many women, and increasingly include gay scientists. There are no obvious associations between the social identities of the researchers and the studies' focus and findings.

Second, this is also not a book about whether differences are "good" for women, sexual minorities, or anyone else. Like many other scientists and science watchers, I believe history shows that there is never a clear message to be drawn from scientific findings of "difference" versus "sameness" (Fausto-Sterling 2000; Terry 1997).[4] For the record, I'm a strong supporter of sexual equality—including full rights and justice for sexual minorities—but so are many of the scientists with whom I disagree when it comes to interpreting research on brain organization.

Finally, and most emphatically, this is not an "antiscience" book, and it's also not an "antidifference" book. Questioning brain organization theory is not the same as rejecting either science or biology. To the contrary, the aim of this book is to invite a deeper consideration of the science that underlies the theory—not to foreclose, but to reinvigorate, free inquiry on the subject of sex differences in the brain. I do not reject the idea of sex differences in the brain as either "dangerous" or implausible—to be honest, I'm somewhat amazed that there are not *more* sex differences in the brain, for reasons that I will explain in the final chapters. But the standard stories about male and female brains are both stale and unscientific, often incorporating unfounded assumptions (about how and when differences arise, and so on), and premature leaps (for instance, drawing firm conclusions from small and unreplicated studies). In more than a decade of looking closely into the very interesting world of research on hormonal organization of the brain, I have tried to reopen key questions that many other scientists assume have been closed, asking not just what we know, but how we know it.

Throughout this book, I examine scientists' methods in conducting brain organization research in a way that can be seen as a hybrid between the kind of critique that is the model for scientific peer review and the kind of critique that is more common in science and technology studies (STS). STS scholars have demonstrated that scientific research contributes to, rather than simply reveals, the meaning of phenomena that are studied. Rather than continuing to point this out as though good science can and should be purged of contaminating social factors, science and technology studies over the past few decades have focused on illuminating *how* social or practical factors are woven into the practice of science—*all* science, that is, excellent as well as bad or simply run-of-the-mill science. Dozens of empirical studies demonstrate how factors that the conventional account would suggest are contaminating or irrelevant to the legitimate business of science, including but moving well beyond scientists' expectations or beliefs, shape science every step of the way. These factors range from the choice of what constitutes interesting problems or "puzzles" for research, to understandings of what information in the world gets recognized as relevant data for the problem at hand, to the physical layout of labs or the choice of one technique over another, to the interpretation of experimental results (Gieryn 1999; Haraway 1989; Latour and Woolgar 1986; Mol 2002; Star 1989).

This book is indebted to recent work in science and technology studies, but it is distinct from recent STS work in two key ways. First, there has been a highly productive emphasis on ethnography and "lab studies" over the past few decades, in which analysts take the role of a naïve, curious, and nonjudgmental outsider whose task is to document the "strange native practices" of scientists. By taking scientific practices as "strange" and refusing either to bracket what scientists *say* is irrelevant or to foreground the conventional descriptions and explanations for what scientists do, these studies have yielded refreshing and often startlingly new ways of understanding scientific knowledge, the objects that science purports to know, and scientific ways of working and knowing. This approach also avoids the circularity common in conventional accounts of science (science is objective, because scientists put aside their biases, because science is objective . . .).

While I did not physically observe scientists in their work, I did interview them—and some of the interviews turned into ongoing collegial conversations. Still, I did these interviews as another scientist—a sort of "critical insider," rather than taking a more anthropological approach. The main way my analysis fits into the ethnographic thread in STS is that I focus very concretely on aspects of scientific practice, especially definitions and measures. I looked much more carefully at how scientists do their studies than at how they describe the theory in question. The gap between theory and practice is interesting, and readers will see that it is often informative to notice when practices and the "overarching narrative" of brain organization theory don't match up.

Second, as I noted above, I do not pay much attention to how so-called social factors shape the work, nor do I examine the "culture" of this branch of science. I look, instead, at technical matters that are well within the conventional account of what science is supposed to be about. I look at things like study designs, measures, and statistical practices, and I evaluate them in terms of accepted scientific practices that prevail today. The good news about this is that even hardheaded "realists" who prefer the older view of good science as objective, and who believe that following scientific methods will yield a value-free reflection of nature, should be able to follow my analysis and agree that I am mounting a fair and reasonable evaluation of brain organization studies. The bad news, for me, is that the present study is apt to be seen as rather old-fashioned among STS scholars, because it is, essentially, a critique. In an age where most academic science watchers think it is somewhat passé (or at least boring) to point out that science-in-practice doesn't actually live up to science-in-the-ideal, critique has gone out of fashion.

I chose this method for a simple reason. I am trained as a scientist, and I value scientific method. I believe it is worth holding scientific research—especially high-profile research on a topic that is of great social and political importance—to the highest standards.[5] Brain organization theory is undeniably powerful, and it derives this power from a widespread belief that it is supported by excellent, state-of-the-art science. There is value in showing in great detail how this work does, and does not, measure up to that billing.

A Three-Ply Yarn: Sex, Gender, and Sexuality

The language of sex, gender, and sexuality is more than vexing—it is confused, confusing, and contentious. And I am in the unhappy position, in this book, of writing at the juncture of multiple disciplines who use these same terms in irreconcilable ways. Clarifying my own usage is necessary, but I am going to attempt this in a way that neither casts my usage as "correct" nor places my analysis in firm opposition to the works that I analyze in this book.

Begin with the distinction between sex and gender. Since the late 1970s, many scholars have used the terms in a general way to signal biological *(sex)* versus social and behavioral dimensions *(gender)*. But there has been an important difference in usage among people who followed this distinction. John Money's elaboration of the notions "gender role" and "gender identity," first articulated in 1955 (Money, Hampson, and Hampson 1955b) and expanded in his landmark 1972 book with Anke Ehrhardt, *Man and Woman, Boy and Girl*, is the one adopted by nearly all brain organization researchers:

> *Gender identity:* The sameness, unity, and persistence of one's individuality as male, female, or ambivalent, in greater or lesser degree, especially as it is experienced in self-awareness and behavior; gender identity is the private expe-

rience of gender role, and gender role is the public expression of gender identity.

Gender Role: Everything that a person says and does, to indicate to others
or to the self the degree that one is either male, or female, or ambivalent; it *includes but is not restricted to sexual arousal and response;* gender role is the
public expression of gender identity, and gender identity is the private experience of gender role. (Money and Ehrhardt 1972, 4, emphasis added)

Notice two key things: eroticism, the realm of sexual attractions, fantasies, and behaviors, is entirely subsumed under gender in this usage. And
note, too, that there is nothing in this definition about etiology. Without
positing causation, Money's distinctions between gender and sex are at the
level of location, or possibly measurement: that is, gender is located, perceived, and measurable at the level of behavior and language; sex is located, perceived, and measurable at the level of the physical body.

Money thought of gender as "his" term, and the record does seem to
support his claim to at least the popular rise (if not the actual invention) of
the term *gender* to describe people as feminine or masculine (Ehrhardt
1988; Gooren and Cohen-Kettenis 1991; Bullough 1994). It was a matter
of great irritation to him that feminists, in particular, deployed the term
with a critical difference (Money 1980). Perhaps the greatest credit in this
regard is due to anthropologist Gayle Rubin's (1975, 159) classic identification of the "sex/gender" system, which she identified as a "set of ar-
rangements by which a society transforms biological sexuality into products of human activity." Rubin's distinction placed the emphasis squarely
on etiology. That is, *gender*—the attribution of aspects of human personality, relationships, behaviors, privileges, and prohibitions to the domain of
either "masculine" or "feminine"—is a social effect, rather than the result
of human biology. *Sex,* in this regard, is conceived as the remainder—the
material body, and those bodily interactions that are necessary to reproduce it.

In addition to the disagreements between mainstream psychologists and
medical doctors, who favored Money's usage, and feminists, social scientists, and scholars in the humanities, who generally favored Rubin's, several later developments have further complicated the terminological field.
For one thing, there is the matter of sexual desires and pleasures and
bodily interactions. Money subsumed this domain entirely under gender,
and Rubin initially conceptually distributed it (unevenly) across gender
and sex. That is, male and female bodies must obviously meet (at least historically!) in order to accomplish the biological necessity of human reproduction. But the elaborate and particular rules about the *ways* in which
bodies meet, touch, and interact, and the designation of some forms of de-

sire and pleasure as "masculine" versus "feminine" as well as "natural" versus "unnatural," belonged to the realm of the social rather than the "raw material" of biology. Anthropologists, philosophers, sociologists, and historians increasingly designated this latter domain as "sexuality" (Foucault 1978; Weeks 1986; Vance 1991). Rubin's work was especially influential in articulating that sexuality and gender overlap and yet need to be distinguished as two "distinct arenas of social practice" (Rubin 1992). Central to Rubin's argument was a growing body of research that undermined the usual assumption that sexuality is grounded in and emerges "naturally" from the biological drive to reproduce. The best-known work on social construction of sexuality is that of the late historical philosopher Michel Foucault, who argued that *sexuality as a domain is itself a social product.* Two key claims in his work are, first, that certain sensations, bodily parts, desires, and forms of interaction are not "naturally" sexual, but *become* sexual (or get designated as sexual) through specific social practices that vary across time and place. The second point, which flows from the first, is that the relationship between the social world and sexuality is not merely one of regulation or imposing limits (a "negative force"), but one of production and instigation (a "positive force"). These points, supported by an ever-growing body of anthropological, sociological, and historical studies, suggest that sexuality is fundamentally contextual, rather than comprising a universal set of desires, sensations, and acts (D'Emilio 1983; Mosse 1985; D'Emilio and Freedman 1988; McClintock 1995; Katz 1995; Stoler 2002).

A classroom exercise developed by Carole Vance, a medical anthropologist who has been extremely influential in feminist sexuality studies, may help bring this discussion down to earth. Suppose a Martian shows up at your home and asks you to explain what sexuality is. What do you say? Reactions to this question can be quite varied, and in my experience, students generally begin with a great deal of confidence that they can arrive at a good answer, but their certainty breaks down. Someone offers "Activities that are related to human reproduction," but that is met with counter questions like "What about oral or anal sex?" Someone else might note that most heterosexuals go to a lot of trouble to make sure that most of their sexual behavior does not result in reproduction. (Of course, some people might answer that sexuality *should* be about reproduction, but as a matter of definition, it is no defense, because if one simply discounts all other activities from being sexual, then there would be no complaints about all the "wrong" or "sinful" expressions of sexuality! These things simply would not count.) Predictably, some other brave soul will offer the definition "Touching genitals" as sexual (hmm: gynecological exams?

touching breasts?) or "Becoming physically aroused by touching someone else" (masturbation? fantasy? a sexual encounter that doesn't really arouse one of the partners?). I could go on, of course, but presumably you begin to see the scope of the difficulties. And this is without even entertaining differences across cultural groups or historical periods. Try this yourself, and I guarantee that if you come up with a really solid answer that can withstand all the counterexamples that any three or four smart adults can offer, you have an immediate career in sexuality studies waiting for you.

But back to my dilemma. For some time, I have thought of sex, gender, and sexuality as a three-ply yarn. I find the metaphor appealing because it suggests three strands that are simultaneously distinct, interrelated, and somewhat fuzzy around the boundaries. As a double entendre, "three-ply yarn" also suggests the narrative aspects of the domains and their relations with one another (Young 2000). In other words, the *perceived* relationships among bodies, desires, and a wide range of social norms governing roles and interactions are central to the stories that we tell ourselves about human nature and the meaning of maleness and femaleness. Following dominant usage in the humanities and social sciences as well as feminist biology (for example, Vance 1989; Butler 1990; Laqueur 1992; Oudshoorn 1994; Kessler 1998; Fausto-Sterling 2000), I have used the term *sex* to refer to characteristics of the physical body, *gender* to refer to psychological attributes and social behaviors that are associated with masculinity and femininity, and *sexuality* for the realm of erotic desires and practices.

The key way that these distinctions have been useful is to enable critical questioning of the relationships among phenomena that are usually taken to be inseparably fused or causally related. Without questioning these assumptions, we cannot probe the boundaries among them, nor perceive shifts in these boundaries. Nor can we investigate causal processes if the theory of causation is already built into our definition of the phenomena. But the slippery nature of the relationships among sex, gender, and sexuality has made it difficult to pin them down long enough to systematically study them: scholars have encountered ongoing difficulty with assigning phenomena definitively to one domain versus the other.[6] An example that is relevant to the story at hand, and the particular dilemmas of terminology that I face in telling it, is that empirical research has revealed the relation between sex and gender to be even more complex than originally thought.

Works by Nelly Oudshoorn and Suzanne Kessler have been especially important for my own understanding. In a text that deserves much wider audience than it has received, Oudshoorn (1994) advanced the somewhat astonishing claim that "sex hormones" were "invented" by twentieth-century endocrinologists. Oudshoorn does not dispute that steroid hormones like estradiol, testosterone, or progesterone are actual material substances, nor does she claim that these substances were not already being produced and used by bodies long before scientists had any notions about them. Her argument specifically concerns the idea that certain steroid hormones are fundamentally *about* and *for* sex. She documents how preexisting ideas of masculinity and femininity caused scientists to look for, create tests for, classify, and perceive steroid hormones in a way that fit them into a dualistic system of sex. This might not be particularly compelling if she didn't also document how this commitment to a fundamentally sexual classification of hormones systematically blocked some kinds of information. For example, scientists had repeated difficulty assimilating the information that both males and females produce and use *both* androgens ("male sex hormones") *and* estrogens ("female sex hormones"). Likewise, biologist Marianne van den Wijngaard (1997) has documented the great difficulty mid-twentieth-century experimental psychologists had with the information that estrogen—the quintessential "female" hormone—was even more important than testosterone for the development of certain "masculine" characteristics. Van den Wijngaard and Anne Fausto-Sterling (2000) have both shown how a dualistic notion of "sex hormones" has been a repeated stumbling block in recognizing the great variety of functions that steroid hormones accomplish, apart from those related to development and function of reproductive capacities. In this vein, a colleague and I recently confirmed that attachment to the term and concept *sex hormones* is associated with systematic misinformation in widely used high school biology texts (Nehm and Young 2008).

Classifying some substances as "sex hormones," then, presumes that we already know the most relevant functions, divisions, and even ultimate "purposes" of these substances. That is not the strongest perspective from which to launch a rigorous scientific program to explore and elucidate the mechanisms of hormonal action. This insight has obvious relevance to my own evaluation of how early steroid hormone exposures may shape the developing brain. The works by Oudshoorn, van den Wijngaard, and Fausto-Sterling form an important backdrop against which I have worked. But I have aimed to take the studies I analyze here on their own terms as much as possible. Thus, the book is littered with the terms *sex hormones, androgens* (literally, "that which creates men"), and *estrogens* ("that which induces estrus," the fertile period in some female mammals). I often place the term *sex hormones* inside quotation marks as a reminder to readers that this is a philosophically invested way to conceptualize the substances I am discussing. I do not put quote marks around *androgens* or *estrogens,* mostly because I found that doing so made the text too clunky and conveyed a less open, less neutral attitude about specific brain organization studies than I had during this research.

A second example of work that problematizes the now-conventional distinctions among sex, gender, and sexuality concerns studies on intersexuality, especially Suzanne Kessler's crucial *Lessons from the Intersexed* (1998). The term *intersex* typically describes a person who is born with "mixed" male and female bodily characteristics: for instance, the female (XX) chromosome pattern but a penis and no vagina, or the male (XY) chromosome pattern but a vulva (including labia and clitoris) and vagina. The most common intersex conditions involve steroid hormone atypicalities, such as a genetic male who produces but cannot respond to testosterone and so develops "female" structures, or a genetic female who produces unusually high levels of androgens, which masculinize her own developing genitalia. An enormous proportion of brain organization research has been conducted with intersex subjects, because many scientists believe intersex people offer an opportunity to study the effects of hormones that "disagree" with gender socialization.

Kessler's groundbreaking analysis showed how ideas about gender and sexuality are called upon by medical doctors and psychologists, both to determine the male or female status of bodily structures and to reshape those structures so that they can be seen as fitting with "normal" male versus female bodies. For example, Kessler explores the centrality of the idea that "boys urinate standing up" in determining whether a phallic structure can, or cannot, count as male. She and, more recently, Katrina Karkazis (2008) have also demonstrated how doctors who conduct "feminizing" surgeries on intersex infants deploy a definition of "normal" female sexual function that is built around vaginal penetrability rather than around the capacity for sexual pleasure and orgasm. Together with analyses of biology and endocrinology, these studies add up to a take-home message that gender causes us to perceive the natural world (the body) in a particular way, and thereby to impose upon it the dichotomous category "sex." Sex, then, is no longer the raw material from which culture produces gender. Instead, sex is in some important sense an *effect* of gender.

Here is another interesting twist. The sex/gender distinction has broken down on the other side of the disciplinary divide as well. While feminist scholars (to oversimplify a bit) increasingly see sex, gender, and sexuality as all proceeding from the same interrelated processes of materially based social relations, an increasing number of biologically oriented thinkers, especially psychologists, see the whole complex as proceeding from biological processes. In some sense, the entire foundation of brain organization theory takes this idea as its starting point, and researchers devise studies that are intended to demonstrate a causal trajectory from the sexed body to gendered behavior, under which most of them subsume sexuality. This fundamentally biological process is in turn often taken to be the basis for broader social structures related to gender. There will be much more on this throughout the book, but for now the point is merely to note that the sex/gender/sexuality distinction has been questioned in the very realm that forms the basis for this study. What, then, is the best way to use these terms in an analysis of brain organization research?

I have settled on this. I am returning to my three-ply yarn of sex, gender, and sexuality, but with an additional twist. It is necessary to explicitly state that my usage, at least in this book, is agnostic as to etiology. In particular, I will not presume that gender is the result of social relations, because to do so would be to discount the studies I review at the outset. Nor, however, will I presume that (bodily) sex is simply given by nature and is not a result of socially inflected classifications and commitments. Finally, I will not presume that sexuality, the realm of pleasures and desires and practices as well as erotic self-concept, flows neatly from either the exigencies of reproduction or the structure of masculinity and femininity. There are two key phrases for which I make an exception and use "sex" where I might, in another work or for another purpose, use "gender." *Sex differences* is a phrase used widely in both popular and scientific work, and I have chosen to use that term rather than *gender differences* much of the time, reserving the term *gender difference* for those situations that are clearly restricted to aspects of psychology and/or behavior. Although the popular and scientific usage of the term *sex differences* does often refer to psychology and behavior, there is often an implicit or potential physiological claim involved, as with the idea that behaviors are a direct reflection of brain structure or function. Likewise, I have chosen to retain the term *sex-typed interests* for a core domain of psychology that some scientists link to brain organization theory. I keep the term largely because I found it created too much confusion to jump back and forth between the term common in the studies and the term I preferred *(gendered interests)*. There will probably be other times when I am describing the positions and theories of brain organization researchers where the usage I describe here may seem to falter. But this does reflect my fundamental intentions in usage for the terms throughout this book.

Structure of the Book

The discussion in this book is organized in three parts. I begin with introductory material concerning emergence of the theory, and my own analytic approach. Chapter 2 explains the theory itself in more detail, and provides some historical context to help readers get a better feel for where this theory came from and how the human research fits with animal studies. It also explains, crucially, why brain organization research in humans has focused so often on particular clinical syndromes and on homosexuality. Chapter 3 covers some basic information about research design, particularly elaborating the importance of understanding brain organization research as comprising a "quasi-experimental network." In this chapter, I introduce several symmetry principles that guide the core analyses in the remainder of the book.

Next I present specific symmetry analyses. Chapters 4 and 5 examine, in turn, the two broadest sets of brain organization studies in humans, those that begin with people for whom scientists have some information about early hormone exposures (cohort studies), and those for whom scientists don't have information about hormone exposures, but instead have information about sex-related characteristics ("case-control" studies). Within each of these designs, studies can be broadly grouped to see the overall pattern of findings for studies that have both similar "inputs" and "outputs" (the most basic sort of symmetry that is described in Chapter 3). The analysis in these chapters is extremely straightforward. I take the findings of particular studies almost completely at face value, and simply highlight how the various studies do and do not support one another, and in some cases show how they set up important internal contradictions. From these chapters, certain kinds of claims about brain organization can be put aside as unsupported, but others appear to merit closer investigation. Chapters 6, 7, and 8 take a closer look at three specific domains in which initial evidence suggests that human brains may be "organized" by early hormone exposures: feminine or masculine patterns of sexual response and behavior, sexual orientation, and sex-typed interests. Each of these chapters constitutes a detailed examination of the studies providing evidence for one specific domain, with special attention to how the "outcomes" of brain organization—that is, the traits that are purportedly influenced by hormones—are actually defined and measured. Chapter 5, for example, examines studies used to support claims that prenatal hormones create permanent "templates" for masculine or feminine sexuality. Although the researchers were mostly interested in what *causes* masculine or feminine sexuality, I am mostly concerned with what they think masculine or feminine sexuality *is*. (I am sure most readers will find the extent of scientific disagreement on this point as surprising as I did!)

In the third section, I begin to look forward to a new kind of sexual science, and indeed, a new science of human development. In Chapter 9 I extend the critique of previous chapters by offering some detailed consideration of how crucial elements of context, both biological and social, have been disregarded in nearly all brain organization studies. Chapter 10 offers some thoughts for moving toward a more sophisticated, scientifically accurate theory of embodiment, including but going beyond issues of gender and sexuality. Here I describe some critical experiments in genetics and developmental biology that suggest how context and processes matter, and how we might do a better job of studying this.

Toward the Open Door

The body unquestionably matters in shaping who we are, what we want, and how we behave. The devil is in the details. One frustrating aspect of brain organization research is that its prominence has gotten in the way of serious and sophisticated research on how biology and the social world interact to shape behavior. I hope this book will point to some potentially more fruitful directions for research on the material and social bases of temperament, skills, desires, and self-concept. Many people have asked me, as I've discussed brain organization research over the years, to comment on how I think psychosexual development *really* works—how do we become the gendered and sexual beings that we are? I must confess ahead of time that I will not present the real and true process of human development. This will be a disappointment to some readers who are seeking a fully developed alternative theory. But for those of you who find open doors inviting rather than vertiginous, I have a number of ideas. Toward the end of the book, I begin to sketch some of the methodological approaches that I think will be more promising. But first I want to open doors that are closed by the "brain organization" story—and that requires painstaking work to move the giant pile of data that sits in front of the door, looking very solid indeed. In the chapters ahead, I show that the evidence for hormonal sex differentiation of the human brain better resembles a hodgepodge pile than a solid structure, and it is a pile that blocks from view the complex and fascinating processes of biology. Once we have cleared the rubble, we can begin to build newer, more scientific stories about human development, stories that narrate genuine exchanges between matter and experience, and that incorporate random events.

Bodies matter—genes matter, hormones matter, brains matter. But how? Even though I'm not entirely certain about the answer, I am certain that we have enough information to eliminate the simplistic story of brain organization as it's currently understood. Let me show you why.

CHAPTER TWO

Hormones and Hardwiring

I T ALL STARTED WITH guinea pigs, and, of course, with sex. Brain organization theory was officially born in 1959, when William Young and his colleagues at the University of Kansas published their groundbreaking article "Organizing Action of Prenatally Administered Testosterone Propionate on the Tissues Mediating Mating Behavior in the Female Guinea Pig" (Phoenix et al. 1959). It rests on a very simple idea: the brain is a sort of accessory reproductive organ. Males and females don't just need different genitals in order to have sex, or different gonads that make the eggs and sperm necessary for conception. Males and females also need different brains so they are predisposed to complementary sexual desires and behaviors that lead to reproduction. This theory suggests that the same mechanism is responsible for both kinds of development—that is, for sexual differentiation of "both sets" of reproductive organs: the genitals and the brain.

Brain organization theory quickly moved past its humble beginnings with the mating behaviors of guinea pigs to become the "grand theory" for sexual differentiation of behavior in mammals, and then even more generally extending to vertebrates. Sexual differentiation is not restricted to those behaviors that are directly involved in reproduction, or even in courting. Instead, brain organization theory is used to explain a very wide range of differences related to gender and sexuality—in humans these include everything from spatial relations, verbal ability, or math aptitude, to a tendency to display nurturing behaviors, to sexual orientation. Still, at its heart this is a story about development of sexuality.

William Young and his junior colleagues considered their research preliminary when they reported it in 1959, but they were spurred to publish it because they considered their studies to be "so much more in line with current thought in the area of gonadal hormones and sexual differentiation" of the genitalia and reproductive system "than the earlier experiments on behavior" (Phoenix et al. 1959, 370). In fact, this new theory suggested a direct parallel between the development of male or female reproductive physiology, and male or female mating behavior: "Attention is directed to the parallel nature of the relationship, on the one hand, between androgens and the differentiation of the genital tracts, and on the other, between androgens and the organization of the neural tissues destined to mediate mating behavior in the adult" (369). Although these scientists described the theory as emerging from experiments on physiology and behavior that had been done within the preceding twenty years, to understand the context of their new model it is helpful to know a bit about the beginning of endocrinology itself.

Endocrinology, the study of hormones, emerged at the turn of the twentieth century, and from the beginning, endocrinologists cast an eye toward understanding how sex differences develop in both bodies and psychology.[1] Rejecting the idea of "nervous regulation" that dominated nineteenth-century physiology, the new endocrinologists focused on how organs could affect distant tissues by secreting "chemical messengers" that are carried to targets throughout the body by the bloodstream. These messengers were also known as "internal secretions," but the simpler name *hormones* quickly took hold (Sengoopta 1998; Oudshoorn 1994, 16–17; Fausto-Sterling 2000, 149–151).[2] Early endocrinologists thought of "sex hormones" as "the chemical messengers of masculinity and femininity," responsible for the development of sex differences throughout the body, as well as for the distinctions in how the sexes think, feel, and act (Oudshoorn 1994, 17). The internal secretions were even explicitly recruited to weigh in on important social struggles over sex, especially the drive for women's suffrage, access to higher education and the professions, and employment outside the home. Writing in 1916, British gynecologist William Blair Bell explained that "the normal psychology of every woman is dependent on the state of her internal secretions, and . . . unless driven by force of circumstances—economic and social—she will have no inherent wish to leave her normal sphere of action" (quoted in Fausto-Sterling 2000, 157).[3]

But it hasn't always been easy to bend hormones to such interpretations, and Oudshoorn and others have documented how scientific evidence about so-called male and female hormones quickly caused trouble. There were three key assumptions about hormones that drove the early research paradigm. First, it was assumed that the hormones would be sex-specific, each hormone appearing and relating to proper functioning only in one sex, while generally causing malfunction in the other sex. Second, it was assumed that the functions of these hormones would be limited to, or at least most important for, functions that were directly related to sex and reproduction. Third, scientists believed that the hormones would be "antagonistic," meaning that "male" hormones would actively counter "female" characteristics, and vice versa. Each of these assumptions was assaulted by data as early as 1921, when a research report showed that "male" hormones seemed to induce the same sorts of changes in the uterus of a rabbit as did "female" hormones.[4] In 1927, Laquer and colleagues, biochemical endocrinologists in Amsterdam, reported that "female sex hormone" was present not only in women, but in both the testis and the urine of "normal, healthy men" (quoted in Oudshoorn 1994, 25). The idea that each sex had its own "sex hormone" was so entrenched that even this report didn't cause much of a stir, but a pair of articles by German gynecologist Bernhard Zondek, appearing in 1934 in the journal *Nature*, did the trick.[5] Zondek reported "mass excretion of oestrogenic hormone in the urine of the stallion" and noted "the paradox that the male sex is recognized by a high oestrogenic content" (Zondek 1934, 209). That is, the so-called female sex hormone could be reliably found in substantially *higher* levels in that quintessentially male animal, the stallion, than in females, including both humans and various equine species (horse, zebra, ass). In contrast, large amounts of the "female hormone" estrogen were not typically present in females except during pregnancy, and even then the level didn't approach that found in the stallion.

The other early assumptions about hormones also didn't stack up well against data. For instance, the notion that the *functions* of sex hormones were restricted, or sex-specific, was challenged on several fronts. Biochemists discovered the very close chemical relationship between the "male" and "female" sex hormones, and evidence accumulated that the hormones have broad-reaching functions in multiple body systems that are not sex-specific, such as liver growth and nitrogen metabolism. This led some scientists to call for dropping the term *sex hormones* in favor of more neutral terms. But of course, even as the huge array of processes known to be affected by these hormones has expanded—now including bone growth, blood cell formation, carbohydrate metabolism, diurnal rhythm, and more—the concept of "sex hormones" seems as entrenched as ever (Nehm and Young 2008; Fausto-Sterling 2000).

One reason for reluctance to drop the concept and the term *sex hormones* might be that so much work in endocrinology involved experiments on sexual differentiation itself, so the focus almost always remained squarely on sex-typed physiology and behavior. In the first few decades of endocrine research, scientific understanding of the roles hormones play in sex-differentiated physiology zoomed ahead of the analogous work on behavior. A series of experiments spanning the late 1940s to mid-1950s resolved many of the central questions about hormonal contributions to physiological differentiation of the sexes. The most important of these were the experiments conducted by the French embryologist Alfred Jost; indeed, the "Jost paradigm" is still the classic view of how sex-differentiated genitalia and internal reproductive structures develop from initially "sex-neutral" tissues in mammals. Being male or female is not simply determined at one moment for any individual organism, but is the result of a multistep process. First, at conception the X chromosome from the egg is joined by a sex chromosome from the sperm, either a Y chromosome (yielding an XY genotype, which is the typical male pattern) or another X chromosome (yielding an XX genotype, the typical female pattern). Next, regardless of whether the sex chromosomes are of the male or the female type, at the earliest stages of development all embryos have the same sexual structures.[6]

The first stage of differentiation of these sex-neutral structures is under control of the sex chromosomes, causing the "bipotential" gonads to become either testes (in XY fetuses) or ovaries (in XX fetuses). Depending on the hormonal environment during the next critical phase of development, the rest of the reproductive tract and genitalia differentiate either as male or female. The hormonal environment is largely a matter of whether the gonads have become testes or ovaries, which are usually the main source of fetal hormones for males and females, respectively. But hormones from other sources, such as the embryo's own adrenal glands or hormonal medications ingested by the mother, can change the usual course of events.[7]

According to the Jost paradigm, "the testes are the body sex differentiators; they impose masculinity on the whole genital sphere which would become feminine in their absence. The presence or absence of ovaries is of no significance" (Jost 1970, 121). Testes are important because they secrete androgens: regardless of an animal's genetic sex, the presence of a threshold level of androgens will push differentiation in a male direction (fetal ovaries typically secrete only a very tiny amount of androgens). Using complex surgical techniques to manipulate fetal rats and rabbits, Jost concluded that the female pathway is the *default direction,* which happens in the absence of *any* gonadal hormones. (Newer evidence indicates that female development is not actually a passive process [Yao 2005; Hughes 2004], but the idea of female development as the passive, default pathway persists, even in scientific literature.)

Jost's experiments increased scientists' confidence that they understood the timing and mechanisms of this process—at least as far as sex-differentiation of genitals and internal reproductive organs was concerned. Yet, as the 1950s drew to a close, much of the data regarding hormone effects on *behavior* seemed disjointed and even contradictory. In humans, a great deal of work had gone into investigating hormones as an explanation for homosexuality. The first formal studies in this vein were conducted by Viennese physiologist Eugen Steinach, whose innovative experiments with rats and guinea pigs "marked the beginning of modern experiments on the role of hormones in sexual differentiation" (Fausto-Sterling 2000, 158). In the prevailing understanding at the time, homosexuality was an "inversion," meaning that the mind or soul of homosexual men was feminine and that of lesbians was masculine (Kraft-Ebing 1930; Ellis 1925). Although others before him had proposed a biological basis for homosexuality, Steinach suggested that male homosexuality could be traced to aberrant cells in the testes that produced "female" hormones. Steinach's role as a physician as well as an experimental physiologist made it fairly simple to test his theory via "therapeutic" interventions with humans. Under his guidance, seven homosexual men each had one of his own testicles removed and replaced by a testicle from a heterosexual man (the latter having been removed because it was undescended). Initial enthusiasm for the transplants was short-lived, as the "heterosexual testes" failed to stimulate heterosexual desire in the recipients (Sengoopta 1998; Oudshoorn 1994; Fausto-Sterling 2000).

Steinach's transplants did not work, but the idea that gay men had more "feminine" hormone profiles continued to be quite popular with scientists (for instance, Glass and Johnson 1944; Neustadt and Myerson 1940; Heller and Maddock 1947). Proponents of the hormonal inversion idea were again thwarted by the dismal failure of gay men to become heterosexual in response to testosterone injections, and by the gradual accretion of evidence that studies seeming to support a "feminine" hormone profile among gay men were badly flawed (Kinsey 1941; Meyer-Bahlburg 1977; Kenen 1997). (The parallel theory that lesbians had "masculine" hormone

profiles was also popular, but lesbians were generally the object of less social and scientific concern. I know of no similar early experiments that
sought hormonal "cures" for lesbianism.[8])

Things weren't going any better with the animal research. Specifically,
the same hormone administered to different animals would sometimes induce "feminine" behavior and sometimes induce "masculine" behavior
(how behaviors are classified as masculine or feminine is an important
point that I take up in later chapters). Scientists increasingly suggested that
hormones should be thought of as "catalysts" that "bring latent mating
behavior to expression" (Young, Goy, and Phoenix 1965, 179). That is,
the hormones themselves, although powerful and sometimes capable of inducing sudden and dramatic behavioral changes, did not automatically determine the *type* of behavior that would be induced, or the frequency or
timing of such behavior. For example, scientists were not able to predict
the lag time between testosterone injection and mounting behavior in different male animals, and could not understand why castration did not immediately stop mounting behavior in males, whereas castrated females
seemed to immediately lose interest in sex.

A conceptual breakthrough came, according to a later report by some of
the researchers involved, with "the realization that the nature of the latent
behavior brought to expression by gonadal hormones depends largely on
the character of the soma or substrate on which the hormones act. The
substrate was assumed to be neural" (Young, Goy, and Phoenix 1965,
179). In other words, perhaps male and female animals had different behavioral responses to the same hormones because their brains were different. If so, then rather than looking to the hormonal status of adult animals
to explain individual differences in behavior, scientists needed to begin investigating the factors that make male and female brains different in the
first place.

In the 1930s Vera Dantchakoff had exposed female guinea pigs to testosterone in utero and found that these animals were behaviorally masculinized, but she had failed to use untreated control animals to contrast
with the treated females. Two decades later, a team of researchers at
the University of Kansas reconsidered Dantchakoff's experiment in light
of the new consideration of the "soma." William Young and his graduate
students Charles Phoenix, Robert Goy, and Arnold Gerall repeated the
experiment with control animals, which was especially important, they
said, because among guinea pigs, "most normal females display male-like
mounting as a part of the estrous reaction" (Young, Goy, and Phoenix
1965, 182). They observed, as Dantchakoff had before them, a marked in-

crease in so-called masculine behavior. They also observed a finding that Dantchakoff had not reported, namely, a lower incidence of lordosis, an arching of the back typically performed by fertile females that allows males to mount and penetrate them.

The report of this research suggested a new interpretive framework for understanding previous studies of hormones and sexual behavior: prenatal testosterone had "organized" the brain for masculine behavior. As biologist and historian of science Marianne van den Wijngaard has explained: "Here, organizing means that the effects are structural and permanent" (1997, 29). The theoretical implications were exciting, because the findings suggested a dramatic parallel between sexual differentiation of the genitals and the brain (Phoenix et al. 1959, 369).

Young and his junior colleagues concluded that the key to predicting the effects of certain hormones was understanding the timing of exposures. Whether the object of interest was genital development or the development of the brain and behavior, they suggested that "the rules of hormonal action are identical," namely: "During the fetal period the gonadal hormones influence the direction of differentiation. During adulthood they stimulate functioning" (Young, Goy, and Phoenix 1965, 183–184). Just as fetal development of male or female genitalia was an irreversible step, sealing the permanent sexual character of the animal, so the organization of the brain was permanent, determining once and for all "whether the sexual reaction brought to expression in the adult will be masculine or feminine in character" (182). And, as with genital development, they concluded that the key substance for sex differentiation of behavior was testosterone.

The new framework significantly shifted the direction of future experiments. If hormones permanently organize the brain areas that mediate adult behavior, changing adult hormone levels would no longer be expected to bring about actual "sex reversals" of behavior. For example, a particular guinea pig might be essentially prone to mount other animals; this would be the latent behavior determined by prenatal hormones. The hormones present at any time in adulthood would determine when and how often this guinea pig actually mounted other animals, but would not create (or remove) the basic behavioral capacity for mounting. Likewise, reducing the androgen levels in a male guinea pig wouldn't suddenly cause the animal to perform lordosis—the receptive position for intercourse. Manipulation of adult hormone levels might affect the timing and frequency of sexual behavior, but the nature of an animal's so-called latent behaviors was "locked in" early in development.

Not Just for Rodents: Hormones and
Human Sex Psychology

Young and his team had immediately proposed that the organization hypothesis should be relevant for humans (Phoenix et al. 1959, 370), but there were hurdles to be overcome. First, there were no actual experimental data in humans, and there was no possibility of generating such data, because experiments would involve risky and unethical manipulation of human fetuses. Second, there was the problem of emerging psychological research from a small group of clinicians at Johns Hopkins University who studied "hermaphroditism," now called intersex syndromes. At Hopkins, John Money and John and Joan Hampson were interested in the fact that intersex people, though born with ambiguous genitalia or with "mixed-sex" characteristics, seemed to have relatively normal gender identity despite their "confusing" anatomy: they felt themselves to be *either* men or women, not in-between or sex-neutral. Money and the Hampsons suggested that the key element in gender identity was how the child was reared (Money, Hampson, and Hampson 1955a, 1955b). Money's dissertation research was especially important. He compared intersex people with similar physiologic conditions and found that no aspect of biology was consistently related to gender identity—not chromosomes, not gonads (ovaries or testes), and not even the appearance of the external genitals. The only variable that reliably predicted whether the intersex person would feel male or female was whether the child had been reared as a boy or a girl (Money 1952).

Young's team was intrigued by findings from Money and the Hampsons that, among intersex people, "gender role or psychologic sex can be independent" of every *physical* aspect of sex: chromosomes, gonads, hormones, internal reproductive structures, and the appearance of external genitalia (Young, Goy, and Phoenix 1965, 178). The two teams were already corresponding before the definitive paper on brain organization theory came out, and the mutual implications of the two research programs were already clear.[9] If the research on intersex patients held up to scrutiny, the applicability of brain organization theory would seem to be limited to nonhuman animals.

Before proceeding further, it is important to pause briefly to consider what intersexuality signified for mid-twentieth-century scientists, both those working with humans and those who studied other animals. Intersex conditions (which scientists at that time still called hermaphroditism) have been of central importance for researchers studying brain organization theory, because genital development is a marker of prenatal hormone exposures—the same exposures that may affect the developing brain (albeit possibly during a different time frame). Fetal ovaries don't produce much testosterone, but genetic females can develop male-type genitals if they are exposed to androgens from some other source. In some cases there are intermediate levels of androgen, so the genitals look ambiguous rather than distinctly male or female. Without androgens (and tissue sensitivity to androgens), the genitals will look female even if the fetus is genetically male. So ambiguous genitalia or genitals that don't "agree" with genetic sex always signal that there has been an unusual hormonal environment during fetal development.

Articles and conference reports from the period indicate that both research groups hedged their bets in comparing their own evidence with the opposing theory: Young and his colleagues repeatedly mentioned that they accepted the importance of learning and experience in shaping behavior (Young, Goy, and Phoenix 1965), and Money routinely asserted his appreciation of the principles of phylogeny—the evolution-derived perspective that predicts continuity between developmental processes in humans and those observed in other animals (Money 1965a, 1965b). Both sets of scientists clearly felt the need to resolve the contradictions between the apparent psychosexual adaptability documented in humans and the brain organization process that suggested that other mammals are "programmed" for femininity or masculinity from prenatal or perinatal life.[10] Their discussions quickly led to a direct attempt to assess the influence of hormone exposure on psychosexuality in the same intersex patients from whom Money and the Hampsons had so recently concluded flatly that "sexual behavior and orientation as male or female does not have an innate, instinctive basis" (Money, Hampson, and Hampson 1955a, 308).[11] A decade later Money had subtly modified his position, walking a fine line between his earlier strong claims for the primacy of sex assignment and the possibility of a psychological predisposition stemming from prenatal hormone exposures.

Suggestive Evidence: Money and the Hopkins Studies

Raised in New Zealand and trained in psychology at Harvard University, John Money was one of the most important and probably the most colorful figure in twentieth-century sexology. Simultaneously charming and abrasive, Money often overpowered his rivals with his riveting narration of case studies, his gift for spinning grand theory, and his delight in coining new sexological vocabulary.[12] In recent years, especially in the enormous controversy over the famous "John/Joan" case, Money has often erroneously been characterized as believing that *only* social factors influence gender and sexuality, while stubbornly refusing "obvious" evidence of biologically based predispositions, especially when it challenged his pet theory.[13]

It is perhaps ironic, then, that Money was actually the *first* to apply the new brain organization theory to data from humans and was easily the most prolific researcher in this field. His influence on brain organization theory research in humans cannot be overestimated. Over the duration of his career, he was author or co-author of literally scores of reports suggesting that early hormone exposures created masculine and feminine sexual predispositions in humans.[14] These suggestions first appeared in a review article, "Influence of Hormones on Sexual Behavior" (Money 1965a), in which Money wove together observations that he and his colleagues had made about intersex patients, on the one hand, with nearly three dozen recent animal studies, on the other, in an effort to formulate a synthetic description of the role of hormones in the development of mammalian sexual behavior. This article deserves detailed consideration because in spite of his initially tentative tone, Money introduced in this article the line of reasoning he would later pursue with more confidence in arguing for the organizing effects of hormones on the brain. The article clearly captured Money's thinking in a way that satisfied him, because certain passages reappeared literally verbatim in a number of his subsequent research reports on brain organization theory.

Money began the piece with caution toward applying brain organization theory to humans, noting as he had in 1955 that "psychosexual differentiation takes place after birth" and "is not automatically preordained by genetics or other factors in intrauterine development" (68). After recounting the original experiment that gave birth to brain organization theory and several subsequent experiments that further refined the theory, he emphasized the known interspecies differences and cautioned that "the leap from guinea pigs and rats to human beings is a broad one, and there may, in fact, be no similarity" (69). Nonetheless, he ventured to place some observations from his own clinical experiences with intersex patients into the context of the new theory. He felt that a "partial human parallel to the animal experiments" might be found in intersex women with "adrenogenital syndrome" (now called congenital adrenal hyperplasia, or CAH). This syndrome, a genetic disorder that causes overproduction of androgens from the adrenal glands, is the most common cause of genital ambiguity.

Androgens are elevated throughout fetal development, which is an especially unusual situation for female fetuses. Because of the hypothesis that high androgen levels may masculinize the brain as well as the genitals, people with this disorder—especially girls and women—have been much studied by scientists interested in brain organization.

Contrary to Money's earlier work, which emphasized the social world, including the cascade of social reaction and reinforcement that might follow birth of a girl with masculine genitals, Money suggested in this piece that women with CAH may have masculine sexuality because of the brain-organizing effects of prenatal androgens. Where he had earlier emphasized that this group of intersex women had developed relatively normal gender identities as female, he now pointed to aspects of their sexuality in which he perceived subtle but telling differences from "ordinary" women. He noted, for example, that "most women" were aroused by touch, whereas CAH women may be aroused by "visual and narrative perceptual material." A key difference, he suggested, was the intensity of arousal, which was "more than the ordinary woman's arousal of romantic feeling and desire to be with her husband or boy friend" (69–70). While it seemed that women with CAH were aroused more easily and more frequently than other women, Money was careful to underscore that "imagery of the erotic thoughts and desires was all suitably feminine," meaning that they were aroused by men, and not by women. In modern terms, their *sexual orientation* did not seem to be male-typical.

This article gave the first indication of the sorts of human traits and behaviors that Money and other scientists would examine in the coming years as evidence that people have been either *masculinized* or *feminized* by hormone exposures in utero. Although a great deal of brain organization research later focused either on broader personality or psychological traits, or conversely, more narrowly on sexual orientation, this passage is a good reminder that brain organization theory is, at its heart, about masculinity and femininity. In the case of sexual orientation, for example, desire for men is thought to indicate a "feminine" brain organization, while desire for women indicates a "masculine" brain organization. In terms of sexuality, the question of whether the brain had developed as masculine or feminine could also be answered by probing such issues as how easily and often she or he becomes aroused, and what kind of stimuli might induce sexual feelings.

In addition to the discussion of women with CAH, Money described clinical observations he had made on patients with testicular feminizing syndrome, a condition in which genetic males have masculine internal organs but female external genitalia. The etiology was not yet well understood in 1965, but it was generally presumed that "the necessary masculinizing principle from the fetal testes either fails to be produced or to be properly utilized" (70). Later named "androgen insensitivity syndrome," or AIS, this syndrome does in fact result from an inability to use testosterone, although it might be produced at normal levels.[15] Money had already reported that people with (complete) AIS develop psychosexually as "fully feminine." That finding was in accord with the theory that socialization after birth determined gender, because these children were assigned as girls from birth and raised in accordance with that assignment. In the context of the persuasive new evidence that early hormone exposures have lasting effects on nonhuman animals, however, Money shifted his interpretation of his findings on these cases. He speculated that "the failure of a masculinizing principle in late fetal life in some manner enhances the capacity of the nervous system to comply with the demands of sex assignment and rearing" (70).

Money held to this position through the rest of his career, suggesting that hormone effects were real, but subtle and limited. He cautioned that "human sexual behavior is not automatically preordained by neurohormonal events at a critical prenatal or neonatal period" (Money 1965a, 71), but he had come to believe that gender and sexuality would develop most seamlessly when sex assignment and rearing were in accord with the hormones that had "primed" the nervous system during fetal life: "masculine" hormones would facilitate masculine gender, and "feminine" hormones would facilitate feminine gender.

Once his interest in the organizing effects of hormones in humans was piqued, Money enthusiastically cultivated this same interest in his colleagues and students. At about the same time as the above article was published, Anke Ehrhardt, a young doctoral candidate who had come from Germany to work with Money at Johns Hopkins, began a study of several intersex syndromes, including the early-treated CAH patients as well as children whose mothers had been treated with progestin during pregnancy. Ehrhardt and Money's collaboration resulted in the first systematic investigations of brain organization theory in human subjects. Their first study, on "psychosexual identity" and IQ in ten girls who were born intersex because of prenatal exposure to progestin, was published in the *Journal of Sex Research* in February 1967 (Ehrhardt and Money 1967). They concluded that these "masculinized" girls had higher IQs than would be expected—which was consistent with the then-current idea that general intelligence, reflected in IQ score, was higher in males than in females. (As I note in Chapter 4, subsequent studies indicated that this result was probably due to sampling bias. In any case, once it became clear from population studies that IQ does not differ between the sexes, this became a dead end for brain organization research, and most subsequent researchers have not bothered to compare IQs for those who have, and have not, been exposed to "cross-sex" hormone environments.)

For the first six years of brain organization research in humans, all of the published work was conducted by John Money and various junior colleagues, especially Anke Ehrhardt but also including Daniel Masica, Viola Lewis, and others, all at Johns Hopkins University Hospital. Hormone-exposed children and adults were compared to either unexposed groups (like "normal" children in their schools, or siblings without clinical conditions or exogenous hormone exposures). In each of these reports, the researchers found that girls and women exposed to high levels of androgen exhibited sexual traits and behavior that were more masculine than was expected. Conversely, they found that unusually *low* exposure to androgen among boys and men resulted in behavior that was more feminine than expected. In addition to sexuality, some of the other sex-typed behaviors that scientists looked at included subjects' interest in careers versus interest in "marriage and motherhood," preferred patterns of games and playmates in the children, manner of dress, extent of expressed satisfaction with their sex, specific cognitive skills like verbal or math ability and spatial relations, and occupational patterns among the adults. The particular ways in which intersex individuals differed from either control groups or the "expected" behavior for their gender varied from study to study, though, and often the differences that showed up in one report didn't show up in another. Still, each report suggested enough differences between hormone-exposed and unexposed groups that the idea of brain organization in humans began to look very convincing.

The clinician-researchers who did these studies described their intersex patients as "analogues" that "simulate [the animal] experiments" (Money, Ehrhardt, and Masica 1968, 105). Interestingly, the researchers were not just studying the same *syndromes* from which Money and the Hampsons had earlier concluded that "psychosexuality" (that is, gender and sexuality) is entirely based on socialization. In many cases they were studying the *same actual patients* in the two sets of studies. What had changed? The idea of brain organization had introduced a theoretical mechanism whereby hormones might influence the brain directly in a process distinct from genital differentiation. The animal experiments had also separated out early effects, when hormones "organize" the brain for permanent masculine or feminine patterns of behavior, from later effects, when hormones "activate" the underlying predispositions into actually expressed behaviors. Intersex patients at midcentury typically had unusual hormone profiles well into adolescence and even adulthood, so it was not possible to separately analyze the "organizing" and "activating" effects of hormones. By the late 1960s, though, diagnosis and treatment of intersex patients were generally occurring much earlier, so most of these patients were presumed to have relatively sex-typical hormone exposures after birth. The physiological conditions of some human patients were henceforth more symmetrical to the experiments in animals, in which hormone anomalies were limited to prenatal or perinatal critical periods of development.

After their first study, Ehrhardt and Money produced clinical research reports pertaining to the organizing effect of hormones in almost rapid-fire succession. In the five years following their initial study, they produced more than a dozen new research reports and released their classic book on human sexual differentiation, *Man and Woman, Boy and Girl*, in 1972. Over time their tentative language dropped away and the team began to assert that there was growing evidence of a real, though limited, organizing effect from prenatal hormone exposures on the human brain.

Two chapters in *Man and Woman, Boy and Girl* may be read as the two ends of the tightrope that Money, Ehrhardt, and their other early collaborators were walking: "Fetal Hormones and the Brain," on the one hand, and "Gender Dimorphism in Assignment and Rearing," on the other. In the first, Money and Ehrhardt explained in great detail the various intersex syndromes they had studied, with special attention to both congenital adrenal hyperplasia (CAH) and androgen insensitivity syndrome. In regard to the former, they paid particularly detailed attention to the many aspects of childhood behavior that were apparently masculinized in these girls, not only concluding that the effects were most likely due to hormones but speculating on the specific neural pathways that were likely implicated in particular behavioral shifts. For example:

> The most likely hypothesis to explain the various features of tomboyism in fetally masculinized genetic females is that their tomboyism is a sequel to a masculinizing effect on the fetal brain. This masculinization may apply specifically to pathways, most probably in the limbic system or paleocortex, that mediate dominance assertion (possibly in association with assertion of exploratory and territorial rights) and, therefore, manifests itself in competitive energy expenditure. (102)

Yet they were still, at this time, convinced that the "big picture" of development in what they called "gender dimorphic behavior" was best seen by focusing not on prenatal hormones but on postnatal rearing experiences. Most distinct from their later work, especially Ehrhardt's research, was their conclusion that the choice of sexual mate—not yet commonly called sexual orientation—was *not* under the influence of prenatal hormones: "Evidently, the deciding factor as to the characteristics of the sexual mate as male or female operates postnatally and not prenatally" (102). In general, they concluded that the organizing effects of hormones are much less pronounced in humans than in other animals:

> In the lower species, fetal androgenization may automatically reverse gender dimorphic behavior by prenatal hormonal decree, so to speak. In human beings there is no such automatic decree. So much of gender-identity differentiation remains to take place postnatally, that prenatally determined traits or dispositions can be incorporated into the postnatally differentiated schema, whether it be masculine or feminine. (102)

Their very next chapter, focusing on gender assignment and rearing, made these points even more dramatically. The first detailed example presented in that chapter was a five-page review titled "Rearing of a Sex-Reassigned Normal Male Infant after Traumatic Loss of the Penis." This review concerned the "John/Joan" case, which is familiar to many readers from the immense popular coverage it received. Here, Money and Ehrhardt gave detailed attention, not only to the child's apparently blossoming femininity, but to the efforts of the parents, particularly the child's mother, to convey consistent "gender messaging," considered essential for the child to develop a feminine identity. This was especially important because the child had been reared as a male until he was 17 months old.[16] They noted approvingly, for example, that "the family was relatively open in regard to matters of sex and reproduction, so that one can study particularly well the differences in treating a girl and a boy regarding sex and their future adult reproductive role" (120).

Clear, no-nonsense information and modeling of distinct male and female roles was seen as the key to this child's successful gender reassignment. Money and Ehrhardt, like most psychologists at the time, were deeply concerned with "successful adjustment to roles." Though in his writing Money in particular frequently showed that he understood gender roles to be matters of convention, rather than somehow determined by natural laws, he nonetheless believed that accepting socially ordained roles was a critical component of psychological health and happiness. It's worth noting, as plenty of critics have, that Money and Ehrhardt's ideas of the *content* of male and female roles were decidedly mainstream, if not downright old-fashioned. "Of course," they explained, "girls and boys are not only prepared differently for their future reproductive role as mother and father, but also for their other different roles, such as wife and husband or financial supporter of the family and caretaker of children and house"

(120). (To give some context for the historically challenged, when Money and Ehrhardt's book was published in 1972, popular support for sexual equality and gender change was at a high point in the United States. It was a watershed year for political gains for women: Title IX (the Women's Educational Equity Act) was passed; the Equal Rights Amendment barring discrimination on the grounds of sex was passed in Congress and looked like it was on course to be ratified by the states; and Shirley Chisholm, the first African American woman in Congress, mounted a bold campaign as a Democratic presidential candidate.[17])

Unsurprisingly, Money and Ehrhardt were in for some fierce criticism from feminist critics (such as Bleier 1984; Fausto-Sterling 1985) for their stereotypical presentation of gender and sexuality. They also caught it from the other side—psychologists who were more deeply committed to a strong role for biology in sexual development (for example, Diamond 1974; Imperato-McGinley et al. 1979). In brain organization work that followed these early studies, both unabashed gender conservatism and preference for seeing postnatal social experiences as the major player in gender and sexuality gave way. Normative statements about "women's roles" and girls' and women's behavior being "appropriately feminine" were replaced with more-neutral statements about what women and girls versus boys and men do and think and say they want. In this way, conventionally gendered behavior was taken out of the context of *prescription* and presented as simple *description*. This had the possibly unanticipated consequence, though, of taking these behaviors out of the context of the social world. The descriptive approach significantly deemphasized the role of norms, social structures, and modeling in developing gendered traits. Instead, disembodied as "naked facts" of sex differences, they began to look more and more like simple reflections of male and female nature.

But this is getting ahead of the story. In 1972, brain organization theory was all very new, and still somewhat tentative. That was all about to change.

Brain Organization Hits the Big Time

Money and Ehrhardt's research was the start of something big. As Marian van den Wijngaard (1997, 35) has noted, by the early 1970s, "results of investigations concerning masculinity, femininity, or sex differences in behavior were no longer convincing without referring to the organization theory." Its popularity has only continued to grow, and brain organization theory has generated progressively more studies in each decade since it was first proposed that early hormone exposures influence masculine and feminine behavior in humans, as well as in other animals.[18] I have analyzed over three hundred scientific papers that explore the link between early hormone exposures, on the one hand, and sex differences in sexuality, personality, and cognition, on the other; these studies represent the great bulk of research on brain organization in humans. Identifying studies related to the theory has become more difficult over time, because the idea of hormonal organization of the brain is now only rarely identified as a theory and is usually simply incorporated into research as a "background fact" of development.

Perhaps reflecting the training of Money and his various protégés, brain organization research has been dominated by psychologists from the start. But the past three decades have seen endocrinologists, neurobiologists, neurologists, and interdisciplinary specialists like psychoendocrinologists enter this work. Study designs have multiplied, too, because scientists recognize that the best way to answer complex scientific questions is to take multiple approaches and see where the answers from various studies converge. Designs also vary because of the different disciplines of the scientists involved, and because the original theory of brain organization has been elaborated to add an organizing role for estrogens, in addition to androgens (a change I describe in later chapters).

Given the many varieties of studies, it is helpful to apply concepts from epidemiology to group them into two main sets: (1) cohort studies and (2) case-control studies. In epidemiology, "cohort research" generally refers to studies in which scientists begin by knowing that certain people have had an exposure (to a substance, an event, or some other factor that is believed to affect health) and other people have not (Mayrent, Hennekens, and Buring 1987). Scientists compare the health of people who did and did not have the prior exposure, in order to determine whether the exposure itself led to any particular outcome. In cohort studies on brain organization, the exposure of interest is prenatal hormones, especially hormone levels that are unusual for the individual's genetic sex. People with these unusual prenatal hormone exposures—whether from a clinical condition or because their mothers were prescribed hormones while pregnant—are later observed and/or interviewed to see whether their sexuality or other sex-linked traits differ from people whose prenatal hormone exposures were normal.

Cohort research on brain organization includes the many studies of people with intersex conditions, because ambiguous genitals, or genitals that don't "agree" with genetic sex, are almost always a signal of unusual prenatal hormone exposures. Other studies have followed non-intersex people whose mothers were prescribed hormones during pregnancies that were thought to be at risk for miscarriage (a practice that was very common in the mid-twentieth century). In recent years, a variation on the cohort design has been added, in which scientists have some information about prenatal hormones among a "nonclinical," or "normal," cohort. One such design involves studying girls with same-sex or other-sex twins (because a genetic female developing with a male co-twin may have a higher than normal exposure to androgens). Another involves inferring fetal hormone levels from samples of maternal cord blood or amniotic fluid. In these latter cases, there are not extreme exposed versus unexposed groups to compare. Instead, comparisons are made within the cohort to see whether prenatal levels of steroid hormones correlate with levels of sex-typed behavior.

While cohort studies begin with information about exposures, case-control studies begin from the opposite direction, grouping people according to outcomes and then looking for information on exposures. (In epidemiology the outcome of interest is usually, but not always, a disease; increasingly one sees studies of "positive" conditions like resistance to disease.) For case-control studies, scientists must have fairly distinct groups to compare, and in brain organization research this means scientists begin with people they consider to be sexually different from the majority— including gay men, lesbians, and sometimes bisexuals, as well as transsexual or transgender people. Some investigators approach the difference among these various sexual minorities as a matter of degree, and others treat them as categorically distinct, but virtually all studies consider bisexuals, gay men or lesbians, and transgender people to have at least partial "cross-sex" psychosexual differentiation, a result that would presumably follow prenatal exposure to "cross-sex" hormones. (I don't mean to endorse this notion. In Chapter 7 in particular, I'll show that it is a troubled proposition that creates a lot of tension within brain organization studies. My point here is simply to explain that it is a core assumption of brain organization research.)

The working hypothesis in all such studies involves two parts: (1) the hypothesis that male-typical sexual orientation or gender identity will correlate with other male-typical physical or psychological traits (and vice versa for female-typical sexual orientation and gender identity); and (2) the hypothesis that traits are correlated because both are influenced by hormones during the critical period of development. For example, a researcher may hypothesize that being sexually attracted to women ("male-typical sexual orientation") will be associated with a "male-typical" pattern of cognition (scoring high on a spatial test like mental rotation ability). A correlation like this is taken as support for brain organization theory. (Note, though, that failing to find such a correlation doesn't necessarily negate the theory, because the two traits may develop during different time periods, and hormone levels may fluctuate during gestation.)

Case-control research, strictly defined, requires concrete information about whether cases and controls had prior exposure to the agent that is being investigated (in this case, unusual prenatal hormones). In case-control studies of brain organization, no one has that information.[19] Instead, investigators *infer* prenatal hormone exposures by looking for some other exposure (like prenatal stress) or outcome (like left-handedness) that is also thought to be associated with unusual hormone exposures. They use statistical tests to determine whether the "cases" (sexual minorities) are appreciably different from the "controls" (those with conventional sexuality and gender) in terms of this second exposure or outcome.

Some examples of case-control studies of brain organization include comparing gay men and straight men on features like left- or right-hand preference, spatial relations, or finger-length ratios—all of which differ, on average, between men and women and are thought to be affected by early hormone exposures. These studies also include the familiar "gay brain" research (where gay men's brains are compared with those of heterosexual men and sometimes women), as well as similar studies that compare transgender and conventionally gendered people. Why do scientists focus on sexual minorities for these studies? Given that masculine interests and sexuality are quite obviously cultivated and encouraged in developing boys, while feminine interests and sexuality are cultivated and encouraged in girls, it would not be particularly informative for scientists to simply compare males and females for these studies. There would be no way to decide that the male–female differences were due to different hormone exposures rather than to lifelong differences in rearing and socialization experienced by girls and women versus boys and men. For this reason, scientists do *intrasex* comparisons of people whose gender and/or sexuality seem sufficiently distinct from one another to suggest that their brains may be differently organized for masculinity or femininity.

Through all of the changes and elaborations of research on brain organization, the basic idea has remained consistent. This theory suggests that regardless of chromosomal sex, having a male-typical hormonal milieu in utero leads to male-looking genitals and "masculine" psychological traits, including erotic orientation to women, as well as broadly masculine cognitive patterns and interests. Likewise, a female-typical hormonal milieu leads to feminine-appearing genitals and "feminine" psychology, including erotic orientation to men, and female-typical cognition and interests. From a division of labor in reproductive intercourse itself, to a reproduction of labor in families, and so on, the core assumption of brain organization theory is that masculinity and femininity are package deals with reproductive sexuality at the core.

What has this vast body of research shown? Before looking at the studies themselves, it is necessary to take a bit of a detour to consider how, exactly, the studies can and should be analyzed. Not haphazardly, as if it didn't matter *where* in the great pile of data any particular study fit in, but systematically. In short, scientifically. The next chapter will explain how.

CHAPTER THREE

Making Sense of Brain Organization Studies

Through the gradual accumulation of modest but generally posi-tive research results, the theory of brain organization by hormones has been increasingly accepted as a fact rather than a theory. Yet a key as-pect of the existing evidence for the theory is often overlooked: individual studies can never be decisive in supporting the theory. Neither can results of multiple studies simply be evaluated in an "additive" fashion. Rather, because all the studies are quasi experiments instead of true experiments, the overall credibility of the theory depends on how well studies fit to-gether. The basic principles behind experimental research are familiar to anyone who has gotten as far as middle school science. What happens to fish in similar tanks when we vary the temperature of their water? How does changing the amount of moisture in the soil affect the growth rate of mustard sprouts? In each case, the experimenter changes one thing at a time to see how that one change makes something else happen. Experi-ments that proceed in this way, by varying just one condition when every-thing else is controlled, are preferred over other kinds of studies because they provide strong evidence that any change observed can be traced to a particular cause. Quasi experiments, on the other hand, are a catch-as-catch-can proposition. Some things, like human development, can't be de-liberately manipulated, so scientists have to improvise creative ways of studying the issue without the systematic and controlled information an experiment would yield.

This chapter covers some basic issues of design and measurement in quasi-experimental research that should guide an evaluation of studies on hormonal organization of the brain in humans. Readers who are familiar with quasi-experimental methods and measurement theory may wish to skip this chapter, or skim selected parts. There are three main sections to the discussion. In the first, I consider how quasi experiments are different from true experiments in terms of the strength of evidence that any individual study can provide. To illustrate these points, I consider, among other things, research on how hormones affect the developing physiology of humans and other animals. In the second section, I introduce measurement theory as a method scientists use to evaluate how well their concrete research practices fit with their theoretical interests and claims. Third, I build on measurement theory to develop three simple symmetry principles that can help both scientists and readers of science evaluate how well studies, especially quasi-experimental studies, fit together to form strong scientific evidence.

Experiments and Quasi Experiments

Studies of how early hormone exposures affect the development of genitalia and reproductive organs in small mammals are good examples of true experiments. In all mammals, reproductive morphology is initially identical in males and females; that is, all mammal embryos have some bipotential structures (those that might develop as either male-typical or female-typical) and also have both male and female duct systems—of which one will regress and the other will develop (leaving aside, for now, the more complex and relatively infrequent intersex cases, in which reproductive development is mixed or ambiguous). Experimental embryologists and other developmental biologists had established early in the twentieth century that hormones play a key role in the process by which animals developed from this "sex neutral" state into males or females, but conflicting results and arguments over research designs had left important questions unanswered. Which hormones controlled which part of the process? Were estrogens and androgens both important? In the 1940s and 1950s, the renowned French embryologist Alfred Jost developed innovative techniques that allowed him to meticulously vary just one factor in development at a time (Jost 1953). Among his most striking experiments were the fetal castrations he performed on rabbits. By operating very early in development, Jost could observe how development proceeded in the absence of endogenous hormones normally produced by the animals' own testicles or ovaries. Jost found that removing the testes in male animals produced the most striking results: rather than a penis, scrotum, and male duct system, the animal developed nearly normal female structures, including vagina, clitoris, uterus, and oviducts. On the other hand, removing the ovaries in female animals did not seem to fundamentally alter the usual course of development: the animal did not develop any typically male structures and did have all the usual female structures. In additional experiments, Jost administered androgens and estrogens to the castrated animals and found that both genetic males and genetic females developed male-typical structures if androgens were administered during the right time in development (Jost 1970; van den Wijngaard 1997; Fausto-Sterling 2000, esp. 199–205).

It will immediately be clear that such truly experimental research on how prenatal hormone exposures affect human development is completely out of the question. That is the case whether one is interested in human genital development, physical development of the brain, or human psychology and behavior. (Importantly, though, as I'll show later in this chapter, the distance from the experimental gold standard is much greater for the development of brain and behavior than for genitals.) True experiments on how hormones affect human reproductive structures, as with studies in other animals, would need to proceed by systematically altering the hormonal environments of human embryos— -removing testes or ovaries or otherwise blocking the embryo's endogenous hormones, introducing external hormones, modifying the maternal hormones that contact the embryo, and so on, then observing the structure and functions of the experimentally manipulated infants at birth. This is abhorrent, of course, even as a thought experiment. But bear with me as I draw the case out a bit further, because it is crucial to the arguments that will follow.

Controlled experiments on human brain organization are even more difficult to conceptualize. This is partly because the brain takes much longer to develop, and environmental inputs—including interactions with other people—are required in order for neural connections to be made, strengthened, and/or maintained among cells within the brain, as well as between the brain and the rest of the body. That is, the physical architecture of the brain (the size and shape of particular brain areas and how various areas connect), as well the function of key cells such as hormone receptors, all depend to some extent upon the particular stimulation and interaction the person experiences. So a true experiment on human brain development would have to subject a substantial number of infants to rigidly controlled stimuli through their development into toddlers and beyond, most especially including early interactions with parents and other caretakers when brain development is happening most rapidly. Even if this were not a nasty idea, it's not plausible that caretakers could be made to follow such a rigid program of interaction and socialization.

But other, less-controlled experimental methods have been devised for understanding the role of single factors in processes that are multiply determined and develop over a long period of time. Think about complex diseases with long developmental periods and multiple risk factors: cancers and any of the chronic diseases like heart disease or diabetes fit the bill. Experimental research on human brain organization might theoretically look something like the clinical trials used to test prevention methods or treatment for such diseases. In a classic clinical trial, people with similar medical conditions are randomly assigned to get either a new drug or a placebo. In addition to using random assignment, these studies also use a technique called "blinding," meaning that the information about who is getting the drug versus the placebo is not revealed. Ideally, the study is "double-blinded," meaning that not only the patients, but also the medical staff who are evaluating patients' progress, must be unaware of who is getting the drug.

So, following this model, one could imagine a clinical-trial-type study of human brain organization. There would need to be hundreds or even thousands of participants, because the statistics required for handling multiple inputs and complex outputs are trustworthy only with large numbers. Such an experiment would entail randomly assigning a large number of pregnant women to different groups and systematically varying the hormone exposures across these groups. This would result in cohorts of children whose fetal hormone exposures could be (eventually) compared. As long as the study was going on, though, no one must know which kind of exposure any particular child had—not the families into which these children were born, nor the children themselves, nor their doctors, their teachers, or anyone else. When the offspring of these experimentally manipulated pregnancies are adults with enough sexual experience to evaluate, scientists could then unseal the record of their hormone exposures and see whether there are correlations between fetal hormone exposures and adult sexuality.[1] This design is obviously, again, in the realm of dark fantasy, not reality. So what is a curious scientist to do?

Because they can't do experiments on how hormones affect human development, scientists rely on a network of quasi experiments. Unlike true experiments, where the controlled circumstances of the research can give strong evidence about cause-and-effect relationships from even a single study, quasi experiments become convincing only when multiple studies, related in specific ways, all point to the same conclusions. Neither individual studies nor even a small set of replications can ever be decisive in supporting the theory of brain organization in humans. Nor can the results of even a vast number of quasi experiments simply be evaluated in an "additive" fashion. Rather, because all the studies are quasi experiments, the overall credibility of the theory depends on how well the studies fit together. The rest of this chapter is devoted to considering what it means that quasi-experimental studies fit together well or poorly, and the rest of this book, in turn, applies these principles to evaluate the specific studies that have been conducted on sexual organization of the human brain.

Because quasi experiments derive their strength from their relationships with other studies, a reasonable evaluation must consider all—and only—the studies that examine the same hypothesized cause and effect. If you can't actually do an experiment to see if X causes Y, then you consider all the quasi experiments about the effect of X on Y. Obviously, it wouldn't make sense to throw in some studies about the effect of X on Z rather than Y, nor would it be fair to include only some of the studies that examined the effect of X on Y. (This is the simplest kind of symmetry, and that is the analysis I present in Chapters 4 and 5.) So far, so good. But what if the hypothesized outcome, Y, isn't something simple and concrete, but is complex, difficult to define, or changeable? In other words, what if the outcome is something like human sexual desires and behaviors? In this case, the issues of definitions and measurement are critical, because we must be sure we are grouping studies properly. Before we can conclude that certain kinds of hormone exposures "organize" the brain for feminine sexuality, we must be clear about what constitutes feminine sexuality (the subject of Chapter 6); the same goes, of course, for homosexuality (Chapter 7), "masculine" or "feminine" interests and priorities (Chapter 8), and so on.

At the most basic level, studies fit together when they have both similar inputs and similar outputs. To get a sense of what this means, consider the example of cancers and chronic diseases again. Although clinical trials are used for developing prevention programs and treatments, it is not possible to conduct clinical trials for the purpose of pinpointing risk factors. That is because once a scientist has a hypothesis about some possible risk factor that is firm enough to warrant investment in a clinical trial, that very same hypothesis also makes a trial unethical: the ethics of research with humans dictate that you cannot subject people to exposures that you suspect are harmful. Instead, scientists do "observational studies," a kind of quasi experiment in which scientists don't actually intervene at all, but simply observe and compare what happens to people under various conditions. Studies linking cigarette smoking to lung cancer are a classic example. These studies employ a variety of designs, but all of them link smoking with lung cancer. One common approach is to compare smokers to non-smokers and see which group develops lung cancer more frequently; though nonsmokers do develop lung cancer, the smokers, as a group, develop lung cancer far more often. Another design is to identify everyone who is diagnosed with lung cancer within a particular geographic area and time period, get cigarette-smoking histories from these people, and then compare the smoking rates among the people with cancer to smoking rates in the general population. These studies show that, compared with the general population, people diagnosed with lung cancer, as a group, have smoked many more cigarettes and have smoked over a longer time period. Even though these designs are different, both kinds of studies "fit together" if they are using similar criteria for the input (smoking) and the output (lung cancer). It is very important to note here, though, that studies of slightly different inputs (smoking marijuana instead of cigarettes, for example, or chewing tobacco instead of smoking it) are not part of the network linking cigarette smoking to lung cancer—such studies are treated as irrelevant for the link between cigarette smoking and lung cancer. Obviously the same thing is true for different outputs: studies linking cigarette smoking to heart disease, for example, or even to breast cancer, do not provide any evidence about the link between cigarettes and lung cancer.

The evidence tying cigarette smoking to lung cancer is among the strongest epidemiological evidence that exists for any risk factor. This is partly because of the numbers of people that have been studied, and because the link between cigarettes and lung cancer is so strong—a more subtle contribution to lung cancer would have been harder to detect and confirm. It is also because the findings have been repeated with so many different designs looking at the same input and output. A further very strong indication that cigarettes actually cause cancer, and are not simply related to it, is the fact that there is a dose–response relationship, meaning that the incidence of lung cancer consistently increases as the amount and duration of cigarette smoking increases. Finally, these quasi experiments in humans fit very well with experiments on other animals, in whom cigarette smoke also induces lung tumors.

A similar network of quasi-experimental research links early hormone exposures to human genital development, and for this reason this network also provides a good example of robust quasi-experimental research. First, controlled experimental research in small mammals, such as the studies by Jost described earlier in this chapter, provided a strong foundation for understanding the basic connection between androgen exposures during a critical period of development (the input) and male-type genitalia (the output). Second, the principles established in those studies seem to hold across all mammal species studied, including primates (an important consideration, because for many purposes primates are considered a better model for human development than are other animals). Given that there are very broad similarities in genital structures across all mammal species, such cross-species data are strongly relevant. Third, the experimental findings accord well with the results of observational research in humans. The human studies that are most important for understanding genital development involve a group of intersex conditions, those cases in which people are born with genitalia that either are not typical for their genetic sex or cannot be clearly classified as female or male. The most common cause of intersex births in genetic females, a condition known as congenital adrenal hyperplasia, or CAH, provides a good example. CAH affects the adrenal glands such that (among other things) the fetus has unusually high androgen levels during the period when the genitals are differentiating. Many genetic females with CAH are born with genitals that range from slightly masculine (for instance, having an unusually large clitoris or partially fused labia) to being indistinguishable from the usual male pattern, with a fully developed penis and fully fused (though empty) scrotum rather than labia. Further, as with the example relating cigarette smoking to lung cancer, there is a dose–response effect: more-severe alterations of adrenal function are correlated with greater virilization of the genitalia (Forest 2004).

Particularly dramatic quasi-experimental evidence for the association between early androgen exposures and the development of masculinized genitalia comes from hormone-treated pregnancies. One popular class of steroid hormones used in "anti-miscarriage" regimens during the 1950s and 1960s were the progestogens, which include both natural progesterone and synthetic steroids that bind to progesterone receptors. Progesterone is critical for physiological processes related to pregnancy (hence the name), but the progestogens, including progesterone, are also androgen precursors, meaning that the body can convert these steroids to androgens. Interestingly, the jury still seems to be out on whether treatment with progestogens does anything to prevent miscarriages (Wahabi, Althagafi, and Elawad 2007). However, one effect that is not disputed is that treating pregnant women with progestogens resulted in the birth of some genetic females with masculinized genitalia (Wilkins 1960; Ehrhardt and Money 1967). The degree of masculinization was correlated with both the timing and the amount of progestogen administration, which helped pinpoint the precise period during which hormone exposures affect sex differentiation of the genitals in humans and also gave dose–response evidence that the hormone itself was what caused the genital changes (Schardein 1980).

Taken together, the nonhuman experiments and the multiple kinds of observational data in humans form a strong quasi-experimental network that supports the role of androgens in the development of masculine genitalia.

But we should be very cautious before extending this very well-established connection between androgens and genital development to conclude that androgens also masculinize the human brain. Genital development provides a much simpler experimental model than the development of sexual "personalities" and desires, for a variety of reasons:

1. The developmental period is much shorter and earlier for the genitals than for the brain. In humans, genitals and reproductive structures develop their basic morphology early in fetal development. Even though genitals go through a second developmental phase at puberty when they grow and mature, the structures are formed long before birth. This is in marked contrast to the development of the human brain, which is extremely undeveloped at birth. Genital development is certainly also different from psychological and behavioral masculinity or femininity. These develop over a much longer period, and even the most diehard biological determinist would concede some effects of culture, rearing, and experience on them.

2. It is much more problematic to draw on animal evidence for understanding the development of brain and behavior than for understanding the development of genitals. First, there is less similarity in brain structure across mammal species than there is in genital structure. Second, human behaviors, personalities, and desires are so different from what we observe in other animals as to complicate or even defy comparison in many domains, including sexuality. Much as genital structures are similar across mammalian species, one core function of genitals—reproduction—is also similar. Reproduction in all mammals requires a male to insert a penis into a female's vagina during her fertile period and to ejaculate (fairly recent variations and technological innovations available to humans notwithstanding). But sexual behaviors are another thing altogether. For one thing, many "sexual" behaviors, both in humans and in other animals, often are not directly associated with reproduction. For another, when reproductive behavior is involved, all animals have complex complements of behaviors that precede, accompany, and follow the act of insemination. Many of the experiments linking sexual behaviors in rodents to early exposure to androgens, or lack thereof, look at "feminine" behaviors like ear wiggling (an invitation to sexually approach) or lordosis, a characteristic arching of the back that allows another animal to mount from behind and

achieve penetration. What would the human counterparts to these behaviors be?

The difficulty is only compounded when we move outside of the domain of sexuality. Some human traits that are believed by some scientists to be affected by androgens do not have clear nonhuman correlates: verbal fluency, for example, or any sort of math skill. The point here is not to claim that there is no possible way to relate complex human behaviors to behaviors in other animals, but to underscore the fact that this is a much more ambiguous undertaking, with much more room for argument and error, than comparing genitals.

3. Human brains, unlike genitals, cannot be "sexed," meaning that they cannot be sorted reliably into "male-type" and "female-type" by observers who don't know the sex of the person they came from. As a thought experiment, imagine that one were to take scientific photographs of the genitals of a thousand human adults and present these photos to a team of judges without any other contextual cues as to the sex of the individual to whom the genitals belong. It would be possible to sort the photographs into male versus female piles with almost 100 percent accuracy. This is not to suggest that there is no intrasex variety in genital size and shape, nor to ignore the existence of intersex people whose genitals might not be so easily categorized, but simply to underscore that in a group of only one thousand people, it will be possible to clearly place almost all human genitals into one of two main types. Human brains are another matter entirely. In spite of much trumpeting that there exist "female brains" and "male brains," the extent and nature of physical differences in the brains of human females and males is highly controversial, with some scientists claiming that there are no clear-cut differences, others claiming that there are some subtle average differences, and still others claiming that the differences are dramatic.

The absence of scientific consensus on this point is not for lack of effort. Indeed, the German physician Franz Josef Gall (1758–1828), the founder of phrenology, which some neuroscientists consider the forerunner of contemporary "brain localization theory," claimed two centuries ago that the differences between male and female brains were so great that he could identify the sex of a brain presented to him in water (Russett 1989, 19). And Paul Broca (1824–1880), a pioneer in neurology who was the first scientist to link a particular function (in this case, speech) with a particular brain area (Broca's area), was passionately interested in the idea that group differences in intelligence and temperament are related to physical differences in the brain. Broca devoted more of his craniometric research to male–female differences than to any other comparisons between groups (Gould 1996). (And though he, like Gall, was quite confident that he had found such differences, history has not been kind to this subset of his "discoveries," which were based on exceptionally biased methods.)

Fast-forwarding, the most extreme claims about sex differences in the human brain that I've seen in some time come from the neuropsychiatrist Louanne Brizendine in her popular book *The Female Brain* (2006). She claims that the "sex-related centers in the male brain are actually about two times larger than parallel structures in the female brain," even likening the male brain's "hub for processing thoughts about sex" to O'Hare Airport while likening the female brain's "hub" to "the airfield nearby that lands small and private planes" (91). This may be an amusing comparison, but from the standpoint of neuroanatomy it is pure imagination, beginning with the assertion that there are "sex-related centers" in the brain. Though neuroscientists have identified a tiny cell group in the hypothalamus that is larger in men than in women, and some of them speculate that this area may be related to some aspect of sexual function, no one really knows yet what the area, called the third interstitial nucleus of the anterior hypothalamus, or INAH3, does (Allen and Gorski 1992; Byne et al. 2000; LeVay 1991). It may even be related to something as mundane and "nonpsychological" as menstruation. At any rate, there's certainly no evidence that it has anything to do with "processing thoughts about sex."[2]

In fact, aside from this little nucleus of unknown function, and an overall average size difference (which seems to be basically a function of body weight), there continues to be great controversy about whether there are any other sex differences in human brain structure. The biologist Anne Fausto-Sterling has conducted an extremely illuminating analysis of various studies on sex differences in the corpus callosum, the band of tissue connecting the left and right hemispheres of the brain. Although many studies claiming sex differences in the corpus callosum have been widely reported in the popular press, Fausto-Sterling finds that, on balance, the studies do not show either an absolute sex difference in corpus callosum size or shape or any relative differences (such as differences in specific brain areas, or differences in size or shape corrected for the overall weight or volume of the brain) (Fausto-Sterling 2000, 127–135). Sex differences in various other areas of the brain (such as the anterior commissure and the bed nucleus of the stria terminalis) have been announced to enormous publicity, but none of them has yet held up to independent replication. And it's worth noting, again, that the identification of sex differences in the brain is one of the longest-standing projects of neuroscience. After more than two centuries of effort, surely any "obvious" differences would have emerged by now. This is not to claim that there are no sex differences in the human brain; indeed, the sex difference in the volume of INAH3 does seem to be real. Given the basic differences in male and female reproductive physiology, and the fact that this has to be represented in the brain somehow, it is somewhat amazing that this is the only sex difference identified to date.

The lesson here is that sex differences in the human brain are subtle and elusive, not obvious. To return to the analogy with our hypothetical experiment on sex-typing human genitals, how successful would a similar experiment be if we were attempting to sort human brains into male and female? Not very; even the most discerning neuroscientists would find it difficult, if not impossible, to sex-type human brains based on their structures.

4. What about function? Contrary to popular claims, there is great controversy about whether there are distinctively gendered patterns of brain function. Do men and women, on average, use their brains differently when engaged in similar tasks? Quite a few psychologists claim that they do. Larry Cahill (2003), for instance, has reported brain-imaging tests of people experiencing emotional memory that indicate more activity in the left hemisphere, on average, for women than for men. But a look at the studies indicating these sex typed patterns of hemispheric activity show broad overlap in the "maps" of activity for women and men (see Cahill 2003, 1239). Likewise, Richard Lippa (2005, 100) suggests that "men's right hemispheres may be more exclusively devoted to visual-spatial tasks and men's left hemispheres to linguistic tasks, whereas women may have more diffuse areas devoted to both kinds of tasks." But the broadest review to date recently contradicted this claim. Reporting meta-analyses of four different reflections of cerebral dominance for language, Iris Sommer and colleagues (2008) found no evidence of greater lateralization related to language in men than in women.[3] A meta-analysis of spatial abilities also found a more complex pattern than Lippa and others have suggested. For example, only some kinds of spatial tasks are processed preferentially in one hemisphere or the other, left-handers seem to be lateralized differently from right-handers, and the gender differences found in studies so far are small enough that a few unpublished studies that do not show gender differences could reverse the overall results (Vogel, Bowers, and Vogel 2003). And even the modest findings to date might overstate the case of gender differences. As Kaiser and colleagues (2009) recently reported, the general preference for publishing "positive" findings in scientific journals is exacerbated in sex/gender research by defaults in the design and analysis

of studies that "inevitably lead to the detection of differences rather than to the detection of similarities" (49).

5. Finally, it is true that some human behaviors and personality traits, unlike brain anatomy, can be roughly sex-typed, but no behaviors or traits are so close to dichotomous as are genitals. For another thought experiment, imagine that you have available to you as test subjects a thousand people whose gender is unknown to you, and you give each of them a test for verbal fluency and a test of spatial relations. It is likely that, on average, the women in the group will score slightly higher than the men on the first test, and the men will score slightly higher on the second. But these group differences will be small, and there will be great overlap in the scores, meaning that the range of scores for women will be almost identical to the range of scores for men. For any given score, you would have a very hard time guessing whether it came from a man or a woman—you could not assign a gender to that score like you can assign a sex to genitals. What if you gave your subjects a test for mental rotation ability, which is considered to be the largest and most reliable gender difference in cognitive ability? For any randomly selected score, you are only likely to be able to guess the sex of the test taker roughly 60 percent of the time. Even if you take a mental rotation score on the extreme end (which, by definition, is not a score that most people of either sex would get), your chances of guessing the test taker's gender would improve only to roughly 75 percent.[4] That's not terribly impressive for something that is supposed to signal a distinctively masculine cognitive skill.

THE POINT here is not to say that there are no sex differences in behavior or cognitive skills; on average, when large groups of people are involved, there do seem to be some small but real sex differences (though this does not mean such differences are innate, as I will consider at length in Chapter 10). The point, instead, is that the nature of differences between the sexes is such that it is misleading to talk about "male-type" or "female-type" cognitive patterns, even though it is currently popular to do so. There is simply too much overlap between the sexes, and too much variation in traits and skills within each sex, for that sort of categorical reference to be meaningful.

The one domain where human behavior arguably more closely approximates dichotomous male and female patterns is sexual partnering. Although significant minorities of both men and women sexually desire and have sex with partners of their own sex, and some people have sex with both sexes, most people most of the time seem to desire and have sex with partners of the other sex. So, sexual orientation seemingly offers more promise than other domains for studying whether hormones shape humans' brains to be "masculine" (oriented to women) or "feminine" (oriented to men). Scientists working on brain organization believe that human sexuality, especially sexual orientation, is the most strongly sex-typed of all human traits and behaviors. It is also the domain where the quasi-experimental network is potentially the closest to the true experiments done on nonhuman animals, where most of the studies focus on aspects of mating behavior.

But "close" is a relative term, even in horseshoes and hand grenades. Returning to the example of the relationship between cigarette smoking and lung cancer, imagine that various studies used entirely different definitions for lung cancer. It would be hard to conclude that smoking caused lung cancer if there was little agreement on what lung cancer is in the first place. So what kind of agreement is there on what constitutes masculine sexuality, feminine sexuality, homosexuality, or other sex-typed characteristics that are thought to result from different kinds of hormone exposures on the developing human brain?

Consider how homosexuality is defined in research on nonhuman animals. "Gay sheep" were coincidentally on the front page of the *New York Times* (Schwartz 2007) one day when I was drafting this chapter; they join the lesbian seagulls, gay penguins, and other much-discussed animals who seem to mate or have sex with partners of the same sex. The "gay sheep" study mentioned in the headline is of particular interest, because it explores how rams' sexual behavior may relate to early hormone exposures and subsequent brain alterations—that is, it's based on brain organization theory (Perkins, Fitzgerald, and Moss 1995; Resko et al. 1996; Pinckard, Stellflug, and Stormshak 2000). But, as I note in later chapters, it's important to dig under the headlines and examine very closely what is meant by "gay" even when it is applied to diverse groups of humans, let alone when applied to nonhuman animals. It rarely means the same thing across species. In the case of these sheep, "homosexual" rams met a three-part test for "preferentially" mounting other rams, but never mounting ewes (Resko et al. 1996, 121). In the case of rats, it has most often meant males who show an increased frequency of either displaying lordosis or being mounted by other males (for example, Dörner et al. 1976). Note that these criteria aren't just different, they are totally contradictory, even though both of them are entirely behavioral.

Human sexuality is particularly complex because it involves additional layers of intention, self-concept, and elaborate social constraints and rewards beyond that in other animals. Even keeping the focus on behavior, human sexual behavior has, by most sociological accounts, undergone seismic shifts in the past five decades. The degree of sexual interest, desire, and activity that is currently considered healthy and normal for American women, for example, was in the 1950s seen as so unfeminine as to indicate mental problems. As sexuality research becomes more sophisticated, human sexuality only seems to look more, rather than less, complicated. Sexual orientation, for example, isn't one singular thing, but comprises sexual desires, affectionate feelings, behaviors, and even how someone thinks of himself or herself socially. People have complex reasons for having sex, and sexual behavior doesn't always accord with what people actually desire. Thus, while the *New York Times* has no difficulty calling some sheep gay, public health experts have had a tremendously difficult time deciding when to use this term for humans. Sexuality experts generally agree that "men who have sex with men," for example, are not exactly the same set of men as those who should be considered "gay" (Young and Meyer 2005).

So research on sexuality, while a cornerstone of brain organization theory and potentially the most direct link to experimental studies in other animals, necessarily involves the measurement of very complex phenomena. Although it is not impossible to study complex behaviors or "traits" such as those involved in sexuality, it requires special carefulness. Scientists have created a whole body of theory and techniques to address the problems involved in studying complex and/or abstract phenomena. This brings us to the issue of measures, which are the working definitions that scientists use in their research.

Measures as Assumption Containers

Measurement is a border area in research, where technical decisions and cultural questions of meaning most clearly come together. In any study, certain ideas about the causes and effects under study are taken as givens or assumed, and these assumptions are embedded in the way that causes and effects are defined and measured. In a study of how early hormone exposures affect mature sexuality, the point is to discern the relationship between hormones and sexuality—not to explore the nature of hormones or the nature of sexuality. But there is no way to study that relationship without either explicitly or implicitly asserting a great deal about the nature of the things being related. Measures are where these assertions are made, because these are the terms that give concrete meaning to the abstract ideas scientists want to study. Ideally, these assumptions are made explicit, which is why the "materials and methods" sections of a research report should always have a description of measures. Measures can be thought of as vehicles through which assumptions travel in studies without being tested.

Scientists can test only what they do not take for granted. That can make studying familiar phenomena particularly challenging. As Evan Balaban, a neuroscientist who studies behavior, notes, "Behavior is all the more difficult to study scientifically because everyone seems convinced that they know what they are studying" (Balaban 2001, 430). This may be especially true of masculinity, femininity, and sexuality, because certain ideas about gender and sexuality are so broadly shared in our culture, including among scientists (Oudshoorn 1994; Fausto-Sterling 2000; Barres 2006).

When I teach scientific methods, I often use the idea of agnosticism to explain the ideal of objectivity in science: an objective approach to a problem requires letting go of fixed ideas about the issues you're studying. Most scientists and virtually all historians of science would agree that this may be a good *prescription* for science, but it is not a good *description* of science.[5] That's because scientists, like all people, have strong expectations about the way the world works, and perhaps especially about the things that we choose to study. Scientific methods are an evolving set of practices that are meant to counter scientists' intellectual and emotional investments in their own research so that their studies avoid getting "rigged" to find outcomes that would please them or confirm their pet theories. That is, scientific methods should give an equal chance for the scientist's hypothesis to be refuted as for it to be supported. But methods that are created to eliminate "bias" (such as random assignment of experimental conditions or blinding) do not totally eliminate the effect of scientists' beliefs and assumptions on their research.[6] One place where both scientists and those who study scientists agree that assumptions always operate is in the process of setting definitions and measures for the phenomena that are studied.

Because brain organization studies are quasi experiments, looking at measures is important not just for understanding what is assumed about sex-typed traits within particular studies, or even what kinds of assumptions you find in this research field more generally. Attention to measures can—in fact, as I argue below, it must—be a part of deciding how well the theory of brain organization is empirically supported by research so far. Close examination of measures is therefore a key part of analyses in this book, especially in Chapters 4 through 8.

The best-known study that focuses on measurement as a way to evaluate scientific knowledge, Stephen Jay Gould's *The Mismeasure of Man*

(1996), illustrates how scientists' expectations can distort research. Gould demonstrated how extremely competent and well-respected scientists systematically erred in their physical measurements of different human groups (divided by race, sex, and social class), such that the measurements corresponded to the scientists' beliefs about the social value of the groups rather than to the physical characteristics they measured. So the scientists literally saw Africans' brains as smaller than European brains, saw "less-evolved" traits in the skulls of women compared with skulls of men, and so on.

Like Gould, I am interested in how scientific measures reflect scientists' assumptions, but my overall approach is quite different from his. In the analysis for this book, I have not looked for mismeasures, but instead have tried to document the *range* of measurement in brain organization research, especially to see the degree of consistency and inconsistency as well as change over time as the field has evolved. Also, I am not so interested in physical measurements (even in studies that correlate sexuality with physical traits like finger-length ratios). Instead, I concentrate on how scientists measure sexual behavior and psychological traits that are thought to reflect certain patterns of brain organization, because measuring these kinds of phenomena poses very particular challenges.

Researchers in behavioral biology are interested in traits, characteristics, or attributes that cannot be directly observed because they are abstractions and simplifications for the way people are—that is, how they behave or feel or think over time and in a variety of different circumstances. The nonobservable attributes that scientists want to know about are referred to as constructs. Hoi K. Suen has explained that "constructs are literally constructed by psychologists to explain some phenomenon" (Suen 1990, 7). Suen uses "math aptitude" to illustrate what a construct is. Whereas the score that someone receives on a single math test is concrete, singular, and directly observable, the notion of math aptitude is an abstraction used to explain or predict how well someone can solve various kinds of tests and problems they encounter. When scientists are trying to understand human development, a single score or a single instance of any behavior is not that informative. They want to know about the underlying characteristic of the person that creates regular patterns in their behavior (the thing that makes it likely someone will score a certain way on next week's test as well as on this one). In practice, the only way we can determine whether an abstract trait is present or not is to use the concrete measures we have (even though we acknowledge that any concrete measure will be at best an approximation of any abstract trait). So the measure becomes the working definition for the trait.

Consider how this works in brain organization research. Focusing just on studies about sexuality, the constructs that scientists want to assess include such traits as "masculine sexuality" or "homosexuality," or perhaps more limited sexual attributes that are thought to signal masculinity or femininity, such as "sexual arousability." When a scientist asks questions about concrete behaviors or responses, such as how often a person thinks about sex or becomes aroused in certain contexts, those responses are not the thing that the scientist is ultimately interested in. Instead, she or he is interested in using that information to draw conclusions about abstract traits such as "libido" or "arousability." It is this sort of abstract trait that is relevant for the following discussion of constructs.

Measurement principles are scientific rules for exploring whether scientists have actually measured what they intended to measure in their research. A great deal of current work in measurement is very mathematical and abstract, but the basic principles all derive from a simple question: Does the measure indicate what it is supposed to indicate? Imagine that one defines the construct "homosexuality" as "an erotic attraction to persons of one's own sex." Now imagine that homosexuality is measured as "the proportion of sexual partners who are one's own sex." The operational definition is different in important ways from the more abstract definition: the question has redefined homosexuality as sexual behavior, while the construct is about "erotic attraction," which may or may not be manifested in behavior. Further, in the abstract definition, opposite-sex attraction is irrelevant, whereas in the concrete measure, same-sex versus opposite-sex attraction are weighed against each other. This example shows that the abstract definition is not necessarily more satisfactory than the operational definition: many people understand being "homosexual" to mean not only that one is attracted to persons of one's own sex, but also that one is significantly more attracted to persons of one's own sex than to persons of the other sex. Otherwise, it might be more appropriate to think of people who are simply attracted to people of their own sex as bisexual rather than homosexual.[7]

Perhaps paradoxically, measurements are strongest where scientists have been uncertain about how best to capture a particular construct. This is because when scientists are not sure their measures are good, they tend to try out various measures, compare alternatives, and gather information about how reliable and valid measures actually are. In a classic text on quasi experiments that covers many measurement issues, Cook and Campbell (1979) note that it is best to test hypotheses with alternative definitions for particular constructs, as well as with alternative measures for particular definitions. Open discussion and mutual critique among scientists may help bring to the surface assumptions that would remain buried if there were no disagreement. It must be stressed that simply using a great variety of measures is not the same as doing actual measurement analysis, where scientists systematically examine the effects of their measurement choices on the results of research. Testing alternative measures is also quite different from the common strategy in brain organization studies of including a great many variables related to psychosexual development and perusing that long list for individual items that are different between hormone-exposed and unexposed people (see, for instance, Ehrhardt, Epstein, and Money 1968; Masica, Money, and Ehrhardt 1971; Meyer-Bahlburg 1977; Kester et al. 1980; Ehrhardt et al. 1985; Lish et al. 1992).[8]

In measurement as in science generally, uncertainty and disagreement are considered fruitful, even necessary, for developing rigorous methods. Yet key scientists repeatedly assured me that the constructs they work with in the realm of sexuality are "common sense." Thus, measures for most traits, especially in the domain of sexuality, have not been seriously debated in brain organization research, with the single exception of how animal behaviors should be used as models for human sexual orientation. A review of that debate will be useful. Because animal studies provide the only experimental evidence on hormonal brain organization, a strong relationship between animal and human studies is especially important for supporting the theory. But brain organization researchers have been criticized by other scientists for their tendency to equate human homosexuality (a preference for same-sex partners) with "sex-atypical" mating behaviors in other animals, that is, behaviors that are more common in the other sex (Meyer-Bahlburg 1977; Byne and Parsons 1993; Fausto-Sterling 2000). In particular this means that female animals who mount other animals with greater frequency than is typical, or male animals who allow themselves to be mounted more frequently than is typical, have both been characterized as "homosexual"—even when they interact with animals of the other sex rather than the same sex, and even though "normal" animals of both sexes mount and receive mounts to some degree.[9] For example, in a 1975 paper, Gunter Dörner et al. wrote that "male rats castrated on the first day of life show predominantly female-like, i.e., 'homosexual,' behavior following androgen substitution" and suggested that this provided a "possible pathogenesis for inborn homosexuality in human males" (1–2). As Byne and Parsons (1993, 231) noted, such reasoning forces us "to conclude that there is only one homosexual when two individuals of the same sex are engaged in sexual intercourse and that the homosexual individual is obvious from the [sexual] position he or she assumes."

Partly in response to this criticism, scientists have increasingly used alternative ways to assess partner preference in animals. One such assessment is called a T-maze or Y-maze. A test animal is placed in a maze with two arms in a T- or Y-shape layout, in which a male animal is tethered or in a "goal box" at the end of one arm, and a female animal is similarly positioned at the end of the other arm. Researchers observe how often the test animal approaches male versus female animals in the maze, and may also observe how test animals behave once they approach the tethered "partner" of choice (for instance, Paredes and Baum 1995; Bodo and Rissman 2007). By introducing these two-armed mazes and other tests (like designating rams as "male-oriented" if they interact only with other rams, and not with ewes), researchers acknowledged that "sexual preference" must be conceptualized and measured in a way that relates more logically to human preference if these studies are to provide informative data for cross-species comparisons.

Significant effort has gone into improving measures in animal research so that animal behaviors can be related to human sexual orientation, but the models are still problematic. (Problems include, notably, the continued use of lordosis as a homologue for female-typical sexual behavior in humans, and the fact that it requires a great deal of inference to suggest that approach behaviors, like those tested with Y- and T-mazes, are a good model of sexual orientation.) Likewise, there is ongoing, productive debate on how best to measure other aspects of personality and cognition that might be affected by hormone exposures, especially handedness (e.g., Holtzen 1994; Satz et al. 1991) and visuospatial abilities (e.g., Hines 2004, esp. 12–16). In contrast, human sexuality as a broad domain, as well as specific categories or types of sexuality (homosexuality, masculine sexuality, and feminine sexuality) are routinely treated as simple, commonsense ideas. There seem, on the surface, to be no disagreements over definitions, no technical issues of measurement to hammer out.

For anyone familiar with historical, sociological, or anthropological research on sexuality, this quietude seems odd, even suspicious. The history of sexuality has literally exploded with creative research in the past three decades, showing that how people experience and express sexuality, as well as expert ideas about sexuality, have undergone vast changes in just the past two hundred years or so in the United States and Europe. To highlight just a few heavily researched topics, interested readers can consult many excellent volumes that historicize notions that often seem quite timeless and "natural" to us, such as sexual orientation (Foucault 1990; Katz 1995), "normal" men's and women's sexual natures and practices (Mosse 1985; D'Emilio and Freedman 1988; Irvine 1990, 1995; Cook 2004), distinctions and similarities between male and female bodies (Laqueur 1986;

Russett 1989; Laqueur 1992), and the relationships among various sites that "signify" sex in an individual—including the physical body; mental, emotional, or behavioral expressions; and relations with others, especially sexual desires and practices (Oosterhuis 1997; Dreger 1998; Terry 1999; Mak 2004, 2006).[10] In addition to historical change, there are difficult conceptual issues in researching sexuality and gender because of extremely different usage of key terms within as well as across disciplines (Sell 1997).

How, then, is it possible to approach an analysis of definitions and measures in studies that examine the possible influence of hormones on aspects of gender and sexuality? Happily, it isn't necessary to resolve these controversies and examine researchers' measures against some abstract "gold standard" definitions. Instead, it is simply necessary to take a close look at the measures scientists use in their work on brain organization. If concepts like "masculine sexuality" or "homosexuality" are indeed simple, then the definitions ought to be pretty consistent throughout the research. But if, as in the comparison of human research and animal research, "homosexuality" in one study is quite different from "homosexuality" in another study, then you can draw two conclusions. First, "homosexuality" is not an obvious concept. Second, studies using different definitions of homosexuality can't be treated as though they are looking at the same phenomenon—you can't simply string the results together and see whether, on balance, the studies show that "prenatal hormone exposures influence the development of homosexuality."

Symmetry Principles

In science, strong measurement should increase the symmetry of meanings at three levels: within each study, across studies that involve the same construct, and from the scientific realm to the larger world into which studies eventually diffuse. Symmetry requires a yes answer to the following three questions about measures:

1. Within particular studies, do the measures used correspond well to the abstract constructs that scientists say they are studying?
2. Across studies that investigate the same phenomenon (such as "homosexuality" or "feminine sexuality"), are the studies' definitions of the construct compatible?
3. Does the definition of constructs within the network of scientific work correspond well to the broadly understood meaning of these

same phenomena outside the world of science? (For example, how does "lesbianism" in the research correspond to "lesbianism" in the larger world?)

Thinking about symmetry at the first level, within studies, is a very interesting philosophical and technical enterprise, but it is an approach that I mostly put to the side in this book. That is because there is a lot of disagreement among researchers about the best definitions, especially for phenomena like homosexuality or "feminine" sexuality, including serious disagreement about whether such phenomena should properly be considered traits that can describe people. In any case, it is not necessary to resolve this question before evaluating the overall evidence for brain organization. Thus, when I examine measures in the remaining chapters, I generally emphasize symmetry at the second and third level.

Regarding the third level, there are no strict scientific principles that require measures to have any correspondence to some term with general meaning outside of science. However, it is important to think about the point at which studies leave the world of science and are interpreted in the context of broad social knowledge, especially because many of these scientists make those links directly in their research reports. In some areas of research, it takes several steps of translation before it is obvious how any particular study might provide generally applicable knowledge. But brain organization studies are big news precisely because no such translation is necessary: they seem to tell us something immediately important about how fundamental differences in human sexuality and gender come about. Measures are rarely, if ever, the focus of news reports about studies, and measures are especially critical because of their relative obscurity.[11] When research findings are reported, abstract constructs, such as "homosexuality" or "feminine sexuality," are discussed. People who hear about the research have direct experience with sexuality, and have their own ideas and sense of what these terms mean. When science seems to pertain so directly to the real world, especially about issues that are socially important, it is critical to consider how well scientists' constructs represent the real-world phenomena to which they are linked.

Note that definitions should not necessarily be exactly the same across studies. A robust relationship between influences and outcomes should withstand minor changes in definition. In fact, research evidence about abstract traits will be more trustworthy if various different definitions have been used, so that the researcher can be sure that minor idiosyncrasies in measurement are not responsible for the relationships they demonstrate in their research. But the definitions across studies that purport to look at the same thing should not be so inconsistent that it is questionable whether they capture the same construct.

At the second level, comparison across studies, there is another critical aspect of symmetry, which has to do with the overall relationship among variables. Not only must "inputs" (in organization theory, specific hormones) and "outputs" (particular traits or behaviors) be consonant across various studies, but the relationships between inputs and outputs must be generally consonant across studies. Importantly, this must include studies with different designs, as well as similar designs. In other words, it is not enough that two different studies should relate "testosterone" to some second variable like "spatial relations," or that "spatial relations" are symmetrically conceived in the two studies. Most obviously, it is also important that the direction of the relationship should be the same (that is, if testosterone increases certain abilities in one study, but decreases the same abilities in the other, the two studies certainly don't support one another). But other kinds of information must also be considered, like timing of exposures or evidence about dose–response relationships. For example, consider a set of studies that look at testosterone exposures within certain, defined exposure periods (say first versus second versus third trimester of pregnancy). Imagine that one study finds testosterone exposure relates to spatial abilities, but only exposures during the first trimester, and a second study finds that only testosterone exposures during the third trimester relate to spatial abilities. These studies are not, in fact, symmetrical. This is particularly important when studies test a great many relationships, as is the norm in brain organization research, because the higher the number of relationships tested, the greater the odds that some relationships will be "statistically significant" only by chance (this issue is raised for a number of specific studies in the chapters that follow).

This is especially crucial for quasi experiments. Quasi experiments are all in some way "weaker" than true experiments, but in addition, particular study designs have particular limitations. Quasi-experimental principles demand that study designs should be varied to offer a balance of strengths and weaknesses, not constantly reproduce the same strengths and weaknesses (Cook and Campbell 1979). Extending this rule to symmetry principles, it is reasonable that the symmetries among inputs and outputs should hold not only for studies with similar designs, but for studies that are significantly different.

Thinking about brain organization research with this principle in mind requires that we first identify the major study designs in this network, and think about the particular limitations for each of these designs relative to experimental research. This would allow us to then group studies for analysis so that we can be certain that symmetry is not limited only to subgroups of studies that have the same sorts of limitations. (The lack of such comparison is the biggest gap in prior attempts to take stock of brain organization research in humans.) As noted above, there are three criteria that make experiments, such as clinical trials, scientifically powerful: (1) the scientists control the exposure to the drug, so they have clear information about timing of exposures, dosage, and the exact chemical composition of the drug being studied; (2) people are randomly assigned to one condition or the other, so that receiving the drug instead of the placebo cannot have anything to do with any particular characteristics of the people in the study; and (3) a double-blind procedure is used, so that neither those studying the drug nor those receiving it know, for any given participant, whether that individual is receiving the drug or a placebo, so their beliefs (including both hopes and fears) about the drug cannot affect how they perceive it to be working.

Of course, no human study on brain organization meets these criteria. Each of the different research designs comes closer in some respects to these criteria than others, but none of them fully satisfies even one of the criteria used in experimental research. As noted in Chapter 2, the studies can be grouped into two broad sets: those that follow people for whom there is some information about early hormone exposures to see how their adult sex-typed traits develop, and those that work backward from adult sex-typed traits, looking for clues about the prenatal hormone environment. The first set offers the advantage that most of the people studied are actually known to have had unusual prenatal hormone exposures; unfortunately, many factors other than possible brain organization effects may result in people with unusual prenatal hormone exposures being "different" in terms of sexuality or gender—including the fact that many have unusual genitals, and, given that the theory of brain organization was popularized so long ago, just knowing that someone had unusual prenatal hormone exposures might have led parents, clinicians, and others to expect that they would have atypical gender or sexuality. The second set of studies, on the other hand, has an entirely different set of strengths and weaknesses. In these studies, there is no way to obtain direct information about the prenatal hormone environment. On the other hand, this also means that there are none of the confounding variables (unusual genitals, expectation of gender-atypical behavior) that go along with knowledge of unusual prenatal hormone exposures. Further, because research subjects in the second design aren't drawn from clinically unusual populations, the findings might be more applicable to a general understanding of how human gender and sexuality develop.

Summary

These are the principles that guide the analysis in the remainder of this book. To properly evaluate the evidence for brain organization theory in humans, we must remember the following:

The studies in humans are quasi experiments, not experiments.
Studies must therefore be evaluated within specific relations to one another.
There must be symmetry in both inputs and outputs.
The symmetry must hold for study designs that have different limitations and strengths.

The next two chapters present brain organization studies in two broad groups, according to study design. In them, I present a best-case scenario for brain organization evidence to date, since I do not yet closely attend to the measures within particular studies. Instead, the focus is simply on grouping studies to examine the overall shape of evidence for broadly similar sets of inputs and outputs, while mostly taking the measures and investigators' conclusions about their findings at face value.

CHAPTER FOUR

Thirteen Ways of Looking at Brain Organization

A CTUALLY, there are more than thirteen ways of looking at brain organization in humans, but all of them are sideways. Because brain organization experiments in humans are out of the question, scientists have
come up with dozens of creative ways to deal with the fact that they can't
take the full frontal approach. This makes it challenging to assemble the
evidence into a coherent picture. This chapter begins that complex assembly, by arranging and examining the cohort studies. There are many different study designs within this group, but they have one thing in common:
the scientists know something about subjects' prenatal hormone exposures
at the outset. Either they know that the subjects have had unusual hormone exposures (because of clinical conditions or because their mothers
were given hormones while pregnant), or there are some physical hormone
measures available (such as from amniotic fluid).

Within the cohort studies, there are two major approaches. In those cohort studies that concern intersex conditions or exogenous hormone exposures, people who had the unusual prenatal hormones are compared with
other people whose exposures were presumably normal. For example,
children or adults with intersex conditions may be compared with people
who attend the same clinics but do not have any hormone-related conditions, or they may be compared with their unaffected siblings or other relatives.[1] In the studies where there is some direct measure of hormones,
there is a different technique. Instead of using a control group, the sci-

entists see whether, within the entire group, the levels of certain hormones correlate with aspects of sex-typed behavior. They ask, for example, whether children whose samples of amniotic fluid contained more testosterone score in more or less "masculine" ways on measures of sex-typed play.

To comprehend the patterns within these many cohort studies, I use the most basic symmetry principle outlined in Chapter 2, grouping studies so that research based on similar kinds of inputs and outputs can be considered together. In this chapter I won't yet delve into aspects of design that might make the various studies either more or less convincing on their own, or other symmetry issues, such as the similarity of definitions or measures—those analyses are left for other chapters. While I do highlight some disagreements among scientists who conduct this research, my aim here is to present a very straightforward summary of findings, grouped according to how the studies are structured and what investigators report having found. It is, therefore, a best-case scenario for the evidence of brain organization from cohort research in humans.

This chapter and the next will set the stage for the more detailed analysis I present in subsequent chapters. The parade of studies can be somewhat mesmerizing, and if a reader simply focused on the number of studies where some effect of hormones had been found, it would read like a slam-dunk case in support of brain organization theory. But, to anticipate the conclusion of this chapter, a more systematic evaluation yields a very different picture: once studies are grouped to reflect similar inputs and outputs, it is clear that there are very few behavioral domains where there have been even marginally consistent findings. Thus, this first level of analysis can function as a winnowing of all the massive data, allowing us to then look more closely at those domains where evidence does seem to support brain organization theory.

Studies of Intersex Conditions

The story that began with John Money and his colleagues picks up where the first researchers outside of Johns Hopkins addressed the organization hypothesis as an explanation for human sexual behavior. In the early 1970s a clinician experienced in treating patients with intersex syndromes in the Soviet Union published two research reports that were pertinent to the organization hypothesis. Arye Lev-Ran (1974a, 1974b) rejected the organization hypothesis, noting that his own intersex cases agreed with Money's early conclusion that "gender is a postnatal and learned phenomenon"(1974a, 392). Likewise, Lev-Ran reported that his intersex patients developed normal gender identity, heterosexual orientation, and psychological profiles that were normal for their sex, in spite of having had cross-sex hormone exposures. Lev-Ran insisted that such findings were incompatible with the idea that prenatal hormones meaningfully affect later sexual behavior, even though Money himself had begun to champion the influence of prenatal hormones. Others, though, suggested another interpretation. Women in the Soviet Union, where strong taboos against homosexuality and open discussion of sexuality in general prevailed, might be less likely to report same-sex orientation or even to develop it in the first place, "regardless of predisposing factors" (Hines 1982, 64).

Lev-Ran's studies were the last to emphatically hold that postnatal socialization overrides biological factors in gender and sexual development.[2] Instead, the trend was quite the opposite, as scientists asserted a much greater role for hormones than Money and his collaborators envisioned. In 1979 Julianna Imperato-McGinley and colleagues at Cornell University Medical Center in New York, as well as in the Dominican Republic, published a study that would define the opposite end of the spectrum from Lev-Ran, claiming that next to the organizing effects of prenatal hormones, rearing and experience are all but irrelevant for shaping gender and sexuality (Imperato-McGinley et al. 1979). The study described male "pseudohermaphrodites" with 5-alpha reductase deficiency (5-ARD), a syndrome in which androgens are in the normal range but an enzyme necessary for male genital development is not present, so genetic males with 5-ARD are often born with slightly ambiguous, but more female-looking genitalia. When testosterone levels increase at puberty, adolescents with 5-ARD develop masculine secondary sex characteristics (beard growth, deepening voice, increased muscle bulk) and some enlargement of the genitalia (Hochberg et al. 1996). Because fetal androgen levels are not affected by the 5-ARD syndrome, brain organization theory suggests that genetic males with 5-ARD should have masculinized brains in spite of their ambiguous genitalia. Consistent with the theory, Imperato-McGinley and colleagues reported that the subjects they studied had developed adult male gender role (including heterosexual orientation and behavior) in spite of having been reared as girls until puberty.

Imperato-McGinley's was the first research report to suggest not only that hormones have a discernible effect on adult psychosexuality, but that the "natural sequence of events" is for masculinizing hormones to *totally override the social experiences* associated with being reared as female (1233). Imperato-McGinley threw down the gauntlet with this article, directly attacking both Money's conclusions from his studies of matched pairs of intersex patients and the clinical protocol that developed out of those studies. Her later studies of the syndrome reiterate this position, using evidence from an additional case in Italy (Imperato-McGinley et al. 1980), five from Papua New Guinea (Imperato-McGinley et al. 1991), and seven from Mexico (Mendez et al. 1995). Her core claim was that the activating effects of pubertal hormones call forth the gender patterns that were "organized" in the brain by prenatal hormones, and that gender socialization during childhood is largely irrelevant.

In contrast, anthropologist Gil Herdt and physiologist Julian Davidson examined this condition among the Sambia of New Guinea (1988) and reached a completely contrary conclusion. Using an anthropological approach to examine the context of rearing experiences and gender role among Sambia "turnim-men" (as people with 5-ARD are referred to in that culture), Herdt and Davidson show that these individuals are neither unambiguously reared as girls, nor unambiguously "male" after puberty. Instead they occupy a culturally specific gender category, which is characterized by limited formal participation in male roles but is distinguished from the category "men" in terms of language and expectations (for instance, they are referred to with a specific third-gender pronoun, they do not participate in the mature male role of initiation rites, and they are permitted to marry only in rare circumstances where their presumed sexual inadequacy is not considered to be a serious problem).[3]

The 5-ARD studies are especially interesting because they reveal an important schism between brain organization researchers who believe that hormones actually *determine* adult sexual behavior to a great extent, and those who believe that the role of hormones is more subtle. Much of the disagreement over 5-ARD in particular revolves around the extent to which any of the subjects in a given study were reared unambiguously as girls, because the apparently abrupt gender change described by Imperato-McGinley would be less impressive if parents and important others had already anticipated that affected children would likely masculinize at puberty. One psychologist who has studied the condition, Peggy Cohen-Kettenis, has suggested that it is especially likely that children with 5-ARD will be reared as girls in communities where the syndrome is rare, but she also notes that "in isolated communities where consanguinity (intermarriage) is rather common, the prevalence can be very high" (Cohen-Kettenis 2005, 400). The Pidgin term *turnim-men* and the existence of a special Sambian phrase *(kwalu-aatmol)* for people with 5-ARD suggest that the syndrome is well recognized among the Sambia (Herdt and Davidson 1988). Similarly, the terms *guevedoce* (roughly "balls at age 12") and *machihembra* ("first woman, then man") indicate that the syndrome is well known in the Dominican Republic (Ehrhardt 1985; Hines 2004). Cohen-Kettenis's review of nearly thirty studies reporting on a total of 140 individuals with 5α-reductase deficiency found that 56 to 63 percent of those "assigned initially as female" eventually decided to live as men (Cohen-Kettenis 2005)—but note that she collapses the more nuanced gender role of "turnim-man" (Herdt and Davidson 1988) into "men" in order to arrive at these figures. Further, because the figure includes an unknown, but possibly quite substantial, number of cases in which the child was expected to masculinize at puberty, the rate of true gender change may be significantly lower. Another clinical expert on the condition maintains that "almost all children with 5-ARD are assigned a female gender at birth" and "most of these individuals continue living in the female role and gender identity" (Wilson 2006, n.p.). In any case, both estimates and interpretations of the rate of gender change in this group differ sharply even among scientists who generally agree with brain organization theory. Some, echoing Imperato-McGinley et al. (1979, 1991), flatly claim that "exposure of the brain to testosterone during development and at puberty appears to have a greater impact in determining male gender identity than do sex of rearing and sociocultural influences" (Swaab 2004, 302). Others suggest that alternative explanations, including ambiguous gender assignment where later masculinization is expected, economic and social benefits of changing to a male gender in some settings, and the fact that affected children may have a generally masculine appearance regardless of assigned gender role, should also be considered when people with 5-ARD change gender (Herdt and Davidson 1988; Hines 2004; Cohen-Kettenis 2005).

Scientists have raised similar concerns in relation to studies of other intersex conditions, especially congenital adrenal hyperplasia, or CAH. As I've described in earlier chapters, CAH is a genetic disorder that causes overproduction of androgens from the adrenal glands. It is the most common cause of genital ambiguity (White and Speiser 2000) and is more heavily researched than any other intersex syndrome.[4] CAH affects both males and females and can occur in both a "classical" early form or a milder, late-onset form. Scientists interested in brain organization have mostly examined genetic females with classical CAH, because high prenatal androgens constitute an extreme "cross-sex" hormone environment in these cases. In genetic females, high levels of testosterone and dihydrotestosterone cause various degrees of genital masculinization, from a larger-than-usual clitoris to a fully formed penis; labia can be fully fused so that they resemble a scrotum (albeit an empty one). A recent review of published research suggested that nearly a quarter of genetic females with classical CAH are initially assigned as males (Dessens, Slijper, and Drop

2005).[5] Early management consists of hormone treatment and surgery (usually multiple surgeries) to try to create more feminine-looking genitals by removing or reducing the size of the clitoris, opening the labia, and creating or enlarging the vagina.

Beginning almost immediately after Money and colleagues began attributing the "masculine" behavior of women with CAH to prenatal hormones, critics pointed to the many unusual experiences of CAH girls as an alternate explanation, especially the possibility that girls with CAH are treated differently because of their masculine genitals (Quadagno, Briscoe, and Quadagno 1977; Longino and Doell 1983; Bleier 1984; Fausto-Sterling 1985). Nonetheless, scientists tend to agree that there is overwhelming evidence of behavioral and psychological masculinization in genetic females with CAH (for reviews, see Reinisch, Ziemba-Davis, and Sanders 1991; Berenbaum, Duck, and Bryk 2000; Hines 2003; Cohen-Bendahan, van de Beek, and Berenbaum 2005). Interestingly, in spite of the broad consensus *within* the brain organization literature, the most comprehensive recent medical review of CAH (by investigators who do not do psychosexual research) suggests that "the influence of prenatal sex steroid exposure on personality is controversial" (White and Speiser 2000, 253). Breaking down the research findings on women with CAH into the specific domains (categories of outcomes) that are studied will help explain the controversy. (A far smaller number of reports include information on genetic males with CAH, and expectations about how CAH might affect males' brains and behavior are less clear-cut, so these are described after the studies in genetic females.[6])

Early reports suggested that "tomboyism," IQ, assertiveness and aggression, rough-and-tumble play, specific cognitive abilities, and sexuality were masculinized in girls with CAH (Ehrhardt, Epstein, and Money 1968; Money and Lewis 1966; Ehrhardt, Evers, and Money 1968). Later reports, however, have contradicted many of these findings, though most studies do find masculinization in the broad domains of childhood play behavior, aspects of gender identity, and sexuality. Consider first those areas where initial reports have not held up over time, such as most cognitive abilities. Initial reports that girls and women with CAH had higher levels of general intelligence (as measured by IQ) were apparently based on biased samples, as later studies found IQ to be either lower in individuals with CAH (Helleday, Bartfai, et al. 1994; Johannsen et al. 2006) or not different from that of unaffected siblings or appropriately matched community controls (Quadagno, Briscoe, and Quadagno 1977; Dittmann et al. 1990, "Congenital Adrenal Hyperplasia I"; Sinforiani et al. 1994; Kelso et al. 2000; Malouf et al. 2006). Though most reviews no longer hold that IQ is higher in girls or women with CAH, the claim persists that the high early androgen exposures in this condition increase specific abilities, particularly spatial abilities. Yet the data do not support this claim either. In terms of overall spatial abilities, most studies have found females with CAH to have either typical or impaired spatial abilities rather than enhanced spatial abilities (Hines, Fane, et al. 2003; Resnick et al. 1986; Hampson, Rovet, and Altmann 1998; Perlman 1973; Helleday, Bartfai, et al. 1994; Johannsen et al. 2006; McGuire, Ryan, and Omenn 1975; Hurtig et al. 1983; Malouf et al. 2006).[7] If there is any effect on spatial relations from the early androgen exposures experienced by women with CAH, the effect is extremely specific and subtle. But studies to date do not suggest any differences at all, because the results are not consistent even for tests of fairly specific abilities such as mental rotations. For example, while Resnick and colleagues (1986) found enhanced mental rotation abilities among CAH women, three later studies found that women with CAH and unaffected women do not differ on mental rotation abilities (Hines, Fane, et al. 2003; Malouf et al. 2006; Helleday, Bartfai, et al. 1994), and one found significantly impaired abilities in women with CAH (Johannsen et al. 2006). Hines and colleagues (2003) did find that targeting abilities were enhanced in girls with CAH compared with unaffected girls, but this finding has not been replicated.

Similarly, verbal abilities, language lateralization, perceptual speed, handedness, and the discrepancy between performance and verbal IQ—all of which are hypothesized to differ in males and females because of early androgen exposures—do not consistently show the hypothesized shift toward "male-typical" cognition in CAH-affected females, and mostly do not differ between CAH-affected and unaffected girls or women (Resnick et al. 1986; Sinforiani et al. 1994; Mathews, Fane, et al. 2004; Johannsen et al. 2006; Malouf et al. 2006). Nor have the early reports of increased aggression among women and girls with CAH (such as Ehrhardt, Epstein, and Money 1968) been borne out by subsequent studies (Helleday et al. 1993; Berenbaum, Duck, and Bryk 2000; but compare Berenbaum and Resnick 1997, in which older CAH-affected girls and women showed increased aggression on some measures; also see my discussion in Chapter 8).[8] CAH girls are also not "masculine" for other traits that may be associated with aggression, such as competitiveness, assertiveness, or dominance (Hurtig and Rosenthal 1987; Dittmann et al. 1990, "Congenital Adrenal Hyperplasia I").

Childhood play behavior is another story. The domain "childhood play behavior" is a broad composite, including energy expenditure as well as the sorts of toys, games, and playmates a child prefers. When considered *en masse*, the results for childhood play behavior look quite impressive: fourteen of eighteen studies that examine aspects of play behavior find that CAH girls are more masculine in at least some dimension. However, when the specific domains are considered separately, there are more negative than positive findings for most particular aspects of play. The most robust finding of play differences between girls with CAH and unaffected girls is that the former are more likely to prefer playing with so-called boys' toys, such as building blocks or vehicles, rather than so-called girls' toys, such as dolls or cooking implements (Berenbaum and Hines 1992; Pasterski et al. 2005). But there are not consistent differences between CAH-affected and unaffected girls for any other aspect of play, including likelihood to prefer playing with boys (Berenbaum and Snyder 1995).

The results on rough-and-tumble play are especially interesting, in part because this variable has often been understood as lending itself well to cross-species comparisons, and animal studies have shown that administering androgens increases "rough" play in several species, including some primates (see, for example, Goy, Bercovitch, and McBrair 1988). Although parents' impressions are often that their CAH-affected girls play particularly roughly, investigators acknowledge that these reports may be affected by parental expectations. The most systematic study of rough-and-tumble play in humans comes from videotaped observations of play behavior, comparing children with CAH to unaffected children of their same sex. Though the measure was clearly sensitive enough to show differences (there were large sex differences between unaffected boys and unaffected girls in rough-and-tumble play), the CAH girls were not different from unaffected girls on any of the four measures for this construct (Hines and Kaufman 1994). A related concept is the idea of "intense energy expenditure," which was reported in early studies to be increased in CAH girls (Ehrhardt and Meyer-Bahlburg 1981). No studies in the literature have supported this finding with reliable measures of energy expenditure (see, for example, Dittman et al. 1990, "Congenital Adrenal Hyperplasia I," esp. 416).

The second domain in which there are consistent differences is sexuality. The very first report on adult women with CAH suggested that this was an area where prenatal androgens had left a permanent impression: "The incidence of homosexual inclinations, the erotic response to perceptual material and the personal freedom in sex, indicate behavior more often found in males than in females" (Ehrhardt, Evers, and Money 1968, 121). Many reports since then have broadly supported the finding of "altered" sexuality in women with CAH (Money and Schwartz 1977; Dittmann, Kappes, and Kappes 1992; Zucker et al. 1996), but the differences have not always been in the same direction as the early reports suggested. Kuhnle and Bullinger (1997), for example, studied a relatively large sample of 45 adult women with CAH and found that, relative to age-matched controls, CAH-affected women had "impaired" (that is, more conservative and negative) attitudes toward sexuality, fewer sexual relationships, and no increased homosexual preference. But decreased sexual interest in men, lower marriage rates, increased rates of sexual fantasy or behaviors involving women, and various other kinds of sexual differences as compared with other women contribute to a continued general impression that sexuality is "masculinized" in women with CAH (for example, Money, Schwartz, and Lewis 1984; Mulaikal, Migcon, and Rock 1987; Dittmann, Kappes, and Kappes 1992; Zucker et al. 1996; Hines, Brook, and Conway 2004; Wisniewski et al. 2004). Because of the broad consistency of findings in this domain, it is one of the areas to which I devote in-depth attention: Chapter 6 is a detailed symmetry analysis of studies on "masculine" and "feminine" sexuality.

Gender identity is the third area where girls and women with CAH may consistently differ from other women, though most studies suggest that the differences are relatively small. Gender identity includes "core" gender identity, that is, one's basic sense of self as male or female (or, rarely, as neither), and more indirect or peripheral aspects of identity that might reflect one's degree of certainty, comfort, and happiness with one's gender, as well as any possible fantasies, wishes, or plans about being the other gender. Although all studies to date show that the vast majority of genetic females with CAH who are reared female develop an unremarkable "female" gender identity, many studies do indicate higher rates of "masculine" gender identity, as well as gender identity disorders, in this group (Ehrhardt, Epstein, and Money 1968; Meyer-Bahlburg et al. 1996; Slijper et al. 1998; Dessens, Slijper, and Drop 2005; Reiner 2005; Gupta et al. 2006; Meyer-Bahlburg et al. 2004 and 2006). Most studies, though, find that effects of CAH on gender are subtle, with no higher rate of actual gender dysphoria (changing, or wishing to change, one's assigned gender) (Zucker et al. 1996), even if there are slight differences in such indicators as "reduced satisfaction with the female sex" (Hines, Brook, and Conway 2004) or being "more likely to question their female gender" (Wisniewski et al. 2004). Two fairly large studies using standardized measures of gender identity find no difference between CAH-affected and control women (Kuhnle and Bullinger 1997) or affected and control girls (Berenbaum and Bailey 2003).

Most studies of males with CAH have come up empty-handed: there are clearly no effects from the hyperandrogenization in CAH males on sex-typed toys, playmates, childhood activities, or gender identity (Berenbaum and Hines 1992; Berenbaum and Snyder 1995; Hines, Brook, and Conway 2004; Pasterski et al. 2005). One study found that boys with CAH had *reduced* rough-and-tumble play relative to other boys, which is counter to the prediction of brain organization theory (Hines and Kaufman 1994). Two studies also counter the idea that the increased androgens in this syndrome would give these boys an extra edge in spatial abilities, instead finding them to have lower scores on spatial tests than control males (Hampson, Rovet, and Altmann 1998; Hines, Fane, et al. 2003), though one study does report that CAH boys have higher rates of left-handedness and a more masculine cognitive pattern, both in accord with the theory's prediction (Kelso et al. 2000). Two studies have reported that CAH males show a more masculine physical pattern, the relative length of the second to the fourth digit on the hand (this is called the "2D:4D ratio," and it is described in detail in the chapter on case-control studies, because many such studies examine this ratio) (Okten, Kalyoncu, and Yaris 2002; Brown et al. 2002). Finally, two studies show that "masculine" personality traits or disorders are no more common in this group, in spite of their very high androgen exposures (Berenbaum et al. 2004; Knickmeyer, Baron-Cohen, Fane, et al. 2006).

In addition to 5-alpha reductase deficiency and CAH, one other intersex condition has been studied extensively in regard to brain organization: androgen insensitivity syndrome (AIS). As the name indicates, people with this condition are "insensitive" to the androgens in their bodies. Testosterone and other steroids responsible for masculine differentiation of the genitalia and (later) secondary sexual characteristics are present, but due to malfunction of the hormone receptors, the tissues do not respond. Thus, these genetic males develop entirely female-appearing external genitalia. Because they appear to be unambiguously female, people with AIS are usually not diagnosed until adolescence, when failure to menstruate causes concern (Wisniewski and Migeon 2002). John Money and colleagues treated women with AIS as a "perfect contrast" to those with CAH—the former are genetic males with the functional equivalent of no androgen (it's present but can't be used), and the latter are genetic females with an overabundance of androgen (Money, Schwartz, and Lewis 1984). Adolescents and women with AIS have repeatedly been shown to be absolutely female-typical in their behavior and psychology, in spite of having the male XY chromosome pattern (Money, Ehrhardt, and Masica 1968; Wisniewski et al. 2000; Hines, Ahmed, and Hughes 2003). (In keeping with these women's self-identification and social and physical presentation, I generally refer to them simply as "women with AIS" rather than the more cumbersome phrase "genetically male women with AIS.") Although this does indicate that there is apparently no direct effect of the Y chromosome on sex-typed behavior, it is less informative in terms of the theory that androgens "organize" the brain to be masculine. The difficulty, as brain organization scientists themselves agree (Cohen-Bendahan, van de Beek, and Berenbaum 2005) is that women with AIS have been reared unambiguously as females, so the effect of androgen insensitivity cannot be analyzed separately from the effect of female socialization in these cases. The experience of women with AIS does provide some insight into potential organizing effects of estrogen, so I return to this syndrome later, when I consider evidence about the role of estrogen.

Normally Androgenized Males Reared as Females

There are several conditions in which genetic males have penises that are absent, extremely small, or damaged beyond repair. Nearly all children with these conditions presumably have had normal androgen exposures, and most have normal sensitivity to androgens. These include congenital conditions such as cloacal exstrophy (a rare, grave birth defect in which the bladder and/or intestines are outside the abdomen, and the penis is either absent or severely malformed), congenital micropenis (often caused by a partial insensitivity to androgens, or PAIS), and the rarest of the three, traumatic accidents (ablatio penis). In earlier chapters I described the case of David Reimer (aka "John/Joan"). Reimer was born a normal male child, but was reassigned female after a botched circumcision destroyed his penis. He later reassigned himself as male. The enormous publicity given to this case, both as it was initially reported by John Money (1975) and after it was revised (Diamond and Sigmundson 1997; Colapinto 2001), has given the situation of sex-reassigned children without hormone disorders extraordinary weight in both public and scientific evaluation of brain organization theory (see also Money and Ehrhardt 1972; Kessler 1998; Fausto-Sterling 2000; Butler 2001; *NOVA*, "Sex: Unknown," 2001; Pinker 2002; Bailey 2003; Bostwick and Martin 2007).[9] The prominent psychologist Steven Pinker goes so far as to call cases of sex reassignment in cloacal exstrophy and ablatio penis "the ultimate fantasy experiment to separate biology from socialization," because such children have had normal levels of androgens in fetal development but are reared as girls (2002, 348). Thankfully, there are very few such cases.

Many scientists and laypeople interpreted David Reimer's situation as showing that Reimer's inability to adjust to living as a girl was caused by the prenatal hormone exposures he experienced as a normal male fetus (for instance, Diamond and Sigmundson 1997).[10] Some scientists have been more cautious, noting that the late timing of the reassignment, as well as parental ambivalence about the gender change, may have played a significant role. Others have pointed to additional possibilities, such as rebellion in response to the "heavy-handedness" with which parents and clinicians enforced femininity for this child (B. L. Hausman 2000). As Susan Bradley and colleagues noted, "Ultimately, all one can conclude is that the experiment of nurture eventually failed, but why it did cannot be determined" (1998, e11).[11]

Ever since David Reimer's self-reassignment to male was publicized, claims and counterclaims about the relative importance of "neurological programming" by prenatal androgens have flown back and forth, even among scientists who generally accept hormonal brain organization as adequately proven. Milton Diamond, Keith Sigmundson, and William Reiner are among the most vocal proponents of a clear and determining role for hormones. In their report of Reimer's self-reassignment, Diamond and Sigmundson suggested an immediate revision of the clinical management protocol for infants with ambiguous or damaged genitalia. That protocol had previously favored female assignment in most cases where a penis was deemed inadequate. This is because a penis has been understood to be indispensable for developing a male gender; constructing a functional penis had until recently been impossible, and it is still considered more difficult than constructing a functional vagina (Kessler 1998; Migeon et al. 2002).[12] Challenging this protocol, Diamond and Sigmundson wrote that genetic males "born with a normal nervous system, in keeping with the psychosexual bias thus prenatally imposed, should be raised up as a male" (1997, 300). Likewise, based on a review of gender outcome in various children with "disorders of sexual differentiation," William Reiner (2005, 549) concluded that "prenatal androgen effects appeared to dramatically increase the likelihood of recognition of male sexual identity independent of sex-of-rearing. Genetic males with male-typical prenatal androgen effects should be reared male." He even hints that male rearing may be preferable in genetic females with CAH, "if prenatal virilization is moderate to severe" (551). These views are echoed in several additional case reports and reviews (Ammini et al. 2002; Reiner and Kropp 2004; Bostwick and Martin 2007).

Other clinicians and researchers, however, maintain that the balance of evidence, including that from cases of gender change after traumatic loss of a penis, tells a more complicated story. In another case of a penis destroyed by a botched circumcision, the details were remarkably similar to those of David Reimer, except that gender reassignment happened much earlier, at 7 months. At 26 years old, she had a clear female gender identity, even though she had been aware since age 12 of her birth sex and reassignment history. But there are caveats. The clinicians reporting this case consider it of "importance" that she "had a strong history of behavioral masculinity during childhood and a predominance of sexual attraction to females in fantasy," going on to compare her with girls and women with CAH, saying that both have experienced "prenatal androgenization of the central nervous system" (Bradley et al. 1998, e12). Still, they caution against simply assuming that her gender behavior and sexual orientation, even if not gender identity, were set in stone by early androgen exposures, because this woman had pursued sexual relationships with men and indeed had sought additional vaginoplasty to make heterosexual intercourse easier. In sum, the only conclusion they felt comfortable making from this case is that "it is possible for a female gender identity to differentiate in a biologically normal genetic male, which supports the original conclusion of Money et al. that sex of rearing may be the most important determinant of a person's gender identity" (Bradley et al. 1998, e12).

Two recent reviews comparing gender outcomes in genetic males with "inadequate" penises, some assigned as female and others as male, came to similar conclusions. Tom Mazur (2005), for example, found that the vast majority of individuals with partial androgen insensitivity or congenital micropenis maintained their initially assigned gender, whether male or female. Of those who did change gender (9 of the 99 partial androgen insensitivity cases), the switch was as likely to be from male to female as in the other direction. Mazur firmly concludes that "the best predictor of gender identity outcome in adulthood is the initial gender assignment" (419). A second review, published by clinicians at the Pediatric Endocrinology unit at Johns Hopkins, likewise concluded that "either male or female sex of rearing can lead to successful long-term outcome for the majority of cases of severe genital ambiguity in 46, XY individuals" (Migeon et al. 2002, 1).

Heino Meyer-Bahlburg, a longtime associate of Anke Ehrhardt and one of the most prolific researchers of brain organization in humans, comes down somewhere in middle of this controversy. He has been among those who have complained of serious methodological problems in reviews of gender outcome to date, including the fact that too many conditions with dissimilar clinical profiles are reviewed together, statistical analyses are biased by either unknown or incorrect baseline estimates, and cases of gender change may be overreported, given that some investigators (especially Diamond) actively seek out such cases (Migeon et al. 2002; Zucker 2002;

Meyer-Bahlburg 2005). In a more stringent test of brain organization theory, Meyer-Bahlburg recently conducted a comprehensive analysis of published cases of genetic males born with "severe genital abnormalities despite a presumably normal-male prenatal sex-hormone milieu" (2005, 423). Seemingly directing his analysis at those who would see hormones as all-important, Meyer-Bahlburg pursued a novel approach, pointing out that if gender is entirely determined by biological factors, then "there should not be any difference in gender identity outcome between female-assigned and male-assigned patients of the same syndrome" (424). Because he found significant differences in gender outcome between male-raised and female-raised patients, he wrote that "one must conclude that gender assignment and the concomitant social factors have a major influence on gender outcome." But he also affirmed a substantial role for hormones, since the female-reared individuals had a higher rate of gender change than those reared male; in the latter cases, in fact, there was no record of patient-initiated gender change (432).

Brain organization studies of intersex syndromes and of other genetic males reared as female are summarized together because they share several interpretive problems. The controversy recounted above highlights the difficulty in deciding whether psychosexual differences among intersex people are due to the direct effect of hormones on the brain, or to other factors like the indirect effects on behavior via the development of atypical genitals, or the experience of illness and multiple surgeries.[13] There is another, very specific problem with those cases in which a normal male child is reared female, though no published critiques that I have seen have raised it. There is a serious ontological and epistemological difficulty with assuming that parents and others can willfully replicate the "naturalized" view of gender. Briefly, I mean to indicate by this that social science evidence on belief systems (Goffman 1963; Garfinkel 1967), as well as gender theory (for instance, Kessler and McKenna 1978; Kessler 1998; Butler 1993) would suggest that it matters a great deal whether one believes that a child's gender has been changed from its original, "true" form, versus coming to an understanding about what the "best" gender is in cases of uncertainty. I explore this point in detail in Chapter 9.

In any case, some scientists see the many interpretive difficulties as invalidating any conclusions about brain organization in humans from intersex studies, whereas those who conduct brain organization research view the problem more mildly (for instance, compare Ehrhardt and Baker 1974, 48–49, with Bleier 1984, 98–103). Nonetheless, scientists long ago searched out less confounded ways to answer questions about brain organization. Journals have continued to publish a steady stream of reports on intersex people, especially people with CAH, but other study designs now account for a much greater proportion of current brain organization research.

DES: Opportunity from a Tragedy

The drug diethylstilbestrol, or DES, provided just such a new angle. DES, the first synthetic estrogen to be available in pill form, was widely prescribed to prevent miscarriages from the 1940s until around 1970, when it was decisively linked to a rare form of vaginal and cervical cancer in young women who had been exposed to the drug in utero. Given the increased risk of cancer, and solid evidence that the drug did not in fact prevent miscarriage, DES was banned in the United States in 1971 (it was still sold and prescribed in other countries until the 1980s). Before it was banned, some two to four million pregnant women received DES. For researchers interested in the possible effects of hormones on the developing brain, the DES tragedy also presented an opportunity, because the popularity of the drug resulted in an unprecedented number of children who had unusual hormone exposures before birth. Even more importantly, DES did not cause any obvious malformations of the external genitals.[14]

A new development in brain organization theory made the study of prenatal DES exposure especially important. Brain organization theory originally suggested a simple model of brain differentiation involving androgens like testosterone. Recall that in this model, a fetus of either genetic sex would develop both physiologically and behaviorally as female unless it was exposed to sufficient quantities of testosterone during the critical organizing period. Animal studies in the 1970s and 1980s suggested that the theory needed to be modified to take into account the chemical conversion (aromatization) of testosterone to estrogen in the brain. Bruce McEwen's group at Rockefeller University and Roger Gorski's group at UCLA, among others, reported that blocking the conversion of testosterone to estrogen during the critical period of brain differentiation prevents the development of male-typical behavior in the rat (McEwen, Lieberburg, Chaptal, and Krey 1977; Clemens and Gladue 1978; Davis, Chaptal, and McEwen 1979; Gorski 1979). The upshot of the modification was that, although the substances are thought to affect distinct aspects of sex-related behavior, both androgen exposures *and estrogen exposures* are believed to lead to more male-typical brain organization.[15]

It may seem puzzling that all fetuses aren't masculinized by maternal estrogens and, in the case of females, also by estrogens produced by the developing ovaries. However, circulating estrogens are generally blocked from reaching the brain, though the mechanisms of this blocking are different in different species (Hines 2004). DES, a synthetic estrogen, has slightly different properties and is able to enter the brain, similar to the effect of estrogen that is produced in the brain by the conversion of testosterone.

These findings about estrogen effects on the brain emerged around the same time that prenatal DES exposure was recognized as a risk for rare cancers and some other serious health problems. Special DES registries and screening clinics were set up around the country to help track the effects of DES exposure and improve the chances of detecting cancer and other problems at early, treatable stages. Thus, DES-exposed men and women became a very attractive study population for understanding the effects of prenatal hormone exposures: millions of people were exposed in utero, and the registries made it possible to locate subjects and even, in many cases, to determine specific information about dosage and duration of hormone exposures. Crucially, for all the problems that DES did cause, it did not cause ambiguous or obviously atypical genitalia.

The first study to examine the offspring of DES-treated pregnancies appeared in 1973, a few years *before* studies in small animals suggested that early estrogens, like androgens, would increase male-typical and decrease female-typical behaviors. The investigators, Irvin Yalom and his collaborators Richard Green and Norman Fisk, expected that DES—an estrogen—would *feminize* the male offspring of treated pregnancies. Though the researchers acknowledged a variety of imperfections in their study design (such as inadequate control groups and the inability to conduct observations blindly as to whether the subject was in the experimental or control group), and dozens of comparisons yielded only a few significant differences between groups, this study is generally taken as evidence that DES feminized the boys who were exposed (Yalom, Green, and Fisk 1973).[16] In direct contrast to this finding, shortly after animal studies suggested that DES would masculinize male behavior, a study comparing various hormone-exposed and unexposed groups suggested that prenatal DES *masculinized* male offspring (Kester et al. 1980).

Like this latter study, most DES research has proceeded from the assumption that DES would shift behavior *away* from the usual female pattern in such sex-related characteristics as sexual behavior and activity, sexual orientation, and cognitive abilities. But, complicating the elegantly simple idea that "masculinizing hormones"—be they androgens or estrogens—masculinize the body and behavior in both males and females, some investigators build upon research showing that early estrogen exposures *inhibit* male differentiation (for example, Whalen and Edwards 1967; Kaplan 1959; Beral and Colwell 1981). As a result, there is a great deal of internal contradiction in the subset of brain organization studies that examine the effect of early estrogen exposures on human sexuality, personality, and cognition. Some studies predict masculinization in both males and females with early estrogen exposures, and others predict masculinization of females but feminization of males. Interestingly, the findings reported in any given study seem to fairly neatly track the preferred hypothesis of the researchers, insofar as there are any apparent differences between hormone-exposed and unexposed groups. This theoretical disagreement would, no doubt, have withered away if decisive evidence emerged to support one theory or the other. In fact, however, there are no behavioral domains that consistently have been shown to be affected by early DES exposures in either sex.

As with the studies on intersex syndromes, most of the studies in the literature seem to suggest that DES does have some effects on psychosexual characteristics, but the various studies conflict regarding both the direction of effects and which specific traits are affected. The findings on cognitive traits such as verbal ability, spatial relations, or general intelligence have been almost entirely negative; most scientists now agree that DES has no effect on these (Hines and Sandberg 1996). On sex-typed interests, such as parenting or childhood play preferences, the findings are contradictory, including a slightly more masculine pattern in DES-exposed women (Ehrhardt et al. 1989), no effects (Lish et al. 1991, 1992), and a more feminine pattern in DES-exposed women (Bekker, Vanheck, and Vingerhoets 1996). On sexual orientation, the findings are just as mixed. For example, studies from the team of Ehrhardt and Meyer-Bahlburg found that, compared with controls, women prenatally exposed to DES reported more bisexuality or homosexuality (Ehrhardt et al. 1985; Meyer-Bahlburg et al. 1995). In contrast, a much larger cohort study (including nearly 4,000 DES-exposed women, compared with less than 100 in all of the other studies combined) found no significant effect on sexual orientation, though DES-exposed women were, if anything, somewhat *less* likely to have any same-sex partners (Titus-Ernstoff et al. 2003).

Another set of measures that scientists use in brain organization research is related to lateralization, because this is also believed to show sex differences. Measures of lateralization include, for example, left- versus right-hand preference, listening tasks to determine whether one ear is more sensitive than the other, and brain-imaging studies to determine whether one side of the brain is dominant when processing certain language or spatial tasks. In addition to more men being left-handed, it is believed that men are more "lateralized," meaning that men are likely to rely on one side or the other for certain tasks, while women use both sides more or less equally for a given task. Recent meta-analyses suggest that men are indeed more likely to be left-handed, but have *not* found the hypothesized sex difference in lateralization for language-related tasks (Sommer et al. 2008), and find only equivocal evidence that men are more lateralized for spatial tasks (Vogel, Bowers, and Vogel 2003). Nonetheless, greater male lateralization is assumed in all brain organization studies that examine aspects of lateralization. Several teams have examined whether DES influences laterality, and again the results have been quite mixed. The only study to report on handedness in DES-exposed men did find a slight increase in left-handedness (Titus-Ernstoff et al. 2003), but this same study—the largest by far in the DES-exposure literature—found no effect on handedness in women. In contrast, the only study that reported other measures of lateralization in DES-exposed males found them to have a "feminized" pattern (Reinisch and Karow 1977). And while the largest study found DES to have no effect on handedness in women, several smaller studies found prenatal DES to be associated with a slight left-shift (an increase in the masculine pattern) (Schachter 1994; Smith and Hines 2000; Scheirs and Vingerhoets 1995).

In sum, though studies of DES are often said to demonstrate organizing effects of hormones in humans, these studies do not, as a whole, suggest any cohesive picture either theoretically or empirically. Neither the existence of effects for any specific domain, nor the direction of effects that should be expected, has been well demonstrated.

Before leaving the issue of evidence for the organizing effects of estrogen on the human brain, it is instructive to return to the syndrome of complete androgen insensitivity (CAIS). Recall that genetically male CAIS patients are insensitive to androgens, so they develop female-typical genitalia. They also tend to have highly feminine secondary sex characteristics (such as full breasts), because the testosterone they produce is converted to estrogen (via aromatization), which stimulates feminine development at puberty. But because people with CAIS produce normal levels of testosterone, which they are capable of converting to estrogen, and because their estrogen receptors obviously function, it is important to consider these cases when evaluating the idea that estrogen masculinizes or defeminizes the human brain. In spite of *higher than typical estrogen levels* during brain development as well as throughout childhood, women with CAIS have been consistently characterized as feminine in terms of gender identity and role, sexual orientation, personality, and cognitive profile. Cohen-Bendahan and colleagues note that the evidence from CAIS indicates that there is a "negligible behavioral role for aromatized estrogens" in human development (Cohen-Bendahan, van de Beek, and Berenbaum 2005, 360).

Hormones in Nonclinical ("Normal") Samples

Moving on from DES, the last variation on the basic cohort design consists of research on nonclinical populations.[17] Rather than comparing people with unusual hormone exposures to "normal" controls, these studies try to determine whether variations in prenatal hormones that are within the normal range can be correlated with normal variations in sex-typed traits. In other words, given that not all genetic males have identical prenatal androgen levels, an interesting question might be whether those males whose prenatal androgens are low-normal are more or less masculine than males with mid-range or high-normal androgens.

These studies seem, at first blush, to be the most direct possible test of the prenatal hormone hypothesis, short of an actual randomized experiment. Yet there are several difficulties with this design. First, it isn't possible to take hormone measures directly from a fetus, so several different proxy indicators of hormone exposures have been used, including hormones in maternal blood, amniotic fluid, or cord blood (which, depending on the sampling technique, may represent the mother's rather than the fetus's blood). Another proxy indicator is the presence of a same-sex versus a cross-sex co-twin, because animal studies have indicated that both physiology and behavior may be altered by the number and proximity of male versus female littermates. There is disagreement about which of these might best indicate fetal hormone exposures, or indeed, if any is a good measure (Cohen-Bendahan, van de Beek, and Berenbaum 2005). Second, hormones may fluctuate during gestation. Other than the sex of co-twins, each of these proxy indicators is generally only available for one time-point, but that time-point may not be relevant for some (or perhaps any) aspects of sex differentiation.[18] Still, these studies definitely avoid the biggest problems with studies of intersex syndromes, namely, the unknown effects of intensive medical and psychiatric management in those syndromes, and the fact that parents, clinicians, and eventually intersex individuals themselves may *expect* cross-sex behavior, due to ambiguous genitals or to widespread knowledge about brain organization theory itself.[19]

To date there have been six different studies of prenatal hormones and later behavior in nonclinical populations, resulting in over a dozen separate published papers. The three earliest papers were from the Stanford Longitudinal Study and linked gender-related traits to hormones in blood samples taken from the umbilical cord at delivery (Jacklin, Maccoby, and Doering 1983; Marcus et al. 1985; Jacklin, Wilcox, and Maccoby 1988). Notably, most of the samples in this study were venous blood, which represents maternal rather than fetal circulation (Maccoby et al. 1979). One finding concerned "timidity," measured via infants' responses to fear-provoking toys. Jacklin and colleagues (1983) reported that girls were somewhat more timid than boys in this 6- to 18-month-old group, and that in boys, lower timidity was associated with higher testosterone in the cord blood, as well as to higher progesterone and lower estradiol. There were no relationships between any of the hormone measures and timidity in girls. A later report on the same sample, this time on moods, again identified a trait that differed between the sexes (boys were more often rated by mothers as "happy/excited," while girls were more often rated "quiet/calm"), and found that this, too, was linked to hormones in boys but not in girls (Marcus et al. 1985). There are no similar findings for timidity or moods in the literature, perhaps because these aspects of temperament have not been shown to actually differ by sex. But the third paper from this group is more easily related to findings from other studies, as it concerns spatial relations. Contrary to the prediction of brain organization theory, investigators found that girls whose (venous/maternal) cord blood had higher levels of the androgens testosterone and androstenedione were, at 6 years old, likely to score *lower* on a test of spatial ability, compared with girls whose cord blood had lower levels of these "masculinizing" hormones (Jacklin, Wilcox, and Maccoby 1988).

Seven other papers come from two different longitudinal cohort studies of the long-term effects of amniocentesis, one in Toronto and one in the Cambridge, UK, region. Interestingly, as with the Stanford Longitudinal Study, the first of the Toronto papers reported that at 4 years, girls with higher levels of testosterone in amniotic fluid had *lower* scores on several cognitive tasks, including block building, a test of spatial relations that typically favors males (Finegan, Niccols, and Sitarenios 1992). Again, this is contrary to the prediction of brain organization theory. They found no significant correlation between any cognitive tasks and amniotic testosterone levels in boys. In the same group at 7 years, and again only among girls, amniotic testosterone correlated *positively* with some aspects of spatial relations (Grimshaw, Sitarenios, and Finegan 1995), a finding that is at odds with the first result.

It's worth elaborating some of the implications of these studies for brain organization theory. First, male superiority in spatial relations, specifically mental rotations, is one of the supposed hallmarks of sex differences in cognition (Voyer, Voyer, and Bryden 1995). From a young age, males tend to use mental rotation to solve certain spatial problems, while girls tend to use other strategies. (These are group tendencies, not absolute differences between the sexes. The individual spatial strategies, as well as test performance, of individual boys and individual girls overlap quite a lot.) As expected, more boys than girls in this study used mental rotation strategies, but when the investigators controlled for sex, children with higher amniotic testosterone were no more likely to use mental rotation (Grimshaw, Sitarenios, and Finegan 1995). Second, they found that among those children who did use mental rotation strategies to solve the problems, girls with higher amniotic testosterone solved the problems faster (as predicted), but there was a trend for boys with higher amniotic testosterone to solve the problems *slower* (contrary to prediction).[20] Because they found contradictory results for this one aspect of spatial relations among boys and girls, the authors of these studies suggest that prenatal testosterone may work differently to influence cognition in males and females. But keep in mind that this requires selectively focusing on just one subset of children (the roughly one-half who use the mental rotation strategy) and one aspect of spatial relations; it does not resolve the discrepancy between the first and the second study on these children.[21]

The second set of amniocentesis studies is directed by Simon Baron-Cohen, a psychologist who is known for his work on autism. In 2002 Baron-Cohen and his students Rebecca Knickmeyer and Svetlana Lutchmaya began reporting various analyses of testosterone in amniotic fluid and childhood behavior (summarized in Baron-Cohen, Lutchmaya, and Knickmeyer 2004). Baron-Cohen's popular book *The Essential Difference* incorporates this research on fetal testosterone into his theory that "the female brain is predominantly hard-wired for empathy. The male brain is predominantly hard-wired for understanding and building systems" (Baron-Cohen 2003a, 1). Like nearly all scientists who do work on "innate" sex differences, Baron-Cohen traces this "essential difference" to fetal testosterone exposures. To date, his team has reported that amniotic testosterone is negatively correlated with the "female-typical" traits of eye contact and vocabulary size in infants (Lutchmaya, Baron-Cohen, and Raggatt 2002a, 2001), sex-typed interests and quality of social relationships (Knickmeyer, Baron-Cohen, et al. 2005), and measures designed to reflect components of empathy (Knickmeyer, Baron-Cohen, Raggatt, et al. 2006). They did not find correlations between testosterone in amniotic fluid and sex-typed play (Knickmeyer, Wheelwright, et al. 2005). It is worth noting that this team's results disagree with the study by Finegan and colleagues, both because Baron-Cohen's team finds that lower amniotic testosterone predicts female-typical cognition in very young children, and also because they do not find, as both studies by the Toronto team did, that testosterone affects boys and girls in an opposite manner. (While the small sample size in Baron-Cohen's various analyses, especially for girls, might preclude finding statistically significant correlations with testosterone, the direction of effects, as well as general effect sizes, should still be apparent, and these show no suggestion of a pattern similar to that in Finegan's study.)

Two studies use blood samples from pregnant women. From these samples, the levels of both androgens and sex-hormone-binding globulin (SHBG) are determined, the latter because it is thought to be an important indicator of how much "unbound" testosterone is available to pass from maternal blood into the fetal sac and influence development (Udry, Morris, and Kovenock 1995; Hines, Golombok, et al. 2002). Udry's group recruited women whose mothers had participated in a large Child Health and Development Study conducted in the San Francisco Bay Area from the 1960s. The investigators predicted that androgens in pregnant women's blood (available from samples collected during pregnancy and frozen), in combination and interaction with androgen levels in their adult daughters, would relate to the daughters' gender-related behavior (measured by a questionnaire the daughters completed). They did not find the predicted relationship with testosterone in maternal blood. But they did find the predicted relationship when they examined the SHBG level, and they reason that this is a better measure of fetal exposure than is the level of testosterone in the mothers' blood (again, because SHBG would indicate not just how much testosterone was present in a mother, but how "free" the testosterone was to pass into the fetal unit).[22]

In an interesting contrast, Melissa Hines's group recently found that testosterone from a maternal blood sample, but *not* SHBG, was correlated with a standardized measure of children's sex-typical games, toys, and activities, and that this effect was apparent in girls but not in boys (Hines, Golombok, et al. 2002). The effect was statistically significant but quite small, accounting for just 2 percent of the variance in girls' scores on this measure. Notably, neither testosterone nor SHBG in a woman's blood was related to the sex of the fetus she was carrying. For these two reasons, Hines's group (like the Toronto team, but in contrast to Baron-Cohen's team and to Udry's team) argues specifically that testosterone may be related to within-sex differences in gender role behavior, but not to between-sex differences.[23]

The final strategy for looking at how prenatal hormones affect behavior in nonclinical populations is the study of children with same- versus other-sex twins. These studies explore "hormone transfer theory," which suggests that androgens from a male fetus may masculinize the behavior of a female co-twin. Resnick, Gottesman, and McGue (1993) found that girls with a male co-twin reported more "sensation-seeking," a trait on which males typically score higher, though they note that the psychosocial effect of having a male versus a female twin cannot be excluded as the explanation for this difference. Cohen-Bendahan and colleagues (Cohen-Bendahan et al. 2004; Cohen-Bendahan, Buitelaar, et al. 2005) found that girls with a boy co-twin had a more "male typical" pattern of cerebral lateralization (in this case meaning that they were more likely to process auditory stimuli in only one side of the brain, rather than both) and showed a more masculine pattern of "aggression proneness." But three other studies found no difference between girls who have a male co-twin and those with a female co-twin in either femininity (Rose et al. 2002) or sex-typed toy preferences (Henderson and Berenbaum 1997; Rodgers, Fagot, and Winebarger 1998), which is a variable that *should* differ if these studies are to be consistent with findings in CAH (for instance, Berenbaum, Duck, and Bryk 2000). To decide how to weigh these positive versus negative findings, it helps to consider how psychologist Celina Cohen-Bendahan evaluates them. While Dr. Cohen-Bendahan's own twin study did seem to support hormone transfer theory, she has abandoned this research approach because of what she perceives as overwhelming, though largely unpublished, negative results for such studies. In the course of her own research, she was told by many different twin researchers that they had examined their twin data for evidence of hormone transfer from a male to a female fetus and, having found no such evidence, never published the results (Cohen-Bendahan interview, October 2007).

The studies of prenatal hormones and sex-typed behavior in nonclinical populations show a great deal of disagreement across different study populations, and in particular across investigative teams, in terms of which behaviors seem linked to prenatal hormones, the particular hormones involved, the direction of relationships, and whether the relationships apply to boys versus girls versus both. At this point, it would be difficult to claim that there is cohesive evidence for any aspect of sex-typed behavior from such studies.

Summary of Cohort Studies

Thus concludes the roundup of cohort research on brain organization, which includes all the studies in which some indication of prenatal hormone exposures is available. Interestingly, although the vast majority of the studies do find some correlation between prenatal hormone exposures and later sex-typed behavior, many of the domains (let alone more finely grained differences within domains) do not show any consistent pattern of effects.[24]

Deeper symmetry analysis will have to wait until the later chapters, but there is one particular kind of methodological problem with these studies that is important enough to mention now. For two reasons, cohort studies bearing on human brain organization are particularly prone to a kind of statistical problem known as a Type II error, meaning that the hypothesis *appears* to be true, but is not. That means that the chances of finding an association between prenatal hormones and later sex-typed traits is inflated. Nearly all such studies look at many aspects of "sex-typed" behavior, instead of just a few, and compare the exposed versus unexposed groups on each of these traits. This creates a multiple comparisons problem. (To illustrate the problem, imagine a simple game that uses one die. If you need to roll a 5 in order to win the game, you have a 1/6 chance to win with every roll. But if you get to roll the die three times instead of once, you have a nearly 50 percent chance of getting at least one 5.[25] Predicting ahead of time that you will roll a 5 under these circumstances does not make you psychic or even unusually lucky—it is just a matter of making the game very easy.) Multiple comparisons are especially troubling in these studies because so many of the sex-typed traits being examined are themselves interrelated, instead of being independent (Glantz 2002). It makes sense to cast a skeptical eye on "scattered" findings that hormones affect various traits in such studies. Unless there is a very clear pattern of effects on certain well-defined domains, the odds are good that the "statistically significant" results we see are actually due to chance. It's also a good idea to look for domains where there are similar effects demonstrated by different study teams, and based on different study populations (that is, not restricted to a single syndrome, such as CAH).

A brief note on the size of samples in the cohort studies is also in order, because it relates to the multiple comparisons problem. There are statistical methods to correct for multiple comparisons (though this, too, has a limit—at a certain point, conducting too many comparisons is considered "fishing" for associations—if you keep at it long enough, something is bound to turn up). In brain organization research, scientists rarely do these corrections for multiple comparisons; they justify this omission on the grounds that the sample sizes are too small for significant effects to show up with the usual more-stringent guidelines (interviews with Drs. N and S). But brain organization researchers should take a leaf from the epidemiologists' book here, because there are better and less biased methods of dealing with small sample sizes, such as increasing the size of the comparison group. Moreover, it would be far better to conduct analyses on a limited, well-defined set of variables, and examine indicators (such as effect sizes) that may suggest real effects, even if the difference between exposed and unexposed groups is not statistically significant. Consistently demonstrating effects under these circumstances would be much more convincing than showing a scattering of "statistically significant" correlations across very broad domains, when the number of variables examined has run into the dozens or more in many studies.

Considering the findings above, which aspects of sexuality, personality, or cognition are affected by prenatal hormones? For traits affected by androgens, are high levels or low levels masculinizing? Is the answer different for males versus females? What is the role of estrogens? Certainly, the studies reviewed so far defy the simple interpretation that "masculine" hormone exposures in utero lead to "masculine" cognition, sexuality, and personality. There does seem to be one fairly clear answer from the studies so far: we can rule out prenatal *estrogen* as having a major impact on human psychosexuality. We see this by combining the very weak evidence of effects from the DES studies with the consistent results seen in XY individuals who have complete androgen insensitivity (CAIS) but are reared female. As noted earlier, if prenatal estrogen has an appreciable influence on masculine psychosexual differentiation in humans, then XY women with CAIS should be in some sense more masculine than other women. These individuals are instead, if anything, more feminine than is typical, but statistical analyses generally show no psychosexual or cognitive difference between them and normal, genetic females.

In regard to the more complex issue of androgens, some patterns of consistent findings emerge within subsets of cohort studies, but there are major discontinuities across different populations and designs. Studies of intersex conditions, especially CAH, indicate differences between hormone-exposed and unexposed individuals in childhood play behaviors, sex-typed interests (usually folded into "gender role"), and aspects of sexuality, including but going beyond sexual orientation. Despite early suggestions that the differences in play may be due to increased energy expenditure, preference for rough-and-tumble play, or aggression among girls with CAH, these hypotheses have generally been ruled out by later studies using better measures of these domains (see Chapter 8). Sex-typed cognitive abilities, however, may indeed be affected in these conditions, especially CAH, but not in the ways predicted by brain organization theory. Instead, it seems likely that there are very subtle but somewhat general effects, with CAH-affected individuals (male and female) being somewhat more likely to have slight cognitive defects and learning disabilities. Two small studies have suggested that left-handedness may be increased in individuals with CAH, but most studies (including all of the larger studies) have found no effect on handedness or other aspects of cognitive lateralization. Core gender identity is *usually not* affected by CAH. There are higher rates of both gender dissatisfaction and/or ambivalence, as well as transgender/transsexuality in genetic females with CAH, compared with the general population, but most women with CAH have typical gender identity.

Cases where genetic males have been reared as females, on the other hand, show that many such individuals seem to reassign themselves as male. Even among the roughly 40 percent who retain the female gender of rearing, it seems that such individuals are more likely to be sexually attracted to women. This could be an effect of prenatal hormones, but that is not the only possible interpretation, as I discuss in Chapter 9. No studies to date have reported any cognitive findings for this group.

Studies in nonclinical populations so far show no appreciable shifts in childhood play behaviors, core gender identity, or sexuality. There have been preliminary suggestions that sex-typed cognitive skills may correlate with prenatal hormone levels, but findings from different study groups have so far reported only contradictory results in this domain (for example, testosterone relates *negatively* to female-dominant cognitive abilities in the report by Lutchmaya and colleagues [2001], but *positively* to similar abilities in the report by Finegan and colleagues [1992], even though both studies were based on analysis of amniotic fluid). The one domain where there is some convergence for studies of nonclinical populations done by different teams is that of sex-typed interests.[26]

In sum, if childhood play behaviors are understood to be another reflection of sex-typed interests, then this leaves three general areas of sex-typed trait where there is some consistent pattern of findings in at least two of the three subsets of cohort studies. These include (in order of the strength of findings): sexuality, sex-typed interests, and gender identity. From the cohort studies, these are the best possible candidates for showing organizing effects from prenatal hormones. In later chapters, I will use the symmetry principles outlined in Chapter 3 to more closely analyze the evidence that prenatal hormones affect human development in these domains. But first let us finish the broad survey of the other set of brain organization studies. The next chapter examines the evidence for the case-control studies, those that begin with presumably distinct psychosexual "types," and work backward to make inferences about early hormone exposures.

CHAPTER FIVE

Working Backward from "Distinct" Groups

A s WITH COHORT STUDIES, there are many different kinds of case-control designs in brain organization research. All of the case-control studies look for some evidence that unusual early hormone exposures lead to what brain organization scientists consider "cross-sex" sexual orientation or gender (for example, being a woman who is attracted to other women, or having male sex chromosomes and genitals but a female gender identity). Beyond that common factor, the study designs are even more varied than for the cohort studies, including investigations of prenatal stress, neuroendocrine function, neuroanatomy (brain structure), sex-typed cognition and personality traits, handedness and other aspects of later-alization, and even aspects of physical morphology like waist–hip ratios or genital size. The traits on which sexual minority and majority (heterosexual, nontransgendered) people are compared are chosen because each of these traits is also linked, either empirically or hypothetically, with prenatal hormone exposures. Often the empirical evidence for a hormonal influence on a trait that is studied comes from other brain organization studies, and may include findings that aren't borne out by the research as a whole. For example, increased rates of left-handedness among women with CAH as well as women exposed prenatally to DES are often cited as one of the justifications for looking at the relationship between handedness and sexual orientation (Green and Young 2001; McCormick and Witelson 1991; Mustanski, Bailey, and Kaspar 2002), even though there is only weak evidence in this regard for the former condition, and a preponderance of evidence *against* such a relationship for the latter.

Neuroendocrine Function:
The "Estrogen Challenge"

The case-control approach was added to the repertoire of brain organization research designs in 1975 by Gunter Dörner, an experimental neuroendocrinologist working in East Berlin. Dörner traces his scientific pedigree directly to the father of experimental endocrinology, Eugen Steinach (Dörner et al. 1991, 142). As you may recall from Chapter 2, Steinach's work with guinea pigs paved the way for the experiments that launched brain organization theory. Moreover, Steinach's "therapeutic" transplants of testes from heterosexual to homosexual men—though ultimately failures—set an important precedent for later scientists who would seek to apply insights from experimental endocrinology to the human condition. Biologist and historian Marianne van den Wijngaard (1997) has persuasively argued that a great preoccupation with homosexuality was the underlying rationale for pursuing brain organization studies in general; certainly there had been discussion of homosexuality or its absence in all of the studies of intersex syndromes or hormone-exposed progeny. Before Dörner, though, brain organization research had placed homosexuality within a larger framework of masculine or feminine sexuality. Dörner's work brought the concern with homosexuality to the foreground.

Given his training, it seems almost inevitable that Dörner would be a major contributor to brain organization research and that his contributions would focus squarely on homosexuality. Like his intellectual "grandfather" Steinach, Dörner's assumptions included the idea that homosexual men are physically and psychologically "intermediate" between heterosexual men and women. Dörner's first brain organization study built upon experiments in rodents that showed how early hormone exposures shape the cyclic pattern of hormone release that underlies fertility in female mammals. One particular element of this cyclicity is called the "LH surge"—a feedback mechanism involving estrogen and luteinizing hormone (LH). In rodents, one of the most dramatic effects of sex differences in early hormone exposures is the fact that after brain organization, females and males have a different biochemical response to estrogen. In a "feminized" animal—one who had very low androgen levels during brain development—increasing estrogen levels (whether from natural production by the ovaries or from an injection of estrogen) causes a sharp increase in the production of LH. In females, this LH surge is followed by ovulation. (Of course, in males whose hormones are manipulated to produce the LH surge, there are no ovaries to produce and release eggs.) Thus, Dörner reasoned that an animal who responds to increasing estrogen levels by producing a surge in luteinizing hormone has a "female-differentiated brain."

Building on the long-standing notion that homosexual men are psychologically "feminine," and that this is somehow related to hormones, Dörner hypothesized that homosexual men would show the characteristic LH surge. In his "A Neuroendocrine Predisposition for Homosexuality in Men," Dörner claimed that homosexual men displayed the LH surge in response to estrogen injections, and concluded that "homosexual men possess a predominantly female-differentiated brain" (Dörner et al. 1975, 1).[1]

This article was the first in a stream of investigations into a possible relationship between biochemical responses to hormone injections (the LH surge response or the positive estrogen feedback effect) and male and female homosexuality and transsexuality. These studies are variously referred to as testing "neuroendocrine function," "LH surge," "gonadostat," or "response to estrogen challenge." Unlike the "Pepsi Challenge" commercials that were made around the same time as these studies, featuring blinded taste tests between Coke and Pepsi, there were no clear winners in the "Estrogen Challenge" tests. They turned out to not differentiate very well between homosexual and heterosexual men, or conventionally gendered versus transsexual people. Between 1975 and 1989, seventeen studies attempted to establish a link between sexual orientation or gender identity and hormone exposure by some variation on this basic study design. In sharp contrast to studies of clinical intersex syndrome and hormone-exposed pregnancies, many of these studies were pursued by skeptics, and there are more outright negative results in this group than in any other subtype of brain organization research. This is probably also due to the fact that these researchers examined only a few outcomes (at most, three or four measures of hormone response), whereas in the cohorts of hormone-exposed and unexposed people, investigators often examined dozens or even scores of measures of behavior or temperament. (As I explained in Chapter 4, the more variables a scientist looks at in a study, the more likely it is that she or he will find some differences between groups just by chance.)

Though LH surge (also sometimes called "positive estrogen feedback") studies are still sometimes cited as evidence for brain organization, this line of inquiry was abandoned by 1990. A fact that played a large, possibly decisive role in dropping this design was the issue that created much of the skepticism about it in the first place, namely, that rodents are not always a good model for human sexual differentiation. As Hendricks and colleagues summed up the problem, "In primates, contrary to rodents, the capacity of the neuroendocrine system to respond with a positive feedback response to injected estrogen may not be a sexually dimorphic characteristic" (Hendricks, Graber, and Rodriguez-Sierra 1989; see also Gooren 1990). Studies showed that "normal" males could exhibit the positive feedback response in adulthood if primed with enough estrogen. Thus, the LH surge design is probably useless for testing brain organization theory in humans.

Prenatal Stress

Dörner's second innovation in brain organization research, studies of prenatal stress and sexuality, has had more staying power. This design, too, built directly on studies of mating behavior in prenatally stressed rats. In the early 1970s at Villanova University, Ingeborg Ward conducted a series of experiments on pregnant rats that she subjected to extreme stress, by such conditions as intense and continuous exposure to bright light. She found that prenatal stress produced male offspring with "low levels of male copulatory behavior and high rates of female lordotic responding" (Ward 1972, 82). Dörner rephrased these findings in a manner that made them directly relevant to the question of human homosexuality: "bi- or even homosexual behavioral patterns were observed in adult male rats which had been exposed to prenatal stress" (Dörner, Schenk, et al. 1983, 83).[2]

In 1980 and again in 1983, Dörner and colleagues reported that homosexual and bisexual men were very likely to be the offspring of "stressed pregnancies" (Dörner et al. 1980; Dörner, Schenk, et al. 1983). The first study was an attempt to compare the relative birth frequency for homosexual men in the German Democratic Republic (GDR) in times of extreme and widespread stress (during and immediately following World War II) compared with the years before and after the war. His method relied on the fact that venereal disease clinicians in the GDR were required to register homosexual patients. Using these registries and broader birth data for the GDR, he computed that there was a higher frequency of "homosexual births" during the highest-stress years, relative to the total number of males born in each period.

The study was and continues to be highly cited, as well as scathingly criticized. It prompted one of the few direct replications in the literature on brain organization: "Does Peace Prevent Homosexuality?" in which Gunter Schmidt and Ullrich Clement (1990) examined various large-scale surveys of homosexual and heterosexual behaviors in the GDR, along with data on birthrates. They concluded that gay men can "go on loving peace and getting involved in the peace movement" because there was not even "the slightest evidence that wartime stress during the prenatal period increases the incidence of homosexual behavior" (Schmidt and Clement 1990, 186). Further, they suggested that a number of biases in Dörner's initial design (all related to the lack of a random sample from the venereal clinic registries) may explain his findings.

Nonetheless, Dörner's initial study struck a chord, and a number of other approaches to investigating prenatal stress and homosexuality followed in short order. Dörner's group published a second study in which heterosexual, homosexual, and bisexual men were queried about "stressful events that may have occurred during their prenatal life." The men were also asked to consult their mothers. (It's critical to note that the mothers weren't asked directly, though almost all reviews and subsequent citations miss this point. Asking sons to report for their mothers by proxy is a highly unreliable method, and would not be accepted as scientifically adequate in most research fields.)

Again, the results were striking: very few heterosexual men reported even moderately stressful events during prenatal life, and none of them reported any major stressful events; over one-third of the bisexual men reported either moderate or severe stressful events, and over two-thirds of the homosexual men reported moderate or severe stressful events in their prenatal lives (Dörner, Schenk, et al. 1983). None of the subsequent studies using Dörner's basic hypothesis of prenatal stress have approached the almost perfectly linear relationship that he found between increasing levels of prenatal stress and increasing levels of same-sex orientation. In fact, the findings of subsequent studies have ranged from mildly supportive (Ellis et al. 1988) to completely negative (Schmidt and Clement 1990; Bailey, Willerman, and Parks 1991). The largest such study (Bailey, Willerman, and Parks 1991), in fact, not only failed to find a relationship between male homosexuality and prenatal stress, but found slightly *higher* prenatal stress among the mothers of lesbians compared with mothers of heterosexual women. Given these investigators' expectation that prenatal stress would lower male-typical hormone exposures, and their further understanding of lesbians as having male-typical sexual orientation, this finding puzzled Bailey and colleagues, who noted that "the primary obstacle to taking this finding at face value is the absence of an etiological theory. The maternal stress hypothesis for human male homosexuality seems reasonable because stress hormones impede the production of hormones necessary for male sexual differentiation. No such scenario exists for human female homosexuality" (290).

Other investigators have suggested a way to reconcile a link between maternal stress and lesbianism. Melissa Hines and colleagues, for example, have noted that although prenatal stress does not alter the sexual behavior of female rats, it does seem to have some effect on "female-typical" behavior in other rodents, including mice and guinea pigs. In a twist on the "cohort" research that uses measures of prenatal hormones among general population samples to explore whether normal variation in hormones relates to normal variation in sex-typed behavior, Hines tested the prenatal stress theory in a prospective cohort study that enrolled nearly 14,000 pregnant women in the United Kingdom during 1991 and 1992 (Hines, Johnston, et al. 2002). A unique strength of Hines's study is the fact that stress was reported *during* pregnancy, rather than years (sometimes decades) after the fact. In addition to the obvious fact that this eliminates faulty memory, this is important because it eliminates the chance that mothers of gay men and lesbians might recall more stress simply because, as Simon LeVay noted in a critique of the Ellis study, they were "racking their brains to figure out what 'went wrong'" (LeVay 1996, 169).[3] Hines and colleagues did find a very small but significant correlation between prenatal stress and gender role behavior in very young girls (42 months), but found no relationship at all between stress and the gender role behavior of boys. The apparent effect of stress was extremely slight, however, explaining less of the variance in gender than at least nine other factors they measured, leading this team to conclude that "prenatal stress does not influence the development of gender role behavior in boys, and appears to have relatively little, if any, effect on gender role behavior in girls" (Hines, Johnston, et al. 2002, 135).[4]

Returning to the case-control research on sexual orientation, there was one other relevant study of prenatal stress. In 2001, Ellis and Cole-Harding published a second examination of prenatal stress and sexual orientation, this time using a complicated scenario involving not only prenatal stress but maternal use of alcohol and/or nicotine during pregnancy. Ellis again found some relationship between prenatal stress and male homosexuality, but with an important difference: instead of stress during the second trimester, which is what most scientists would speculate is the important period and is the timing he found to be important in his earlier study, this time he tagged the first trimester as critical.[5] In a particularly complicated sequence of events, the study also suggested that nicotine in the first semester, followed by stress in the second trimester, was related to lesbian orientation in female offspring. A bit more information on data

from experiments in other animals, regarding the specific relationship between stress and testosterone levels, adds to the incoherence one might already perceive in these findings. In rats, Ward and others have found that
intense and prolonged stress leads to an initial surge of testosterone, followed by a sharp drop; levels tend to remain low for a long period afterward. Given this two-part scenario, it might be plausible to find associations between different gestational periods for lesbians and gay men. If the
prenatal stress scenario is to explain sexual orientation, then lesbians should
indeed have had earlier prenatal stress than gay men. But the model is supposed to apply to both genetic males and females; there is no theoretical
rationale for achieving this result by introducing some variables (nicotine
and alcohol) for females only. Ellis and Cole-Harding (and the extraordinary number of other scientists who have cited them) blithely ignore all
these pesky details that show their findings don't actually fit the theory.
(The fact that investigators continue to find some positive associations between prenatal stress and sexual orientation likely speaks more to the critique raised by Simon LeVay and others: that same-sex orientation is still sufficiently stigmatized that the mothers of gay men and lesbians are more likely
than other mothers to recall negative events during those pregnancies.)

In sum, research has yielded little support for the idea that prenatal
stress affects later gender-related behavior, including sexual orientation.
Several important scientists who are generally supportive of the brain organization hypothesis (such as Simon LeVay and Melissa Hines) have expressed skepticism about the notion that prenatal stress plays a meaningful
role in sexual orientation. Currently some scientists are so disenchanted
with the prenatal stress hypothesis that they do not even mention this line
of research when they review biological research on the development of
gender, sexual orientation, or sex-typed cognition (for example, Berenbaum 1998; Cohen-Bendahan, van de Beek, and Berenbaum 2005; Rahman 2005). The stress studies are still actively cited, though, both to support the general claim that maternal stress in humans alters behavioral and
cognitive development (Bowman et al. 2004) and the specific claim that
maternal stress leads to "increased occurrence of homosexuality" (Swaab
2004, 308).

Neuropsychology Studies

Neuropsychology as a field concerns the relationship between physical and
biochemical properties of the brain, on the one hand, and, on the other,
aspects of human psychological function, including both behaviors and

other kinds of traits (like lateralized information processing, the processing of a certain type of information in one hemisphere rather than both hemispheres of the brain). Neuropsychologists conducting brain organization studies look for ways in which sexual orientation or gender identity correlates with various other characteristics that they believe are influenced by early hormone exposures. But again, they do not have any direct information about hormone exposures. Instead, these scientists use physical or psychological traits that differ, on average, between men and women as proxy indicators of early hormone exposure. Examples of their proxy indicators include aggressiveness, competitiveness, verbal abilities, spatial relations, handedness, hemispheric specialization (the degree to which one side or the other of the brain is engaged to process certain cognitive tasks), aspects of auditory function (dichotic listening and otoacoustic emissions), and motor tasks such as throwing at a target.

Although these traits are often called sex-typed, it's important to recognize two things about them. First, the sex differences for most are small—you couldn't take an aggression score, for example, and predict with very good accuracy whether that score came from a man or a woman, and you'd have an even worse time trying to predict someone's sex by knowing scores on other cognitive tests, or knowing whether they are right- or left-handed. For all of these traits that are used as proxies for early hormone exposure, the male–female differences are not apparent unless you have very large groups to compare. In fact, as noted in Chapter 3, recent meta-analyses indicate that men are more likely to be left-handed, but men and women do not have different lateralization patterns for language-related tasks, and possibly not for spatial tasks either (Sommer et al. 2008; Vogel, Bowers, and Vogel 2003). Second, the evidence that early hormone exposures are responsible for the sex differences that do exist in these traits is still indirect and often inconsistent. This limits the overall symmetry among studies, as explained in Chapter 3. As noted earlier, the decision to choose a certain trait for contrasting heterosexuals and homosexuals is often made on the basis of isolated findings from other brain organization studies, which don't always hold up when the studies are viewed as a group. Thus, the evidence for brain organization that comes from these neuropsychology studies (as well as the neuroanatomy studies reviewed next) is a step further removed from brain organization theory than the evidence that comes from hormone-exposed pregnancies.

Studies of hand preference provide a good illustration of the general shape of findings in neuropsychology research, because they compose the largest subset and are quite interesting and complex in terms of the hypotheses at play. Recall that in brain organization research, the classic prediction for sexual orientation is that higher levels of prenatal testosterone will be followed by greater attraction to women (the "male-typical" pattern), regardless of the individual's genetic sex. But these predictions get messy in the handedness studies. In these, scientists tend to predict that homosexuality in *either* men or women will be associated with higher rates of left-handedness, which is the masculinized pattern (that is, because men in general are more likely to be left-handed, this trait is thought to follow relatively high levels of prenatal testosterone during the critical developmental period when handedness is determined).

While this fits neatly with brain organization research in the case of women, it includes the somewhat surprising suggestion that gay men have had "hypermasculine" androgen exposures, at least during one portion of their early development. One very highly cited study, by psychologist James Lindesay (1987), draws on a finding of a "left-shift" in handedness among gay men to suggest that gay men are hypermasculine in general, and suggests that a link between (male) homosexuality and left-handedness in humans contradicts animal research that links "gender-anomalous sexual behavior in males" to reduced androgens in development (967). Lindesay looks to an unusual authority—the Old Testament Book of Judges—to suggest that there is precedent for linking hypermasculinity, left-handedness, and same-sex preferences among men: "the Benjamite army at Gibeah contained 'seven hundred chosen men left-handed; every one could sling stones at an hair-breadth, and not miss' (interestingly, the men of Gibeah were punished for the same transgressions as the Sodomites)" (968).

Most other studies reconcile a link between increased left-handedness (a "hypermasculine" trait) and sexual orientation to men (a "feminine" trait) by reference to a model advanced by Norman Geschwind and Albert Galaburda (Geschwind and Galaburda 1985a, 1985b, 1985c). Appeal to the Geschwind-Galaburda model is a more conventionally scientific approach, but it's not necessarily less controversial, because the theory has been widely criticized (for example, McManus and Bryden 1991; Berenbaum and Denburg 1995). Nonetheless, most researchers linking sexual orientation to handedness, especially in men, rely on this model. In brief, the Geschwind-Galaburda hypothesis is an extension of the animal research mentioned above suggesting that prenatal stress causes testosterone to surge briefly and then fall to permanently low levels.[6] If handedness develops first, and sexual orientation develops later, then such a two-stage response to prenatal stress might plausibly link sexual orientation and left-handedness in men. Of course, this still leaves the question of why handedness and sexual orientation should follow this pattern in men (especially given the lack of empirical support for the prenatal stress hypothesis) but not in women. Perhaps handedness develops differently in men than in women? That could be the case, but there are no independent data to suggest that this is so, and it seems to contradict animal data.

Another theoretical possibility may be that both gay men and lesbians have atypically high androgen exposures early on, but the pattern persists for lesbians and reverses for gay men. If so, that should result in a pattern of consistent masculinization across traits in lesbians, but a more mixed pattern of hypermasculine, feminine, and possibly "average" traits among gay men. But again, this is not consistent with the findings for other traits among gay men and lesbians. One prominent researcher believes current evidence even shows that lesbians might be "more female-like than heterosexual women," suggesting that scientists and laypeople should "not get bogged down in thinking of lesbians as sort of like guys. Think of them as killer women" (Dr. A, interview with author, February 5, 1999). In Chapter 7 I look closely at the various ways brain organization researchers think about sexual orientation, especially how they think homosexuality relates to sex differences, and how this matters in interpreting the evidence that the human brain is "organized" by hormones to be masculine or feminine, as well as gay or straight. For now it is simply important to note that brain organization researchers have never systematically mapped out either the specific traits that should and should not be correlated, or the specific sequence of events required to produce the pattern of associations that these studies hypothesize. (I begin to do that mapping in this book, especially here and in Chapter 7.)[7]

Even without rigorously tracing the way evidence about handedness and sexuality *should* fit together to support the idea that prenatal hormones influence sexual orientation in men, the evidence to date is less than overwhelming. There are currently slightly more studies that fail to find increased left-handedness among gay and bisexual men (Rosenstein and Bigler 1987; Marchant-Haycox, McManus, and Wilson 1991; Satz et al. 1991; Gladue and Bailey 1995b; Bogaert and Blanchard 1996; Cohen 2002; Mustanski, Bailey, and Kaspar 2002) than studies that find such an association (Lindesay 1987; McCormick and Witelson 1991; Becker et al. 1992; Holtzen 1994; Lippa 2003). One meta-analysis has suggested that the preponderance of negative results is due to the fact that the difference in handedness between gay and straight men is extremely small, so that most studies have not been large enough to find this link (Lalumiere, Blanchard, and Zucker 2000), but this point is undermined by the fact that the largest studies tend to find no significant associations, and the fact that the inability to select truly random samples of gay men and lesbians may subtly bias these studies.[8]

Fewer studies have examined sexual orientation and handedness in women, but at least on the surface, the data are stronger for a link between homosexuality and left-handedness or "decreased right-hand preference" for women than for men. Three studies do find an association between lesbianism and hand preference (McCormick, Witelson, and Kingstone 1990; Mustanski, Bailey, and Kaspar 2002; Lippa 2003), and the review by Lalumiere and colleagues (2000) did find the larger effect size in women compared with men, which they predicted.[9] Two studies found no difference in handedness between lesbians and heterosexual women (Rosenstein and Bigler 1987; Gladue and Bailey 1995b), though the former of these used an exceptionally poor measure of sexual orientation. Studies of people with atypical gender identity have consistently shown this group to be more left-handed than controls (Green and Young 2001; Zucker et al. 2001; Cohen-Kettenis et al. 1998).

Other traits studied in sexual minorities do not follow these same patterns. For example, consider digit-length ratios, which have now been extensively examined in both men and women. The relative length of the second to the fourth fingers (known as the 2D:4D ratio) tends to be roughly equal in women, but lower in men, reflecting a longer "ring" finger than index finger. There have been seven studies comparing 2D:4D in gay and straight men, and the findings are all over the map. Three studies find higher (feminized) ratios in gay men (McFadden 2002; Lippa 2003; Collaer, Reimers, and Manning 2007), but two other studies found gay men to have *lower* (hypermasculinized) ratios (Robinson and Manning 2000; Rahman and Wilson 2003) and two found no difference in digit ratios between gay and straight men (Williams et al. 2000; Voracek, Manning, and Ponocny 2005).[10] Among women, two studies found lesbians to have "masculinized" digit ratios (Rahman and Wilson 2003; and Williams et al. 2000), but one small study (Anders and Hampson 2005) and the two largest studies (Lippa 2003; Collaer, Reimers, and Manning 2007) found no difference compared with heterosexual women. Only one study has reported a link between 2D:4D and gender identity, and this found a "feminized" pattern in male-to-female transsexuals, but no difference between female-to-male transsexuals and other genetic females (Schneider, Pickel, and Stalla 2006). In sum, if there is any link between digit ratio and sexual orientation or gender identity, it appears to be very small, and to be limited to (genetic) men. The evidence is similarly mixed for other indicators.

Neuroanatomy Studies: Is There a Gay Brain?

Last, but certainly not least, it's important to consider evidence from neuroanatomy, because a small group of studies from this field have been among the most influential in the literature. As with neuropsychological studies, scientists who use neuroanatomy as evidence for brain organization do not actually assess or attempt to control for hormone exposures during some early period. Instead, focusing on brain structures that they believe are different between men and women, investigators look for differences in the brain that correlate with differences in sexual orientation or gender identity. The general brain organization hypothesis is that "male-typical" structural features (like the size, cell density, or shape of a particular brain region) will go with "male-typical" patterns of sexuality or gender identity. But brain structures, even more than psychological traits, don't come in neat "male" versus "female" types. Further, as with the neuropsychology studies, scientists typically *infer* that a high correlation between "male-pattern" sex-dimorphic structures and masculine patterns of sexuality indicates that both the structures and the traits developed as a result of early hormone exposures, though this may not be the case, as I describe below.[11]

Given that these studies have had extraordinarily high citation rates and eager popular coverage, they seem to appeal to people as particularly "concrete" evidence of a biological basis for differences in sexuality. Perhaps that is simply because the design involves actual examination of the brain, which, along with genitals, is the alleged "target" of the organizing effects of hormones. But the reality is more complicated, for several reasons, not least of which is the controversy over the extent and nature of sex differences in the human brain. Brain organization research has mostly focused on regions within the hypothalamus, because studies in rats have indicated the existence of a sex-dimorphic nucleus within the preoptic area of the hypothalamus (SDN-POA, or simply SDN). Based on this rat research, the SDN is relevant to brain organization on two counts: (1) an area of the hypothalamus sexually differentiates under the influence of steroid hormones in early development, and (2) the same area is functionally involved in sex-typical mating behaviors (Gorski 1979).

But locating the human homologue to the rat's SDN proved elusive, indeed. The search for a human SDN has centered on various cell groups in the anterior hypothalamus. The Dutch neuroscientists Swaab and Fliers first reported a sex difference in this region in 1985, and designated it the "sex-dimorphic nucleus" to indicate the presumed parallel with the SDN that Gorski had identified in the rat (Swaab and Fliers 1985). Laura Allen, then a graduate student working with Gorski, later failed to confirm the sex difference reported by Swaab and Fliers, so Allen and Gorski suggested the name "first interstitial nucleus of the anterior hypothalamus" (INAH1), for that area (Allen et al. 1989). Allen and colleagues also identified three other nuclei in the anterior hypothalamus, INAH2–4, and reported that two of these, INAH2 and 3, were sex-dimorphic. The sex difference in INAH3 has been replicated at least twice (LeVay 1991; Byne et al. 2000). Two later studies failed to replicate the sex difference in INAH2 (LeVay 1991; Byne et al. 2000), although some neuroscientists continue to claim that the area is sexually dimorphic (for instance, Koutcherov et al. 2002). Why all the confusion? As authors of a recent developmental study of the human brain explained, "cell groups in the adult human hypothalamus are less circumscribed [than in the rat], and, thus, more difficult to identify" (Koutcherov et al. 2002). (This is exactly the point I raised in Chapter 3—that it is difficult to apply a genital model to understand brain development, because interspecies differences in brains are both larger and more difficult to discern.)

At least four studies have reported attempts to link one or more of these nuclei with sexual orientation. The first of these was conducted by Dick Swaab, the scientist who had initially called INAH1 the "sex-dimorphic nucleus," and his junior colleague Michel Hofman. Swaab and Hofman (1990) reported that, in contrast to the sex difference Swaab had earlier found, INAH1 did not vary by sexual orientation in men. They did, however, find a sexual orientation difference in a fifth cell group in the anterior hypothalamus, the suprachiasmatic nucleus. The finding was never replicated.

The next report to emerge is one of the most influential (and controversial) studies in all of brain organization research. Simon LeVay, then working at the Salk Institute in southern California, released his comparison of homosexual versus heterosexual men's brains in a 1991 article in the journal *Science,* receiving a veritable storm of press coverage. Building on Laura Allen's identification of sex differences in INAH2 and INAH3, LeVay reported that INAH3 was smaller, closer to the size of presumed heterosexual women, in gay men's brains (LeVay 1991). As noted earlier, LeVay found no sex difference in INAH2; neither was there a difference by sexual orientation.

A partial replication of LeVay's work came from somewhat surprising quarters, a team headed by William Byne of Columbia University. Byne has been an outspoken skeptic of brain organization research and has offered particularly pointed criticisms of what he has characterized as sweeping and premature conclusions from preliminary, often ill-designed studies of variations in the human brain by sex and sexual orientation (for instance, Byne and Parsons 1993). Like LeVay, Byne found INAH3 to be smaller in women than in men, with some interesting qualifications. Byne's group found the sex difference in INAH3 size to be accounted for by women having *fewer neurons*. In contrast, the difference between homosexual and heterosexual men was not a matter of the number of neurons, but was a near-significant trend for the former to have *smaller INAH3 volumes* relative to brain weight. So, although there was a difference, the sexual orientation difference did not mirror the sex difference (Byne et al. 2001). Byne and colleagues interpret both findings, but especially the sexual orientation-related difference, as consistent with a pattern that arises *after* the critical period of development, because "in humans, the major expansion of the brain occurs postnatally while the individual is in constant interaction with the environment" (91).

In addition to the INAH studies and the (unreplicated) finding of a difference in the suprachiasmatic nucleus (Swaab and Hofman 1990), one other study has reported a difference in brain structure between heterosexual and homosexual men. Laura Allen, "scooped" by LeVay's research on gay men's brains, added her own finding about a sexual orientation difference.[12] She and her mentor, Roger Gorski, examined the anterior commissure, which is a thick band of tissue that connects the two sides of the brain. While not known to have any specific functions related to sexual behavior or reproductive capacity, the anterior commissure has been found by some teams to be larger in women than in men. Allen and Gorski reported that this structure was also larger in homosexual than in heterosexual men, a "feminine" pattern that they believed was due to female-typical hormone exposures in development (Allen and Gorski 1992). Other studies have failed to replicate that sex difference (Byne and Parsons 1993; Byne et al. 2001), and another study later found that the anterior commissure did not vary by either sex or sexual orientation (Lasco et al. 2002).

Though no studies have examined the brains of lesbians, one study has now looked at the brains of male-to-female transsexuals (Zhou et al. 1995; Kruijver et al. 2000). The study, by Dick Swaab's lab at the Netherlands Brain Institute and conducted in conjunction with Louis Gooren, a renowned clinician-researcher who works with transsexuals, concerns yet another portion of the hypothalamus, the central subdivision of the bed nucleus of the stria terminalis (BSTc). In an initial report, Jiang-Ning Zhou and colleagues simultaneously reported a difference in the volume of the BSTc between nontranssexual men and women, and between transsexual and nontranssexual men. The study is particularly interesting because the scientists had sought to identify "a brain structure that was sexually dimorphic but was not influenced by sexual orientation, as male-to-female transsexuals may be 'oriented' to either sex with respect to sexual behavior" (Zhou et al. 1995, 68). (Most other brain organization researchers, especially in the early years, have missed this point about the variability of sexual orientation among transsexuals.) When Zhou and colleagues found that the six transsexuals they studied had, on average, a smaller BSTc vol-

ume than other genetic males, they concluded that they had found "a female brain structure in genetically male transsexuals," which they said
supported "the hypothesis that gender identity develops as a result of
an interaction between the developing brain and sex hormones" (68).
Swaab's team later refined the analysis, determining that the volume differences they observed reflect a difference in cell number (Kruijver et al.
2000). They also ruled out a relationship between cell number and (male)
sexual orientation, because nontranssexual men, regardless of sexual orientation, had almost twice as many of a specific kind of neuron (soma-
tostatin neurons) as women in this brain area, and the number in male-to-
female transsexuals was similar to that of the women.[13] (It's worth emphasizing that the second paper is an elaboration, rather than a new study, because 26 of the 34 subjects in the main analysis were the same as those
studied by Zhou et al. (1995). Thus, the finding of a relationship between
the BSTc and gender identity has been extended but not replicated.)

In part because the spotlight has shined on them, these studies have also
drawn a great deal of criticism for certain aspects of their design. The
study that has been most heavily criticized is LeVay's examination of a section of the hypothalamus, but the criticisms apply equally well to the other
studies. First, because they are autopsy studies, there is very little information on the subjects' sexual behavior, and none whatsoever on their desire,
which is what most sexuality researchers consider a better reflection of
sexual orientation. Almost all of the "homosexual" men in these studies,
as well as a portion of the "heterosexuals," are people who died of AIDS.
So the designation "homosexual" or "heterosexual" reflects information
that was collected for purposes of classifying HIV transmission, not for
sexuality research. This is far from ideal.[14] Second, a number of critics
have raised the issue of how HIV might have affected the brain areas that
the scientists compare. By including some heterosexual subjects who died
of AIDS, as well as others who did not, scientists can conduct various analyses to try to rule out an effect from the disease. But the underlying virus,
HIV, affects different people in different ways, and none of the studies so
far has actually controlled for the specific clinical diagnoses of different
subjects. (This is especially important because the gay men may have been
less likely to be users of intravenous drugs, and more likely to have other
sexually transmitted infections, than the other men with AIDS—and each
of these covariates may have distinct effects on the brain.) In any case,
nothing short of studying brains from homosexual subjects who have not
died of AIDS can actually eliminate this problem. Third, all of the studies
are very small, and the amount of overlap in size or shape of the brain area
in question between the homosexual and the heterosexual groups in each
case is quite substantial. Fourth, there is no temporal information about

the development of the brain structures in question relative to the development of same-sex orientation. Many scientists simply assume that structural differences predate and are causally related to differences in sexual desire and behavior; others, most notably William Byne and colleagues (2001), point to a large body of evidence showing that brain structures are responsive to experience and retain plasticity well into adulthood, if not throughout life.

In the end, there is one replicated finding linking neuroanatomy to sexual orientation, which is that the portion of the hypothalamus known as INAH3 seems to be larger in heterosexual men than in homosexual men. What to make of this finding is a matter of dispute even between the two main researchers involved: Simon LeVay (1996) suggests it means that gay men have likely had less testosterone exposure in utero, so they have undergone a different brain organization, but Bill Byne and colleagues (2001) believe it is just as plausibly a reflection of influences from later in life.

So what is the take-home message of case-control research on brain organization in humans? Taking the studies reviewed here at face value, there seem to be several research areas that we can eliminate as providing reliable evidence for "organizing" effects of prenatal hormones in humans: neither prenatal stress, nor neuroendocrine function, nor any somatic characteristics are reliably related to sexual orientation or gender identity. While there are some data suggesting a link between sexual orientation and handedness, more studies overall suggest no link, and there are major difficulties in reconciling the suggested empirical pattern of increased left-handedness in both gay men and lesbians with the predictions of brain organization theory. There do seem to be some data suggesting that gay men and lesbians are more likely to have sex-atypical interests and possibly also sex-atypical cognitive skills (especially spatial relations) relative to their heterosexual counterparts. But as I have noted above, these data don't fit into any coherent pattern, either on their own or together with data on prenatal stress, handedness, or digit-length ratios. There is also a portion of the hypothalamus, INAH3, that appears to be larger in heterosexual men, though major questions remain regarding the functional significance of this difference, as well as the timing (especially whether it predates or follows the emergence of sexual attraction and behavior).

Moving Beyond the Best-Case Scenario

As I described in Chapter 3, analysis of quasi-experimental studies requires special attention to studies that use significantly different research designs. In this chapter I have not attempted to consider how the findings from case-control studies on brain organization stack up next to the cohort studies. That is a critical next step. Rather than examining every domain for congruence between the cohort and case-control studies, I limit the cross-design analysis to those domains where there seems to be the most evidence of early hormone effects.

Given the overall pattern of findings in the cohort and case-control studies on brain organization, what domains appear to be the most likely candidates for showing an influence of prenatal hormones? First, a variety of studies seem to suggest that early hormone exposures affect aspects of sexuality, both sexual orientation and broader "masculine" or "feminine" patterns of sexuality. Before we conclude that hormones affect masculine or feminine sexuality, though, it's important to consider just what "masculine" or "feminine" sexuality *is,* and make sure that we consider evidence in a way that compares apples to apples. The same goes, of course, for "homosexuality," which is by far the most common characteristic that has been studied in connection with prenatal hormone exposures. Thus, I devote the next two chapters to a deeper symmetry analysis of evidence about sexuality.

Second, there seems to be evidence from both clinical and nonclinical populations that some aspects of sex-typed interests and cognitive patterns may be associated with prenatal androgens. The basic analysis in Chapter 4 showed that some common claims in this regard are at the very least overstated—for example, if there is a connection between androgens and spatial relations, it is neither simple nor linear. Because the evidence is so weak for the effect of prenatal hormones on sex-typed cognition, and because there are already several excellent critical reviews of this subset of research (including Hines 2004; Spelke 2005), I do not give additional attention to symmetry analysis for that domain.

Sex-typed interests are another story. This is a domain where, based on this surface analysis, there seem to be consistent findings across several categories of studies. A closer analysis of evidence regarding interests is especially important in light of the fact that, to my knowledge, there have been no deep reviews of this research to date. Moreover, claims about innate sex-typed interests are central to the way "male brain" and "female brain" arguments are built in relation to important policy issues. Do schools fail boys if reading lists aren't sufficiently action-oriented (Tyre 2005; Brooks 2006)? Should we eliminate "disparate impact" analyses that support discrimination claims with evidence that there is a lack of parity in certain jobs? Some argue that parity should be dropped altogether as a goal, because they interpret brain organization research to show that men and women "self-select" into occupations based on interests (Holden 2000). I don't directly tackle these policy issues, but Chapter 8 takes a closer look at evidence that is recruited for making these suggestions.

Finally, with a few exceptions, most scientists who study hormone effects on gender identity agree that one's basic sense of self as male or female is not "set in stone" by the level of androgens in utero, but that these androgens do seem to have some effect. The evidence that is considered strongest in this regard comes from intersex syndromes, as well as studies of other conditions in which children are reared in a gender that does not fully accord with their genetic sex and/or their presumed prenatal hormone environment. As I have suggested, these studies have been subjected to a number of criticisms, most frequently the inability to control for the effect of unusual genitals on either parental treatment or one's developing sense of self. But there are critical issues about interpretation of these studies that have not yet been deeply explored, issues that signal a broad shortcoming in current thinking about how sexuality, gender, or indeed personality in general might be connected to both physical bodies and to one's social experiences. For example, what might be the effect of repeated questioning about one's gender, coupled with repeated physical exams, surgeries, and the awareness that one's physique is "different" from others'? I do not devote an entire chapter to the evidence on gender identity, but I explore such questions in Chapter 9. There, I consider how a number of relevant and measurable variables are systematically excluded from analyses of gender identity development. There, and in Chapter 10, I also revisit the evidence about gender identity as one of several strategies for imagining a more dynamic theory of human sexual development.

CHAPTER SIX

Masculine and Feminine Sexuality

ONE OF THE MOST apparently obvious features of masculinity and femininity is that they are different from one another—whether conceived of as two poles of a single continuum (where being more masculine implies being less feminine) or as entirely different dimensions—masculinity and femininity are thought to be sharply different and easily distinguished. In my interviews with the top scientists conducting brain organization research, I heard repeatedly that masculine and feminine sexuality are simply "commonsense" ideas. As one scientist said: "Most people . . . don't have any problem understanding that male sexuality is different from female sexuality. It's a no-brainer" (Dr. A interview, February 1999). Like Dr. A, nearly all scientists conducting brain organization research treat masculinity and femininity as commonsense ideas that don't require explicit definitions.

Most people—scientists included—also tend to think of masculinity and femininity as stable across times and cultures, especially when thinking about how biology might matter to sex or sexuality. Things that are "natural" don't change much over time, at least not in the short run. Evolution is a slow process. But even a cursory familiarity with the history of sexuality—or recent social history more generally—brings the obviousness of masculine and feminine sexuality into serious question. Just in the decades since brain organization research began, the sexual revolution, second-wave feminism, and the gay and lesbian liberation movement have brought big questions into public discussion: How important is sexual satisfaction to women? What gives "normal" women orgasms? How many and what sort of sexual partners is it normal to have—and is the answer different for women and men? Brain organization researchers have been asking, "What *causes* masculine or feminine sexuality?" Many activists, scholars, and regular people trying to work out their own sexual lives have been asking, "What *is* masculine or feminine sexuality?" That discrepancy underlies the central question of this chapter: How have scientists conducting brain organization research come up with definitions and measures for masculine and feminine sexuality, when these concepts are politically contested and seem to be in flux?

As it turns out, these scientists' certainty that we all know what we're talking about, so they don't have to spell it out, masks dramatic changes in how they distinguish between masculine and feminine sexuality. In this chapter, I trace the shifts in scientific definitions by focusing more particularly on feminine sexuality, because the changes all involved a broadening of feminine sexuality to include aspects that previously were treated as clear signals of masculinization. Interestingly, no aspects that were originally considered feminine ever became a part of the "normal" masculine repertoire.[1]

Given that ideas about timeless differences in the fundamental sexual natures of men and women form the conceptual structure for brain organization research, it's useful to point out just a few historical changes before turning to close examination of "feminine sexuality" in the studies. For example, the contemporary idea that men have a much greater sexual appetite than women is the exact reverse of the idea in Renaissance Europe that women were generally sexually insatiable and men were much more capable of controlling their "base desires" with their naturally greater rationality (Laqueur 1992). Another example concerns the type of conclusions one might draw about the sexual habits of a man who is extremely concerned with his appearance and dress. According to current American stereotype, the man is likely to be gay; but according to the common wisdom of early modern Europe, he was likely to be a womanizer (van der Meer 1997). Interestingly, being a womanizer in this earlier period was considered effeminate, because it signaled a lack of manly self-control. So much for the popular idea that a desire for lots of sex and multiple partners has always and everywhere been considered masculine!

Closer to our historical "home," it is somewhat amazing to trace the shift in ideas about "normal" women's sexuality—both psychological and physiological—over the twentieth century. In a 1933 lecture, Sigmund Freud articulated a theory of femininity that was dominant for decades:

"mature" female sexuality required giving up the active, clitoral-focused erotics of childhood for the passive pleasures associated with vaginal penetration: "In the phallic phase of girls, the clitoris is the leading erotogenic zone. But it is not, of course, going to remain so. With the change to femininity, the clitoris should wholly or in part hand over its sensitivity, and at the same time its importance, to the vagina" (Freud 1965, 118). Women who "failed" to experience their greatest sexual pleasure from vaginal penetration were viewed as sexually and psychologically disordered, while healthy "adjustment" to femininity was characterized by "passivity in coitus." Freud's followers even went so far as to suggest that femininity entails masochism, which "impels [woman] to welcome and to value some measure of brutality on the man's part" (Marie Bonaparte, quoted in Ehrenreich 1978, 272). Even Freud's critics, like Alfred Kinsey, agreed that women had a naturally lower sexual "capacity" (though Kinsey attributed this to physiological differences rather than the different developmental mandates for males and females) (Irvine 1990, 49–50).

The idea of men's greater sexual *appetite* has carried through the decades, but the idea of men's greater sexual *capacity* was soundly put to rest in 1965, when Masters and Johnson published their paper "The Sexual Response Cycle of the Human Female." Among their most riveting findings, which helped propel their 1966 book *Human Sexual Response* to best-seller status, was their conclusion that men's orgasmic capacity paled beside that of women, whose capacity for multiple sequential orgasms was usually limited by "physical exhaustion alone" (Masters and Johnson 1965, 64). Likewise, Freudian ideas about "mature vaginal sexuality" were trounced by behavioral sexologists, whose work in the 1950s and 1960s emphasized the importance of clitoral stimulation for the sexual satisfaction of most, if not all, women (Irvine 1990, esp. 161–163; Lloyd 2005, esp. chap. 2).

These issues just touch the surface of recent changes in expert knowledge about feminine sexuality. Note also that the discussion above is limited to mainstream ideas in Western Europe and the United States; the range of ideas about "normal" masculine and feminine sexuality expands greatly when one takes a cross-cultural perspective or considers nonmajoritarian and nonexpert views. Several excellent historical studies, *Intimate Matters: A History of Sexuality in America* (D'Emilio and Freedman 1988), *The Invention of Heterosexuality* (Katz 1995), and *The Long Sexual Revolution: English Women, Sex, and Contraception, 1800–1975* (Cook 2004), richly detail the changing sexual landscape for both women and men, before and into the twentieth century. The pace of change was uneven, and a general pattern of rapid liberalization was punctuated by periods of sexual conservatism. But over the course of the twentieth century, the horizon of sexually acceptable behavior moved farther and farther from the nineteenth-century standard of reproductive intercourse in marriage. From our current vantage point, changes in the visibility and acceptability of same-sex relationships may seem to be the most dramatic sexual change of the past few decades. But heterosexual norms and behaviors have been transformed as well. Each time, norms and behaviors shifted first for men, then for women, with more and more behaviors eventually constituting the "normal" repertoire for both (D'Emilio and Freedman 1988; Cook 2004).

In fact, at the very same time that brain organization studies on humans began in the late 1960s, political and social movements were shaking traditional ideas about sexual behavior, and a revolution in birth control and the legalization of abortion increasingly separated sexuality from reproduction. This separation had a particularly dramatic effect on notions of acceptable—and therefore normal—sexual behavior for women. Consider how the definition of "promiscuous" rapidly transformed from the 1950s to the 1970s, as historian Hera Cook has pointed out:

> In 1972, the *Sunday Times* reported on a study in which 3,000 women who suffered unwanted pregnancies were interviewed and then divided into four classes: permanent cohabitees; people in stable liaisons (anything over three months); temporary relationships (under three months or undefined), and lastly 'those who had disordered casual sex with a variety of partners'. Only the last group . . . was identified as promiscuous. . . . In the 1950s, all of these women would have been considered immoral and all of those not in permanent relationships would have been considered promiscuous." (Cook 2004, 294)

It wasn't just opinions that were changing, but also behavior. Citing demographic figures on cohabitation, marriage, and births outside of marriage in the United Kingdom, Cook concluded that seismic shifts in sexual behavior that began in the 1970s actually "accelerated in the 1980s and 1990s" (338). The end result, which she attributes above all to the availability of contraception, is that "women's behaviour has become closer to that long considered acceptable for men" (338).

Similar changes swept the rest of Europe, as well as the United States. Young men's descriptions, from the 1950s and 1960s, of the emotional content of their sexual desires and behavior are a particularly compelling indicator of just how much changed in the second half of the century. The current popular and scientific view holds that male sexual desire is largely, if not entirely, independent of concerns for love and intimacy, but this "masculine" separation of lust and emotional connection was demonstrably less pronounced just a few decades ago. In interviews about "premarital" sex, some young men justified their sexual activities outside of marriage with reference to emotions, especially love: "We didn't believe in petting because of the sex alone, but because we were very much in love and this was a means of expressing our love to each other" (Rothman 1984). Others felt that it was disrespectful to make sexual advances to "nice" girls, even if that meant settling for little or no sex themselves. "As one college male explained: '. . . the fact that I was never sexual with anyone is because I never dated anyone I didn't care for'" (263).

This is not to say, of course, that sex now is or ever was an equal proposition for women and men. To the present day, the consequences of heterosexual intercourse generally remain higher for women than for men, both because of the risks of pregnancy and because of the damage to reputation that can result from being sexual outside the (expanding, but still gender-defined) boundaries of social acceptability. But from this side of the sexual revolutions of the twentieth century, it is easy to lose track of just how much has changed, and how rapidly.

Surprisingly, against this backdrop of change, most brain organization researchers have used the common term *feminine sexuality* through more than four decades as though it is absolutely self-evident and unproblematic. But the ground has been shifting under their feet. While ideas and practices associated with "normal" sexuality changed in the broader world during those decades, the transformation of masculine and feminine sexuality was just as dramatic in the studies that are intended to determine how male and female sexual natures develop.

In brief, from the late 1960s until around 1980, brain organization researchers relied on a model of human sexuality that sharply divided masculine and feminine sexual natures. Their model exhaustively divided aspects of sexual desire and expression into male and female "forms": initiating versus receptive, versatile versus conservative, genitally focused versus diffuse. As posited in the early studies, feminine sexuality looks frankly Victorian: it is romantic, dependent, receptive, slow to waken, and only weakly physical. Scientists thought of female sexual activity not as an end in itself but as a means for fulfilling desires for love and motherhood. Their picture of masculine sexuality was a mirror image of their feminine model: active and energetic, initiating, dominant, penetrating, frequent, intense, and genitally focused. Masculine sexual activity was assumed to be its own end, unsentimental and undiluted by romance. Certain activities, especially masturbation and having multiple partners, or sex outside of marriage, were unequivocally coded as masculine. Above all, sexual desire for women—only and always for women—was understood to be the central feature of masculine sexuality, while exclusive sexual desire for men was central to the feminine pattern.

Things began to change in the early 1980s. In the most general way, scientists continued to assert that early exposure to "masculinizing" hormones makes sexual development either more masculine or less feminine, and their preoccupation with sexual orientation intensified. But a closer look at the specific behaviors coded as masculine or feminine in the later studies shows some surprising and very important differences from the first period. In particular, masturbation, genital arousal, and sex with multiple partners came to be understood as "commonsense" features of feminine sexuality, even though these had earlier been read as clear signs of masculinization.

The remainder of this chapter will provide detailed examples of how sexual patterns have been interpreted as masculine or feminine in specific studies over the years. Scientists' definitions are usually left implicit. But by painstaking examination of research reports and the materials used in conducting these studies—especially the actual questions asked and the rules for evaluating answers—it is possible to identify the operational definitions at work. What follows is based both on what scientists explicitly said female sexuality is, and also on how they interpreted women's actual behaviors as "masculine" or "feminine."

The survey is divided into two roughly chronologic periods when one or the other model of sexuality was dominant among scientists. Perhaps more surprising than the outright reversal in scientists' ideas about some key components of female sexuality is that scientists (those who conduct research related to the theory; those who review studies for journals, funding proposals, and other publications; and those who lean on the theory for broader arguments about sex differences or sexuality) have apparently never noticed these critical changes. After examining the transformation in scientists' interpretations of sexuality, I'll turn to a consideration of how it matters for brain organization theory itself.

Working Definitions, Take One:
The 1960s and 1970s

Brain organization research on human sexuality was officially launched by John Money's team with a 1968 study of women with congenital adrenal hyperplasia (CAH) (Ehrhardt, Evers, and Money 1968). CAH interferes with the synthesis of the steroid hormone cortisol, causing overproduction of androgens, including testosterone and dihydrotestosterone. As noted in earlier chapters, these high androgen levels cause various degrees of genital masculinization in genetic females with "classic" CAH (as opposed to the milder, late-onset form). Until 1950, when it was found that treatment with cortisone would suppress the overproduction of androgens, women with CAH would undergo progressive virilization, so that the initial masculinization of the genitalia was complicated by the development of secondary sexual characteristics.

Money, whose own doctoral research had focused on the psychosexual development of people with various intersex disorders (then known as hermaphrodites), had been studying people with CAH for over a decade by the time of his first brain organization report. Housed in the Department of Psychiatry and Behavioral Sciences and the Department of Pediatrics at Johns Hopkins University School of Medicine, Money's projects were coalescing into a new "psychohormonal research unit." The team included a young German doctoral student, Anke Ehrhardt, who came to the United States to work with Money and would go on to become one of the most distinguished sexuality researchers of her generation.

Ehrhardt took the lead on a paper about women with CAH whose treatment did not begin in infancy, as is now the norm, but at adolescence or even later. These "late-treated" women had come of age prior to the introduction of cortisone therapy, so "the majority of the sample had lived for many years with the stigma of heavy virilization, sometimes uncorrected genital morphology and lack of feminine secondary sexual development" (Ehrhardt, Evers, and Money 1968, 117). Several features of the article recommend it as an excellent anchor for understanding scientists' ideas about masculine and feminine sexuality in early brain organization studies. First, although the title vaguely referred to "some aspects of sexually dimorphic behavior," the focus of the report was squarely on eroticism and sexual practices. As such, it was the first research report concerning the influence of prenatal androgens on frank sexuality in humans. Second, while the authors acknowledged that late-treated women with CAH provide an imperfect window into hormonal brain organization (because elevated androgens have not been confined to the prenatal period in this case), the report was nonetheless interpreted by its authors and others as providing evidence that prenatal hormones shape adult sexuality. It is, in fact, an especially influential study that continues to garner an impressive number of citations in the current period.[2] Third, in this paper Money's team drew a detailed template of masculine and feminine sexuality that psychologists working on brain organization studies would subscribe to for the next dozen years.

Close reading of this influential early study gives a good sense of scientists' working definitions of masculine and feminine sexuality. Consider, in particular, two brief sentences in which the scientists describe how prenatal androgen exposures have affected sexual arousal in women with CAH:

> The arousal was in the women's own genitals and was such as might lead to masturbation in the absence of a partner. It was not the sentimental arousal, more typical of the normal female, which leads to romantic longing for the loved one alone and which will, in his absence, require waiting for his return. (Ehrhardt, Evers, and Money 1968, 120)

This short passage is a rich source of information about scientists' ideas of masculine and feminine sexual natures. By characterizing this arousal pattern as "masculinized," Ehrhardt, Evers, and Money thereby defined masculine arousal as being focused in the genitals and requiring "release" (if no partner is available, masturbation will do the trick). Conversely, they defined feminine arousal as sentimental, romantic longing; it is explicitly *not* located in the genitals and does not reach the threshold of frank sexuality without a partner present; masturbation will not follow arousal. Further, feminine sexuality is tightly attached to love ("longing for the loved one alone"), and the longed-for loved one is, by definition, a male (*his* absence, *his* return).[3] Finally, female sexuality is passive: in the absence of a partner, a "normal" woman simply waits. Elsewhere in this same article, the researchers were even more explicit on this point: "The female role in heterosexual relations is conventionally defined as more passive and receptive in comparison with the more aggressive, active and initiating role of the male" (120).

The contrasts that Ehrhardt and colleagues drew between women with late-treated CAH and "normal" females suggested a number of additional distinctions between masculine and feminine sexuality. For example, women with CAH showed "versatility in coitus," operationally defined as "experimenting in sexual foreplay and coital positions." They noted, for example, that "nine patients were versatile and enjoyed a variety of positions in sexual intercourse rather than just one or two" (120). This versatility, as well as the women's "erotic response to perceptual material and the personal freedom in sex," they suggested, "indicate behavior more often found in males than in females" (121). They also divided sexually arousing stimuli into masculine and feminine, with different senses engaged in these two types of arousal. For example:

> Dependency on touch, including kissing and caressing, for the erotic arousal of most of these patients fits into the expected female pattern. However, almost the same number of women also reported sexual response to visual and narrative stimuli. The latter are generally believed more frequently associated with the male. (120)

Most importantly, Money and Ehrhardt's model divided stimuli into masculine and feminine in terms of *content*. In other words, whether it is evoked by sight, smell, narrative, or otherwise, is the arousing image that of a man or a woman? Setting the tone for literally all brain organization studies that would follow, they treated erotic desire for men as the most dramatic and clear-cut sign of feminine sexuality, and erotic desire for women as the clearest sign of masculinization. Underscoring the fact that sexual object choice was central to them, they embedded information about the gender of partners in *all* data on sexual behavior. For example, rather than simply listing "frequency" of sex, they listed frequency separately as "frequency of homosexual contacts" and "frequency of heterosexual contacts" (Ehrhardt, Evers, and Money 1968, table II, 119). They similarly divided "amount of satisfaction from sex," "age of initiation," and virtually all other aspects of sexual experience into "homosexual" and "heterosexual"—even when there were no homosexual data to report.

Some of Ehrhardt and colleagues' qualifying statements, meant to indicate that these women were not *fully* masculinized, are also instructive. In particular, the report highlighted the fact that the subjects' sexual relationships were all with steady partners. In an earlier commentary on the sexual patterns of women with late-treated CAH, Money had interpreted their arousal as masculine because arousal in these women was "likely to be accompanied by . . . masturbation or the willingness for sexual intercourse with even a transitory partner" (Money 1965a, 70). Continuing that line of thought here, Ehrhardt, Evers, and Money suggested that having sex only with "steady partners" is a feminine trait.

While it is important to not judge forty-year-old studies by current standards, a brief pause to situate this study historically shows that many of these claims were seriously out of step with sexual standards at the time. Ehrhardt and her co-authors had sent the article to the *Johns Hopkins Medical Journal* in December 1967, a few months after the "summer of love" launched the sexual revolution into mainstream consciousness. Masters and Johnson's (1966) *Human Sexual Response* had already been out for a year, which gave wide publicity to their findings that the physiology of arousal and orgasm is extremely similar in men and women. And long-standing evidence was already available from several studies of nonclinical samples of women that bluntly contradicted the prim, almost disembodied picture of normal female sexuality painted by Ehrhardt, Evers, and Money. Their statements on masturbation are especially fascinating in this regard. The notion that normal women do not masturbate ran counter to literally all survey data collected from the 1920s to the late 1960s, showing that anywhere from one-third to one-half of American women reported masturbation, with the proportion generally increasing for each younger age cohort (Kinsey et al. 1953; Davis 1924–1925; Gagnon, Simon, and Berger 1970).

Given this context, it is important to recognize that in this "sexual revolution"–era study, Ehrhardt, Evers and Money compared the women with CAH to outdated, *stereotypical expectations* about normal women, rather than to actual women without CAH.[4] Almost all later studies (including studies from Money's team) did include some comparison or control group, but the neglect of population-based data on the sexual practices of "normal" women has been the rule rather than the exception in brain organization research.[5]

In sum, "feminine" sexuality in this key study followed a sort of romantic, sleeping-beauty model: awaiting the touch of her special prince, a normal woman may experience romantic longing, but will not feel physical, genitally focused lust. The scientists divided sexual arousal into many distinct aspects, some of which had to do with the stimulus to arousal, and some with the response—but all of which came in either a masculine or a feminine form. Does the arousing stimulus, whether evoked by sight, smell, narrative, or otherwise, refer to a male or a female partner? What senses must a stimulus engage for sexual arousal to follow? Where in the body is the physical sensation of sexual arousal located? Is the quality of desire that is aroused specifically sexual, or is it romantic? What about the speed and intensity of response—how much stimulation is required to be sexually arousing?

Following Ehrhardt, Evers, and Money's first report in 1968, the picture of feminine sexuality drawn by brain organization studies remained remarkably consistent until around 1980. During these years, at least nineteen published studies linked prenatal hormones to sexual behaviors and/ or desires in humans (many additional studies looked at other aspects of masculine or feminine cognition and personality). Each of these studies employed definitions of masculine and feminine sexuality that accord with the study I've just detailed.[6]

The Early Years: "His and Hers" Sexualities

The scientific model of sexuality that held sway during the first decades of brain organization research was fundamentally one-dimensional, meaning that masculine and feminine sexuality could be represented as not just distinct, but polar opposites. That is, masculinity and femininity were thought of as the extreme ends of a single continuum, and an individual's sex or gender "profile" was a trade-off between the two: the presence of

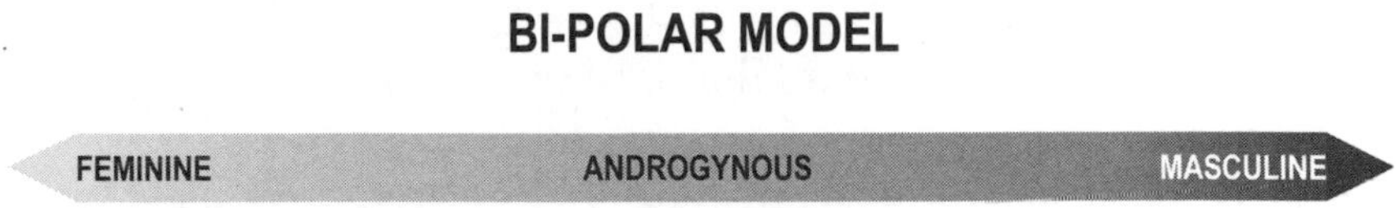

BI-POLAR MODEL

| FEMININE | ANDROGYNOUS | MASCULINE |

Figure 6.1. Bipolar model of gender.

feminine traits automatically made someone less masculine, and vice versa (see Figure 6.1). Twenty years after brain organization theory had first been proposed, Anke Ehrhardt and Heino Meyer-Bahlburg summed up the theory in a way that nicely illustrates the underlying one-dimensional model of masculinity and femininity that had guided research to that point:

> Deprivation of testosterone in the male animal during his species-specific critical time of brain differentiation will result in a female pattern of sex-dimorphic behavior. In contrast, a female rat given testosterone shortly after birth will, as an adult, perform more like a male in the tests for male and female sexual behavior. This principle holds true even though *mammalian behavior is not completely sex dimorphic but is a matter of degree and relative frequency of behavior components that are more typical for one sex or the other.* (Ehrhardt and Meyer-Bahlburg 1979, 417, emphasis added)

Within this logic, femininity can be demonstrated by the presence of certain traits specifically defined as "feminine," but also by the *absence* of traits defined as "masculine." In this model, there is no way to represent someone who is either unfeminine *and* unmasculine, nor *both* feminine and masculine. (As I describe further on, this model would change in the 1980s.) From the beginning, sexual orientation was the most consistently interesting item to investigators. Prior to 1980, every single study that gave any attention to sexuality and eroticism included information on sexual orientation, which was treated as a master element that conveys all (or at least the most important) information about masculine and feminine sexuality. In this scheme, sexual desire for females is the sine qua non of masculine sexuality, and desiring males is likewise fundamental to feminine sexuality.

But given that the hormone-exposed women in these studies were nonetheless largely or even exclusively heterosexual, what did scientists see as the hallmarks of their "masculinized" sexuality? The early investigations concerned sexual arousal, libido, patterns of sexual relationships, specific sexual activities, interest in marriage, and the timing of sexual "milestones" (such as first date or first sexual intercourse) in their search for organizing effects of hormones on the brain.

Arousal and Libido: Male = More

The model of feminine arousal elaborated by Ehrhardt, Evers, and Money (1968), for instance, is closely echoed in a study comparing ("masculinized") women with CAH to ("feminized" or "unmasculinized") women with androgen insensitivity syndrome (Masica, Money, and Ehrhardt 1971). In this report, the scientists again reported "more conventionally male" sexual arousal in some of the CAH women, "the arousal having a strong genitopelvic component with little sentimental component" (138). Likewise, they focused on the fact that more of the "masculinized" women reported "sexual arousal from visual stimuli" or from "narrative stimuli, such as love scenes in novels" (138), in contrast to the more "feminine" women, who were aroused only by "tactile" stimuli—the touch of a partner. Arousal in response to visual stimulation, especially from pornography, was mentioned as an aspect of masculine sexuality in several other studies from this early period (Yalom, Green, and Fisk 1973; Money and Daléry 1976; Money and Alexander 1969).

Investigators consistently defined a strong libido as masculine, and a weaker libido as feminine (for example, Ehrhardt, Evers, and Money 1968; Masica, Money, and Ehrhardt 1971; Lev-Ran 1974a, 1974b; Money and Ogunro 1974; Kester et al. 1980). In many studies, "libido" was implicitly treated as a key difference that in turn explains many other aspects of male–female differences in sexuality. This is apparent from the fact that under the heading of "libido," investigators discussed number of partners, "versatility" versus "conservatism" in sexual positions, being passive and receptive versus active and initiating, and even being orgasmic versus nonorgasmic (for instance, Ehrhardt, Evers, and Money 1968, 118–120; Masica, Money, and Ehrhardt 1971, 137).

In terms of masculine versus feminine relationship patterns, research reports often simultaneously conveyed expectations about the *number* of sexual partners that normal men versus women might have, and the *type* of relationships in which normal men versus women have sex. The idea that women's response is normally directed to "a loved one alone" (Ehrhardt, Evers, and Money 1968, 120; Masica, Money, and Ehrhardt 1971, 138) contains two requirements: an emotional connection (the *loved* one) and exclusivity (alone). The conventionally male pattern, in contrast, was thought to have "little sentimental component" (Masica, Money, and Ehrhardt 1971) but instead to involve a response to thoughts or images of "girls" or "women"—nearly always referred to both generically and in the plural (Money and Alexander 1969, esp. 117; Money and Daléry 1976;

Money and Ogunro 1974, esp. 203; Yalom, Green, and Fisk 1973). In the early studies, especially, masculinity was established by dating many girls or women, as well as having sex with multiple partners. For example, to emphasize the masculinity of three genetic females with CAH who had been reared as male, Money and Daléry (1976) noted that one had "four different girlfriends," another had "four or five different girlfriends" before marriage, and the third claimed "having had 19 different girlfriends" (364–365).

The first study of children exposed in utero to diethylstilbestrol, or DES, provides a detailed illustration of the basic expectations for boys' sexuality, and of the corollary expectations for feminine sexuality (Yalom, Green, and Fisk 1973). Irvin Yalom and his colleagues—who included Richard Green, an important researcher with a special interest in "sissy boys" and who later founded a "gender identity disorder" clinic—were working from the hypothesis that DES, a synthetic estrogen, would *feminize* prenatally exposed boys. What did that mean for them? The DES-exposed boys were expected to have fewer sexual contacts with girls, less sexual activity, less frequent erotic daydreams, and so on. Interviewers rated the boys for, among the many indicators of masculine development, "existence and number of girlfriends, past and current, general amount of social contact with girls, amount of physical contact with girls (coded on a 6-point scale from no physical contact to hand-holding, to sexual intercourse with many girls)" (555). Note that the masculine sexual profile built up with this rating system is a composite of high frequency, high number of partners, and sexual orientation to females.

It's worth pausing here to note that Yalom and colleagues' hypothesis about estrogen "feminizing" the brain was a sort of folkloric extension of brain organization theory, which did not originally include any role at all for estrogens. I will return to this point at the end of this chapter when I describe some troubling trends in how earlier studies are cited even though the specific working hypotheses in brain organization research have changed. For now the crucial point is that throughout the 1970s, various research teams produced studies suggesting that estrogens would *masculinize* behavior in several species (McEwen, Lieberburg, MacLusky, and Plapinger 1977; Simon and Gandelman 1978; MacLusky, Lieberburg, and McEwen 1979). Accordingly, the next study relating DES exposure to human sexuality found that DES-exposed boys were *more*, rather than *less*, masculine (Kester et al. 1980).[7] The findings about DES effects were different, but definitions of masculine sexuality were quite consistent between the two reports.

For example, Kester and colleagues wrote that hypotheses regarding sexuality of "female hormone–exposed" boys and men in this study in-cluded that they "would have a lower sex drive and a higher incidence of sexual dysfunction" as well as a "higher incidence of homosexual fantasy and experience" (271).[8] In contrast to this hypothesis, they reported that the 17 subjects exposed to DES alone were the most "masculine" of the groups, especially in comparison to those exposed to progesterone alone. (Again, note the operative definition of "masculine" sexuality in the study. The finding itself should not be taken too seriously, given the dizzying number of dependent variables and the multiple comparison strategies with very small groups.)

In spite of emerging confusion about whether estrogen would increase or decrease the "masculine" pattern of sexuality, all studies conducted through 1980 were unanimous in characterizing multiple partners and high libido as hallmarks of masculine sexuality. Scientists characterized the feminine pattern, on the other hand, as a "relative weakness of what most people identify, subjectively, as sexual drive and its urgency" (Money and Ogunro 1974, 195).

Specific Activities: "Normal" Girls *Don't*

During these early years, brain organization studies were also based on the idea that certain *activities* are the purview of males, not females. These included masturbation, initiating sex with a partner, having erotic dreams, using pornography, plus the catchall category "versatility." Versatility, for example, reflected the idea that feminine sexuality is "conservative," meaning that ("normal") women would be interested in or willing to have coitus in "one or two positions only," while men are more "experimen-tal," preferring "a variety of positions in coitus" (Ehrhardt, Evers, and Money 1968; Masica, Money, and Ehrhardt 1971, 134–137; Money and Ogunro 1974).[9] Of supposedly masculine activities, masturbation and ini-tiation of sex got the most attention: one or both of these was coded as masculine in each of the ten studies from the early years that include in-formation on particular sexual behaviors (Ehrhardt and Money 1967; Ehrhardt, Epstein, and Money 1968; Ehrhardt, Evers, and Money 1968; Money and Alexander 1969; Masica, Money, and Ehrhardt 1971; Yalom, Green, and Fisk 1973; Money and Ogunro 1974; Lev-Ran 1974a, 1974b; Kester et al. 1980; Money and Lewis 1982). Neither masturbation nor ini-tiation of sex was ever coded as feminine in these studies. For example, in a study comparing women with two different intersex syndromes (Masica, Money, and Ehrhardt 1971), the scientists asked women about their "role in heterosexual relations," meaning the proportion of the time a woman is "initiating" (the male pattern) versus "reserved and passive" (the female pattern) (see also Ehrhardt, Evers, and Money 1968; Money and Ogunro 1974).[10]

As with homosexuality, the idea of masturbation played an important conceptual role in these studies even when there was not much masturbation to report. Along with early signals of sexual orientation, masturbation was the only aspect of frank sexuality that was often assessed in studies of children with "cross-sex" hormone exposures. The association of masturbation with masculinity, and specifically the notion that masturbation is "logically" more connected with male anatomy (the penis) than with female anatomy (clitoris, vulva, and/or vagina) is apparent in this passage from the study of early-treated girls with CAH: "The enlargement of the clitoris was not as important a stimulus to sexual play as it might have been. Each patient had had a clitorectomy within the first few years of life with the latest one being done at age 3-1/2. It is noteworthy that this patient with the latest clitorectomy is the same patient who is neurotic and confused about her femininity" (Ehrhardt, Epstein, and Money 1968, 163). This context is important for recognizing that, even when they were not explicitly framed as "masculine" behaviors, both masturbation and increased "childhood investigative curiosity" about genitalia signaled masculinity in these early studies. For example, in discussing girls whose genitalia had been virilized by prenatal progestin exposure, Ehrhardt and Money (1967) wrote that "there were no parental complaints of masturbation" (93–94). This framed masturbation by girls as something that would be troubling—if not for the investigators, then for the girl's parents. None of the reports ever suggested that boys' parents might "complain" of masturbation. Instead, masturbation among young boys with sexual precocity due to CAH was interpreted as rather typical, "a strictly masculine, heterosexual developmental pattern" that was unusual only in that it "occurred at an earlier age than in nonprecocious boys" (Money and Alexander 1969, 116–117). Likewise, in a report of seven cases of genetic females with CAH, full masculinity among the three who were reared as males was established, in part, by highlighting frequent masturbation: "All three reported orgasm by masturbating and by sexual intercourse . . . O.I. claimed, besides having pleasurable vaginal intercourse two to three times a week, that he masturbated two to three times a day" (Money and Daléry 1976, 367–368).

Other reports followed suit. Echoing the earlier article on late-treated women with CAH, Masica, Money, and Ehrhardt (1971) noted that arousal would "very possibly [lead] to masturbation" in women with

CAH (138). They contrasted the "much higher incidence of masturbation" in CAH women (136) to the *complete absence* of masturbation in the "fully feminine" women with complete androgen insensitivity (see esp. table II, 133). Similarly, other scientific teams approached masturbation as a signal of masculinity, even though they did not find differences in rates of masturbation between hormone-exposed and unexposed subjects (for example, Yalom, Green, and Fisk 1973; Kester et al. 1980).[11]

Though the focus of this chapter is on the definitions used in the research, rather than the findings of studies, some observations about findings are unavoidable, and the data on masturbation provide a good point of departure for discussing the problem of comparison data. As noted above, brain organization researchers have generally avoided introducing population-based data on sexual desire and behaviors, instead preferring to compare hormone-exposed subjects to specific comparison samples. Had scientists used larger and more representative comparison groups, or even included reference to population-based data on the sexuality of healthy men and women, it would have been difficult or even impossible to maintain that masturbation was a completely "masculine" trait. But even if they conceded that "normal" women masturbate, would they have found that women with CAH masturbated *more* than "normal" women? Interestingly, it is impossible to tell from the research reports, because even the most detailed reports do not contain sufficient information to make a comparison.

Table II in Ehrhardt, Evers, and Money's 1968 study of late-treated women with CAH illustrates this difficulty. Instead of frequencies for various behaviors being presented separately, the data on sexual activities are presented in a hierarchical manner, such that masturbation is discussed only for women with no experience in sexual relations with any partners. Furthermore, the data are presented in a dichotomous "ever" versus "never" format, without frequencies. More recent data on sexual behavior among women with CAH suggests exactly the opposite: women with this condition are, if anything, likely to masturbate *less* (Zucker et al. 2004). There will be more to say about that when I describe the later studies. For now, let us return to the remaining ways in which scientists coded sexual patterns as masculine or feminine in the early studies.

The Tricky Bits: Marriage and Milestones

Marriage, while not exactly a component of "frank" sexuality in the same sense as erotic behaviors and attractions, has been treated as a signal of heterosexuality since the very earliest brain organization studies. Yet scientists' approach to marriage as an indication of masculine or feminine brain organization has been complex. In the early studies of adult men and women, being married was frequently used to signal heterosexuality (Money, Ehrhardt, and Masica 1968; Lev-Ran 1974a, 1974b; Money and Daléry 1976; Imperato-McGinley et al. 1979), and even core gender identity: "Marriage, when it is combined with romantic affection and falling in love rather than being a perfunctory arrangement, is one manifest confirmation of gender identity" (Money and Ogunro 1974, 190).

Juliana Imperato-McGinley and colleagues' (1979) well-known study of people with 5-alpha reductase deficiency (5-ARD) provides a good example of how marriage was used to confirm sexual orientation to women, which in turn signaled masculine brain organization. In this intersex syndrome, genetic males have fetal androgen exposures within the normal range, but are born with female-looking genitalia because additional substances required for genital development are deficient. Therefore, while they are born with slightly ambiguous but more female-looking genitalia, brain organization theory would suggest that the brains of children with 5-ARD have been masculinized. At puberty, adolescents with 5-ARD undergo dramatic physical masculinization—the voice deepens, the penis grows significantly, and facial and body hair becomes abundant.

Imperato-McGinley and colleagues (1979) used the fact of marriage among men with 5-ARD as a gloss for information about sexuality, briefly describing subjects as "initially anxious" and "fearing ridicule" from potential sex partners, but ultimately sweeping this aside with the conclusion that "15 of 16 subjects who changed to a male-gender role are either living with women in common-law marriages or have lived with women in a common-law relation" (1234).

Likewise, in the early studies of children, researchers presented both interest in marriage and a "timely" interest in romantic relationships and motherhood as feminine. For example, Ehrhardt and Money described how girls who were exposed prenatally to progestins failed to display what scientists considered to be the feminine attitude: "Five aspired to professional careers; two of these preferred career to marriage and motherhood which they approached in a rather perfunctory way as something that might or might not happen in the future" (Ehrhardt and Money 1967, 93). Rather than this "perfunctory" approach, feminine sexuality was expected to be accompanied by wedding fantasies and prioritization of marriage over a nondomestic career (Ehrhardt, Epstein, and Money 1968). Slightly later studies from Anke Ehrhardt and colleagues showed similar expectations about gender and interest in marriage. While still collaborating with

John Money at Johns Hopkins, Ehrhardt had moved in 1970 to the Department of Psychology and Pediatrics at the State University of New York (SUNY) in Buffalo. There she began collaborating with Heino Meyer-Bahlburg, with whom she has continued working to the present day. Their earliest publications together concerned children exposed to the synthetic progestogen medroxyprogesterone acetate (MPA), and proceeded from the hypothesis that MPA would "antagonize androgen action," resulting in the feminization or demasculinization of behavior in girls and boys (Ehrhardt, Grisanti, and Meyer-Bahlburg 1977; Meyer-Bahlburg, Grisanti, and Ehrhardt 1977). The data showed very few differences between children exposed to MPA and unexposed controls, but the reports are nonetheless informative about the investigators' expectations regarding interest in marriage. As part of a "cluster of behavior data [that] concerns toy preference and role rehearsal of becoming a mother in adulthood," interest in marriage was coded as feminine (Ehrhardt, Grisanti, and Meyer-Bahlburg 1977, 395; Meyer-Bahlburg, Grisanti, and Ehrhardt 1977, 389).

Conversely, scientists never mentioned marriage fantasy as a possibility among the studies of boys, even in studies where they scrutinized boys for evidence of feminization (Yalom, Green, and Fisk 1973). Scientists consistently rated boys as more masculine for having more girlfriends, rather than for having particularly serious relationships. Nonetheless, if boys were *actively hostile* to the idea of marriage, this would be a strike against their masculinity, too. Money and Lewis (1982), for instance, rated boys' "positive orientation toward marriage and parenthood" as masculine (343).

Marriage, then, was a somewhat complicated indicator of masculine or feminine sexuality in these early studies. Boys were expected to not express an active interest in marriage, but as adults, men should be married in order to confirm their heterosexuality. Girls and women, on the other hand, were expected to approach marriage as a priority, actively fantasizing and planning for it, as well as marrying when the time came.

Investigators' hypotheses regarding the timing of sexual milestones is another interesting feature of the early studies. Given that CAH, if untreated, results in "precocious virilizing puberty in the female" (Ehrhardt, Epstein, and Money 1968, 160) as well as conventional though precocious puberty in the male, the investigators seem to have expected that masculinization would lead to *psychosexual precocity* as well: rather than noting that girls were late to develop, they initially tended to comment that fetally androgenized girls were *not precocious*. For example, girls with the early-treated syndrome of CAH "did not show more childhood investigative curiosity about sex than their matched counterparts" (163). In several discussions, Money and Ehrhardt subtly employed data on nonhuman primates to indicate that earlier manifestations of sexual behavior would be expected among masculinized females (Ehrhardt and Money 1967, 98; Ehrhardt, Epstein, and Money 1968, 160; Money and Ehrhardt 1972, 88 and 100).

The hypothesis that "masculinizing" regimens would engender precocity was consistent across sexual as well as nonsexual outcome variables, and for both androgen and progestogen exposures (see, for instance, Dalton's 1968 study of progesterone).[12] By the time Money and Ehrhardt published *Man and Woman, Boy and Girl* in 1972, they already had the "impression that [fetally androgenized girls] are late in reaching the boyfriend stage of development and in getting married" (Money and Ehrhardt 1972, 102). Still, when Yalom and colleagues (1973) examined estrogen-exposed boys, and when Kester and colleagues (1980) looked at males exposed to "female hormones," they looked for delay in reaching sexual milestones as a sign of *feminization*.

The Centrality of Object Choice

The importance of object choice (a concept that was not yet, at the time, referred to as sexual orientation) in this early model is apparent from the fact that investigators often reported early indications of object choice even in studies of children, for whom they otherwise made few observations or speculations regarding sexuality and eroticism. The same year that Ehrhardt, Evers, and Money published the report on sexuality and eroticism among women with CAH whose treatment had commenced at later ages, Money's team also released the first report on gender identity and related behavior among the girls with CAH whose treatment had begun much earlier, usually in infancy (Ehrhardt, Epstein, and Money 1968). The study's main finding was that the girls were more likely to be "tomboys" than were girls in the comparison group (healthy girls chosen from Baltimore public schools). Ehrhardt and colleagues were at some pains, though, to emphasize limits to the girls' tomboyism, especially concerning sexuality and core gender identity. Driving home the point that same-sex interest would be a clear sign of masculinization, they mentioned the *lack* of such interest three separate times in this article—for example, "tomboyism in the adrenogenital girls did not, however, extend to erotic interests or sexual play. . . Their tomboyism did not include implications of homosexuality or future lesbianism, or a belief of having been assigned to the wrong sex" (165). Note, however, that the investigators contrasted the kind of tomboyism observed in these girls with that observed in the late-treated girls and women: "*Their* tomboyish traits included, in some cases, homosexual fantasies and dreams and, in some few cases, frank bisexualism" (166, emphasis added). In other reports, too, investigators' consistent attention to lesbianism even in its absence confirmed their idea that orientation toward men is a critical marker of feminine sexuality.[13]

In their study of 5-ARD, Imperato-McGinley's team explicitly identified "sexual object of choice (the sex of the person chosen as an erotically interesting partner)" as one of the four key criteria for evaluation of "sexual behavior differentiation" into masculine or feminine patterns (Imperato-McGinley et al. 1979, 1234). The other three included gender identity, or self-concept as male or female; "sexual patterns," in which they included sex-related but nonsexual behaviors like aggression or occupation; and sexual mechanisms, those "features of sexual expression over which an individual has little control, which for men include the ability to obtain and maintain an erection and to achieve orgasm." The report's narrative showed that these four aspects were not equally ranked, though: next to core gender identity as male, the number of mentions as well as the specific context indicate that Imperato-McGinley and colleagues approached sexual orientation toward women as the most important signal of brain masculinization among subjects with 5-ARD. For example, "The change to a male-gender role occurred either during puberty or in the post-pubertal period, after the subjects were convinced that they were men (male-gender identity) and were experiencing sexual interest in women." Also, as noted above, this team emphasized sexual orientation through their focus on subjects' marriages and common-law relations with women (1234).

With the advent of the case-control model in brain organization studies, the emphasis on sexual orientation was further amplified. Recall that these are the studies where researchers begin with subjects whom they perceive to be "psychosexually distinct" enough to suggest that they may have been exposed to an early hormone environment that was unusual for their sex. Brain organization researchers interpreted homosexuality as a partial "sex reversal" in psychology, reflecting a cross-sex pattern in the physical organization of the brain. Gunter Dörner was the earliest to state this bluntly, suggesting that "homosexual men possess a predominantly female-differentiated brain" (Dörner et al. 1975, 1).

Oddly, the overarching importance of sexual orientation is clearest in the studies that focus on transsexuals. Although none of the case-control studies reported information about sexuality *other* than sexual orientation, nearly all of the early case-control studies that focus on transsexuality (also defined as cross-sex gender identity) also report on sexual orientation (Dörner et al. 1976; Seyler et al. 1978; Boyar and Aiman 1982; Dörner et al. 1983; Weisen and Futterweit 1983). (The only exception in this regard is Goh, Ratnam, and London 1984.) In keeping with now-outmoded clinical criteria that required anyone undergoing sex-change surgery to demonstrate that they would be heterosexual at the end of the process, investigators in all the early studies used sexual orientation both to establish that the transsexual subject groups in these studies were "really" transsexual, and as a criterion for establishing the "normal," sex-typical status of individuals in the control groups. Describing female-to-male transsexuals in a study of endocrine response to injected estrogens (one of the "estrogen challenge" studies described in Chapter 5), for example, Seyler and colleagues (1978) wrote that "sexual interest is directed toward 'straight' women and they are not attracted sexually to homosexual women" (177). Likewise, they stipulated that the men and women serving as controls were "exclusively heterosexual in orientation" (177). (Interestingly, this study seems to be unique in the brain organization literature for treating attraction to heterosexual women as qualitatively distinct from attraction to lesbians. This suggests that the construct of sexual orientation is more complex than simply attraction to a particular body type. Instead, they imply it should be framed in terms of the ideal sexual partner's own sexuality and perhaps gender, instead of or in addition to the ideal partner's sex.)

The importance of sexual orientation in brain organization research has intensified over time; since around 1980, the overwhelming majority of studies have reported *no* actual sexual activities at all, relying only on information about sexual orientation to indicate masculinizing or feminizing influences of hormones. Definitions changed over time, but sexual object choice has remained the most constant aspect of masculine and feminine sexuality, at least in theory. As Chapter 7 will demonstrate, however, the actual measurement of sexual orientation is so different within various studies that that this can't be considered a stable variable either.

Summary: Masculine and Feminine Sexuality in the Early Studies (1967–1980)

Figure 6.2 summarizes how characteristics of sexuality were interpreted as either masculine or feminine in studies of human brain organization through about 1980. Qualities that could be quantified were put on scales with *masculine* at one end and *feminine* at the other: libido, number of partners, and number of sexual positions enjoyed. Which end of the scale

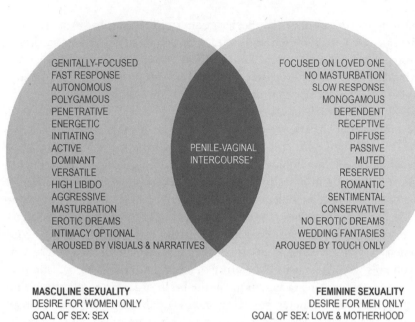

GENITALLY-FOCUSED
FAST RESPONSE
AUTONOMOUS
POLYGAMOUS
PENETRATIVE
ENERGETIC
INITIATING
ACTIVE
DOMINANT
VERSATILE
HIGH LIBIDO
AGGRESSIVE
MASTURBATION
EROTIC DREAMS
INTIMACY OPTIONAL
AROUSED BY VISUALS & NARRATIVES

PENILE-VAGINAL
INTERCOURSE*

FOCUSED ON LOVED ONE
NO MASTURBATION
SLOW RESPONSE
MONOGAMOUS
DEPENDENT
RECEPTIVE
DIFFUSE
PASSIVE
MUTED
RESERVED
ROMANTIC
SENTIMENTAL
CONSERVATIVE
NO EROTIC DREAMS
WEDDING FANTASIES
AROUSED BY TOUCH ONLY

MASCULINE SEXUALITY
DESIRE FOR WOMEN ONLY
GOAL OF SEX: SEX

FEMININE SEXUALITY
DESIRE FOR MEN ONLY
GOAL OF SEX: LOVE & MOTHERHOOD

Figure 6.2. How brain organization researchers divided activities and desires into masculine and feminine in the early studies, roughly 1967 through 1980.

*Penile-vaginal intercourse should be understood as a "meeting point" for masculine and feminine sexuality in this model, not an overlapping element.

is which? Brain organization researchers almost always treated this issue as though it goes without saying, and the simple rule was "male is more." More libido, more partners, more "versatility" in terms of actual positions, and more ways to be aroused were all a part of scientists' ideal for masculine sexuality. In contrast, there was scarcely any conceptual space for autonomous feminine sexuality in these studies. Given that scientists consistently interpreted behaviors such as initiating sex, expressing intense physical desire, or masturbation as masculine, it is scarcely an overstatement to suggest that sexuality itself was seen as a masculine trait. Female sexuality—if not an outright oxymoron—was nonetheless thought to be decidedly responsive rather than autonomous, requiring a masculine sexual partner to move it from mere *possibility* to *expression*.

Around 1980 the sexual revolution began to have an effect on brain organization research. Or rather, that *might* be why things began to change in the studies around that time. The crucial point here isn't *why* they changed, but that they did change—and sometimes quite dramatically. Scientists still believed that androgens were crucial to the development of masculine sexuality and damaging to feminine sexuality. But what went under those general umbrella terms *masculine* and *feminine*—especially *feminine*—wasn't stable. Broadly, scientists began to see "healthy" and "normal" feminine sexuality as a more corporeal, active phenomenon than they had previously thought.

Working Definitions, Take Two: Active Feminine Sexuality in the 1980s and Beyond

What do women want? To this famously perplexing question, brain organization researchers have offered a simple answer: women want men. From the start of these studies, sexual orientation to males has been the most important indicator of a "feminine" sexual profile, and sexual orientation to females has been the most important indicator of a "masculine" sexual profile. The weight of sexual orientation has intensified over time: since 1980 the great majority of studies that include *any* information about sexuality have exclusively focused on sexual orientation, omitting mention of any other aspects, such as libido, arousal patterns, and so on (for example, Money and Lewis 1982; Dörner, Schenk, et al. 1983; Gladue, Green, and Hellman 1984; Lindesay 1987; Hochberg, Gardos, and Benderly 1987; Ellis et al. 1988; McCormick, Witelson, and Kingstone 1990; Swaab and Hofman 1990; Bailey, Willerman, and Parks 1991; LeVay 1991; Allen and Gorski 1992; Bogaert and Hershberger 1999; Lalumiere, Blanchard, and Zucker 2000; Titus-Ernstoff et al. 2003). But even when investigators have conducted detailed sexual histories, their reports have followed the example set in the earliest studies by John Money, Anke Ehrhardt, and their various colleagues, in that sexual object choice has been the clear organizing principle of masculine and feminine sexuality.[14] It is still standard practice for brain organization studies to structure information on sexual arousal and behaviors so that object choice is inseparable from most or all specific information about sexuality (see Money and Mathews 1982; Meyer-Bahlburg et al. 1984; Dittmann, Kappes, and Kappes 1992; Mendonca et al. 1996; McCarty et al. 2006).

Depending on how you look at it, the focus on sexual orientation can seem like either a mark of continuity in the way brain organization studies have approached sexuality, or a signal of change. That's because virtually none of the other aspects of sexuality are treated in the same way during the period before 1980 versus the period afterward. As the examples below indicate, some things—arousal patterns and the timing of sexual milestones, in particular—are simply dropped out of the framework of masculinity versus femininity. To the extent that these aspects of sexuality are mentioned at all in later studies, they are seen as "gender neutral" aspects of the development of sexuality. Other aspects of sexuality have undergone an even more remarkable transformation, and this is where things get really interesting. High libido, multiple partners, and more varied and frequent sexual activities are all treated as part of *normal female sexuality* in more recent brain organization studies, even though each of these was previously interpreted as a *masculine* sexual pattern. As surprising as this is, what is *more* surprising is that the changes have gone unnoticed. Some readers might be tempted to simply say "Phew! Thank goodness they changed that!" but it is not so simple. These changes seriously affect the overall structure of the network of research, such that studies showing a link between androgens and *masculinized* sexuality in the second period are showing a link between androgens and *feminine* sexuality, according to the definitions used in the first period.[15] Like a shell game that has gone on too long, not only the observers but also the people wielding the shells seemed to lose track of what was underneath.

To appreciate the "new female sexuality" that appeared in the 1980s, it is helpful to understand two key ways that brain organization theory had been elaborated. As noted above, researchers had generally accepted by then that estrogens, like androgens, contributed to male-typical sex differentiation. During the 1970s, experiments with rodents began to indicate that some aspects of male-typical behavior are not directly differentiated by testosterone; instead, testosterone must first be converted (aromatized) to estrogen in the brain.[16] The important implication of this elaboration of brain organization theory was that it gave a clear new paradigm for studying estrogens as a potentially masculinizing agent for human sexuality.

Second, most scientists no longer worked from a behavioral paradigm that looked at sexuality as one-dimensional, in which masculinity and femininity occupy opposite poles of a single continuum. Another way of thinking gained currency in the 1970s and is now (at least officially) more favored than the idea that masculinity and femininity are trade-offs. It had long been known that normal female animals routinely display both "male-typical" and "female-typical" behaviors (and the same is true of male animals), but this observation fit poorly with mid-twentieth-century theories about human sexuality and gender, which assumed that "normal" and healthy development involved either *consistent* masculinity or *consistent* femininity. In 1974, psychologist Sandra Bem (1974) proposed the "orthogonal" model, suggesting that masculinity and femininity are two separate domains of personality and behavior. According to the orthogonal model, any individual may be more or less masculine *and, independently of their masculinity,* more or less feminine. The orthogonal view fit well with animal data showing that normal male and female animals regularly display both so-called male and female sexual behavior, but in different proportions that vary by factors like species, breeding strain, and rear-

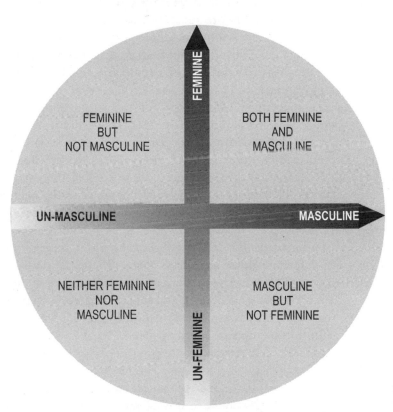

Figure 6.3. Bipolar versus orthogonal models of gender.

ing conditions, in addition to early hormone exposures (Phoenix et al. 1959; Beach 1971; Goy and Goldfoot 1975).[17]

Figure 6.3 contrasts the bipolar (or one-dimensional) model with the orthogonal (or two-dimensional) model. Note that in both models, masculinity and femininity may be seen as completely distinct domains—that is, there are no traits that signal both masculinity and femininity. But the way the properties of masculinity or femininity may be combined or absent in any individual are quite different in the two models.

Bem's model specifically departed from the traditional psychological "assumption that it is the sex-typed individual who typifies mental health" and suggested that "in a society where rigid sex-role differentiation has already outlived its utility, perhaps the androgynous person will come to define a more human standard of psychological health" (1974, 162). Research has tended to support this view, indicating that androgynous people are better able to engage in "situationally effective behavior without regard for its stereotype as more appropriate for one sex or the other" (Stake 2000; Shifren, Furnham, and Bauserman 2003; Lefkowitz and Zeldow 2006; Hunt et al. 2007). Brain organization researchers never picked up the idea of androgyny as the apex of psychological health, and generally didn't credit Bem with the innovation, but both animal and human studies mostly dropped the polarity model for an orthogonal model by the early 1980s (for instance, Arnold and Gorski 1984; Ehrhardt et al. 1984; Gooren et al. 1984; Hines and Shipley 1984).

The orthogonal concept of masculinity and femininity appeared at about the same time that the role of estrogen in sexual differentiation was broadly accepted, and the two concepts were linked through the "two-pathway" model of sexual differentiation. Scientists had observed that hormone treatments did not evenly affect all sex-typed traits, including components of mating behaviors (such as mounting versus lordosis). They reasoned that male-typical sex differentiation required both the development of masculine traits (masculinization) and the active suppression of feminine traits (defeminization). Instead of sexual differentiation happening through one chain of neurochemical events (through androgen directly stimulating masculine trait development and suppressing feminine trait development), the new model suggested that differentiation could happen through either androgen effects or estrogen effects.[18]

The change of model had particularly serious implications for the way in which feminine sexuality was understood. In the old model, where masculine was associated with active expression of sexual desire, higher libido, more activities, and so on, it could seem very much as if moving toward the feminine "pole" of sexuality meant moving away from sexuality itself, toward a sort of sexual "null set." The shift to an orthogonal model required positing female sexuality as a domain *with content*—not an absence of traits that men had, but an expression of traits that were typical of "healthy" and "normal" women.

Let's see how the changes play out in specific studies.

IN 1982 John Money and Dana Mathews published a follow-up on an early brain organization report that had focused on girls (Ehrhardt and

Money 1967; Money and Ehrhardt 1972) who were exposed in utero to progestins. Whereas the earliest reports of progestin-exposed girls had suggested that these girls were "tomboys" (a complicated designation that included erotic masculinization), the overall argument of this later study was that "girls with a history of progestin-induced prenatal androgen-ization" were not, as adults, "behaviorally different from women in general" (Money and Mathews 1982, 81). In fact, the authors emphasized that their findings "correspond to everyday criteria of what to expect from teenaged girls and young adults in contemporary America" (81). This was not especially surprising, because various human and nonhuman studies since the 1967 report had produced complex and seemingly contradictory information on the effects of prenatal exposure to progestogens (Dalton 1968; Shapiro et al. 1976; Reinisch and Karow 1977; Ehrhardt and Meyer-Bahlburg 1979). Further, though Money had been an important pioneer in human brain organization research, he was always anxious to note that the effects of early hormones were subtle and may even decrease over time, provided gender socialization was relatively normal (Money and Ehrhardt 1972).

So the findings themselves were not remarkable. The way in which Money and Mathews arrived at their conclusions, however, was quite another story, because they departed so dramatically from scientists' earlier expectations regarding aspects of sexuality and gender among normal women. One of the clearest changes concerned the way they described sexual arousal. Most of the women were "erotically responsive to the sight or touch of the partner and also to reading narrative material of romantic and explicit erotic content" (Money and Mathews 1982, 79). Further, they experienced arousal primarily in the genital area. Here is where it gets interesting: Money and Mathews characterized these responses as "stereotypically feminine sexual behavior" (82). Recall that just a few years earlier, Money and colleagues had consistently characterized genitally focused arousal, as well as arousal in response to visual or narrative material, as "masculine" (Money 1965a; Ehrhardt, Evers, and Money 1968; Masica, Money, and Ehrhardt 1971; Money and Daléry 1976).

This study was the last to describe subjects' patterns of sexual arousal as masculine or feminine. In fact, very few brain organization studies after 1980 mention sexual arousal at all, but even these do not indicate that some kinds of arousal are feminine and other kinds are masculine. For example, Meyer-Bahlburg and colleagues (1985) listed the relative frequency of vaginal responsiveness as well as "nongenital sexual arousal" in a table comparing DES-exposed women to women with abnormal Pap smears, but did not in any way suggest that either genital or nongenital arousal is associated with a masculine or feminine pattern. What is most telling is that after Money and Mathews's 1982 report, scientists *no longer referred to the previous findings* of "masculinized" or "feminized" arousal patterns among subjects with anomalous hormonal histories, even in review or background sections summarizing studies that explored arousal patterns at length (for example, the review of CAH research in Dittmann, Kappes, and Kappes 1992).

It is always worth noticing when a research finding that is considered important in one period is no longer mentioned at all in a later period, because it signals a change in scientific thinking. Ideally scientists document the changes in their thinking, so that other researchers and consumers of the research can understand why they changed their approach, and to clarify the relationship of early studies to those that come later. In this case, though, students of brain organization research will search in vain for an explanation of why arousal patterns have been dropped as a signal of masculine or feminine brain organization.

Marriage and Milestones: Outside the Frame of Gender

A change in the way sexual arousal was conceived is just one of several scientific "turnarounds" regarding masculine and feminine sexuality patterns that have not been accounted for in the scientific literature. Another aspect of sexuality that was reframed in later studies is the timing of sexual milestones. Recall that scientists had initially expected that "masculinized" children would reach sexual and other milestones earlier (Ehrhardt and Money 1967; Dalton 1968; Ehrhardt, Epstein, and Money 1968; Money and Ehrhardt 1972). The hypothesis that androgens would lead to precocity was gradually dropped as evidence continued accumulating that children exposed to exogenous hormones, as well as children with clinical intersex conditions, were more likely to be *delayed* in sexual development. Again, the 1982 study of progestin-exposed women by Money and Mathews is interesting, less because of the findings than because of the change in how the findings are framed. Earlier studies of "masculinized" subjects had either reported some developmental precocity (Dalton 1968) or had noted the *absence of precocity* (for example, Ehrhardt, Epstein, and Money 1968), but in this report the investigators noted an *absence of delay* in reaching milestones: "There were *no reports indicating slowness* in arriving at the dating age, relative to age mates, and no evidence of difficulty in establishing a dating relationship with a boyfriend" (Money and

Mathews 1982, 77, emphasis added). This is a signal that the hypothesized effect of "virilizing agents" on the timing of milestones had shifted from precocity to delay (see Tenhula and Bailey 1998 for a recent example of this hypothesis, and similarly negative findings).

After that short transition, though, later studies have generally shifted hypotheses about hormones and the timing of milestones to the extent that they no longer fit a paradigm of sexual differentiation. Instead these studies are shaped around the expectation that "unusual" hormone exposures may lead to developmental delay, regardless of whether the hormones were masculinizing or not (for instance, Meyer-Bahlburg et al. 1984; Hurtig and Rosenthal 1987; Dittmann, Kappes, and Kappes 1992). An especially clear example of this comes from Meyer-Bahlburg and colleagues' 1984 report on psychosexual milestones in women who had been exposed to DES prenatally. In this article the investigators did not describe the timing of physiological or psychological puberty in the familiar terms of masculinization, feminization, defeminization, or any of the increasingly awkward variants of those terms. Instead, they framed developmental delay as an implicitly gender-neutral effect of the hormone exposure, effectively removing this aspect of sexuality from the brain organization framework of masculinization or feminization.

By retrospectively reframing all earlier research on the initiation of sexual behaviors in this same gender-neutral language of *timing* and *delay,* Meyer-Bahlburg and colleagues make the findings across time seem more coherent than they are. Their description of Yalom, Green, and Fisk's 1973 study of male adolescents with prenatal exposure to DES and/or progestogens is of particular note. In 1984 Meyer-Bahlburg and colleagues discussed the Yalom group's finding of "decreased sexual interest or activity" in the gender-neutral language of delayed psychosexual development (Meyer-Bahlburg et al. 1984, 360). There are two reasons to question this characterization. First, Yalom and his colleagues had interpreted the delay of sexual development among hormone-exposed males not as a neutral matter of *timing,* but as a sign of *feminization.* Second, in a 1977 article, Meyer-Bahlburg's team had likewise summarized the Yalom group's findings *not* as (gender-neutral) developmental delay but as "decreased heterosexual experiences," which suggested "an anti-androgenizing effect" (Meyer-Bahlburg, Grisanti, and Ehrhardt 1977, 384).

Besides the Yalom group, only one other team had looked at timing of sexual development in men prenatally exposed to DES and/or progestogens, and the fate of that study in later brain organization literature is also troubling. Patricia Kester and colleagues (including Richard Green, who co-authored the study with Irvin Yalom) had found that men prenatally exposed to DES reported *earlier* initiation of sexual activities (Kester et al. 1980). This was in keeping with ideas at the time, because DES had been reconceived by then to be a masculinizing agent, and masculinization was still expected to cause earlier sexual development. The study by Kester and colleagues is still frequently cited, but their findings on sexual milestones have sunk like a stone: I could not identify one subsequent brain organization study that refers to them.

Good for the Gander, Good for the Goose: Libido and Sexual Activities

While arousal patterns and the timing of psychosexual development were dropped from the masculine/feminine framework, other elements were moved from one clear position to another. In each case, these elements shifted from exclusively masculine to neutral—that is, to common elements of human sexuality that are shared by both "normal" men and "normal" women. One study that provides an especially dramatic contrast with the original picture of feminine sexuality is Meyer-Bahlburg and colleagues' 1985 report, "Sexual Activity Level and Sexual Functioning in Women Prenatally Exposed to Diethylstilbestrol."

Citing animal evidence of *defeminization* secondary to prenatal DES exposure, these investigators hypothesized that prenatally DES-exposed women would be defeminized, specifically meaning that they would "have lower levels of sexual interest, activity, and function than unexposed controls" (Meyer-Bahlburg et al. 1985, 498). They indicated that their data generally supported this hypothesis:

> The DES women were found to have less well-established sex-partner relationships and less experience with child-bearing, to be lower in sexual desire and enjoyment, sexual excitability, and orgasmic coital functioning. (497)

It is remarkable that these findings were presented *in toto* as *defeminization.* An emphasis on strong relationships and childbearing had always been seen as components of feminine sexuality, but earlier studies had without exception framed *lower* rates of sexual desire, excitability, and orgasm, and lower number of partners, as the feminine expectancy, while relatively higher rates were interpreted as evidence of masculinization. If anything, consistency with the earlier definitions would suggest that these DES-exposed women had been *unmasculinized.*

Here, a brief pause to consider the context of this study helps to make some sense of the stunning shift in definitions. First, there was the background factor of studies on DES in nonhuman animals. Animal research had shown that in several species, estrogens "defeminized" or "masculinized" behavior. So there were two possible hypotheses available for interpreting differences between DES-exposed and unexposed women in a way that would be consistent with current thinking in brain organization studies: either DES would defeminize women's behavior or it would masculinize it. But once women were found to have fewer partners, fewer orgasms, less desire, and so on, there was only one choice that seemed to fit the cultural, "commonsense" logic of sexuality. In this logic, it would be truly unthinkable to interpret lower libido, fewer partners, and so on as indicating "masculinization." But in the post-sexual-revolution common sense of 1985, it made a great deal of sense to view these elements as showing an impairment of "normal" feminine sexual appetite, that is, defeminization. The problem, of course, is that the interpretation of the evidence that looked perfectly reasonable from a "commonsense" standpoint looks totally inconsistent when it is placed side-by-side with the interpretation that would be dictated by previous studies' definitions of female sexuality.

This same study also decisively shifted the status of two specific sexual activities: masturbation and sexual night dreams. Researchers had previously consistently interpreted both masturbation and sexual night dreams as more masculine activities. By explicitly including these activities among those for which a lower rate indicates *de*feminization (see, for example, table 3, 502), Meyer-Bahlburg et al. (1985) redefined masturbation and sexual night dreams as part of normal *feminine* sexuality. As with higher orgasm rates, higher libido, and so on, the researchers (apparently unwittingly) aligned their interpretations with cultural logic rather than with the logic of scientific precedent.[19]

Very few studies in the period after 1980 assess specific sexual activities other than penile-vaginal intercourse, but those that do follow the example set by the DES studies of Meyer-Bahlburg and colleagues in discarding the old notions of feminine sexuality. Ralf Dittmann's work provides an example of how the definitions of normal feminine sexuality shifted even in studies that do not rely on the concept of defeminization. In his publications on a sample of German women with CAH (especially Dittmann, Kappes, and Kappes 1992), Dittmann characterized women with CAH as "masculinized" in comparison to their nonaffected sisters. While the scientists' interpretation of "normal" feminine sexuality in these sisters includes many features that have been characterized as feminine since early brain organization research (being in love, having long-term relationships with partners, and especially being attracted only to men), it also quite remarkably includes many features that were always characterized as masculine in the early studies. The sisters with presumably "feminine" prenatal hormone histories report *more* initiation of sexual relationships, *more* sexual contacts, *more* orgasmic experiences, and *more* romantic/erotic night dreams than the "masculinized" women with CAH. Yet all of these differences are presented as indication of "feminine" sexuality and taken to support the brain organization hypothesis! Similarly, more recent analyses of sexual functioning and sexual activity in women with CAH either present lower levels of sexual arousal, masturbation, and heterosexual intercourse as "apart from" those aspects of psychosexuality that would be affected by early androgens (Kuhnle et al. 1993), or they include the suggestion or explicit interpretation that lower levels of arousal, masturbation, and heterosexual activity may reflect brain *masculinization* (Federman 1987; Mulaikal, Migeon, and Rock 1987; Zucker et al. 1996; Meyer-Bahlburg 1999; Zucker et al. 2004).

Figure 6.4 illustrates the newly expanded version of feminine sexuality that showed up in brain organization research around 1980. Recall that in the early studies, masculine and feminine sexuality have no overlap—they occupy sharply bounded territories. In the later studies, there is a middle, neutral zone. Certain aspects of the old model were retained in the new model. Sexual orientation continued to intrigue researchers and to be the most frequently investigated aspect of human sexuality. As before, sexual orientation was treated as a master element, from which all important information about an individual's sexuality could be inferred. Finally, when investigators did focus on particular behaviors rather than sexual partners (real or fantasized), penile-vaginal intercourse was always treated as the pinnacle of sexual activity: usually it was treated as the *only* sexual activity of interest.

But a great deal changed, too. As noted early in this chapter, the major change between definitions in the first and second model of sexuality used in brain organization research can be thought of as a migration of certain elements from the exclusive domain of masculine sexuality into the *neutral* zone, meaning that normal women are also expected to exhibit these traits. The new norm after 1980 placed a relatively high value on the quantitative aspects of sex (libido, number of partners, frequency of sexual activity) for *both* men and women. In this version of feminine sexuality, these formerly masculine traits are now considered feminine, as long as the partner or object of desire is male. Finally, masturbation and initiation of sexual activity, once coded as unequivocally masculine, were later treated as the expected norm among women.

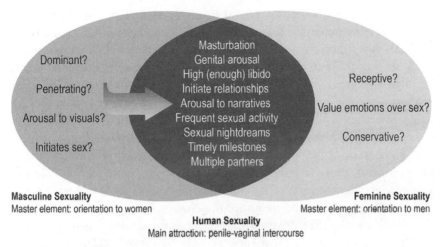

Figure 6.4. Masculine and feminine sexuality in the later studies, from about 1980.

Implications: Bullet-Proofing the Theory

What, specifically, is wrong with the fact that brain organization researchers came to expect relatively high libido, frequent sex, multiple partners, masturbation, sexual night dreams, and initiation of sexual activity from sexually normal women? Updating their assumptions about feminine sexuality might seem to be a fairly straightforward matter of scientific progress—especially when the later model seems to fit better with contemporary women's sexual behavior, and when many early studies of brain organization have been criticized as sexist due to their outmoded assumptions. So why the critique? Aren't the changes relayed in the preceding discussion a good thing? Unfortunately, as I noted before, it's not that simple. Newer and more complex conceptualizations of feminine sexuality may be an improvement in abstract theory, but unfortunately they make it difficult to grasp the meaning of research underlying brain organization theory. We might read that two different studies have shown that prenatal androgens "masculinize" women's sexuality, but we can't be confident that those two studies are talking about the same thing—in fact, there's a good chance the studies directly contradict one another. My point is not that updating the definitions is bad, but that updating the definitions has serious consequences for how brain organization studies can be compared to see if the theory is supported, on the whole, by the evidence.

It is important to be entirely clear about one thing: I do not believe that scientists shifted definitions on purpose in order to make their favored theory look better. From the studies and from the interviews I conducted with scientists, I am convinced that they are totally unaware of the transformation this key concept has undergone over time in their studies. As I noted early in this chapter, these scientists—like most people, probably—tend to think that masculine and feminine sexuality is a no-brainer. My sense is that over time, as sexual-revolution ideas about women's sexuality became increasingly mainstream, scientists' definitions changed without their even realizing it. This was an especially comfortable and "natural"-feeling change because of the ways that data on DES-exposed women seemed to fit hypotheses with the newer definitions of feminine sexuality. But regardless of *why* the definition of feminine sexuality changes or whether scientists are aware of it, once we notice the change, we have to acknowledge that studies with different definitions of feminine sexuality generate irreconcilable evidence about the theory.

According to the symmetry principles reviewed in the previous chapter, studies with definitions that are starkly different or even contradictory can't be lumped together as evidence of how hormones affect feminine sexuality. The general term *feminine sexuality* is used to gloss divergent— and in some cases diametrically opposed—concepts in the research. This level of departure from earlier definitions veers dangerously close to making brain organization literally unfalsifiable. It begins to look as though *any* difference from a comparison group could be interpreted as in keeping with the idea that CAH-affected women are masculinized, or DES-exposed women are defeminized, and so on. Further, because scientists have not noticed the changes and confronted them directly, they have both reinforced the notion that "masculine" and "feminine" sexuality are universal, timeless constructs and created the illusion of a seamless line of evidence supporting human sexuality as hardwired by hormones.

In addition to the dramatic shifts in definitions for key constructs, scientists have also sometimes failed to notice serious discrepancies between specific studies and the actual predictions of brain organization theory. One of the clearest examples is the study of DES-exposed boys by Yalom, Green, and Fisk (1973). Yalom and colleagues' hypothesis that estrogens would have a feminizing effect was based on the notion that androgens and estrogens are antagonists, and thus would have *opposing* effects on development. As I noted earlier, this "sex antagonism" paradigm was inconsistent with brain organization theory. (Recall that the theory at the time concerned the "active" role of androgens in differentiating as masculine; development as feminine was believed to be the *default*, not the result of exposure to estrogens.) Moreover, the antagonism hypothesis had been empirically refuted in many experiments from as early as the 1930s (Beach

1971; Oudshoorn 1994). Yet as historian Marianne van den Wijngaard (1997) has documented, sex antagonism held sway with a great many biomedical researchers through the 1960s and beyond. In fact, van den Wijngaard has convincingly argued that the "consensus among scientists" that androgens and estrogens were "carriers of biological maleness and femaleness, respectively" was a potent factor in scientists' readiness to accept brain organization theory in the first place (37). It should not be surprising, then, though it is indeed disappointing, that in spite of its inconsistency with brain organization theory at the time the study was done, and its subsequent outright contradiction by studies that found DES to be a masculinizing agent, Yalom and colleagues' study continues to be widely cited as supporting brain organization theory.[20]

The issue of citations, of which studies have "staying power" and which do not, as well as which specific findings continue to be mentioned while other findings—even in otherwise highly cited studies—disappear, is likewise important. For example, I noted that Kester and colleagues' study is still actively cited, but that the specific findings on the timing of psychosexual milestones have literally disappeared from the literature. The troubling possibility is that those findings have disappeared because both Kester's hypothesis and her findings of *early* sexual initiation among DES-exposed men contradict both the theory and the evidence presented in later studies. But instead of directly confronting and disputing the older model (as is supposed to be the scientific norm for advancing knowledge), scientists have replaced the previous conceptualization of masculine precocity/feminine delay by selective reference and reinterpretation of earlier research. Again, I do not think that this practice has been deliberate; instead I think it is a function of researchers' searching the literature for findings that seem "salient" when they are exploring their current hypotheses. But the "cleansing" effect that it has on the scientific record is unhealthy, because to eyes trained solely on the current literature and its characterization of the research trajectory, it makes brain organization research appear seamless, coherent, and ever-advancing—rather than characterized by many dead ends, reevaluations, and continuing contradictions.[21]

To sum up the effect of shifting definitions on masculine and feminine sexuality, scientists' superficial certainty about these concepts masks both an enormous amount of disagreement among them and a lot of change in the concept of feminine sexuality over time. Their cavalier approach to definitions of their key variables has led them to be conceptually sloppy, and the result is devastating for the existing network of evidence about brain organization. As we shall see in the next chapter, they also stub their toes on their "commonsense" approach to homosexuality.

CHAPTER SEVEN

Sexual Orienteering

Orienteering, n. 1. The action of determining one's position correctly. 2. A competitive sport in which runners have to find their way across rough country with the aid of a map and compass.

WHY ARE SOME PEOPLE gay and other people straight? This question is quite intriguing to some scientists, judging by the amount of time and money they put into looking for evidence that hormones influence sexual orientation. And the public seems very interested, too, because studies about the possible origins of homosexuality always get great news coverage. But sexual orientation is notoriously difficult to define. So when a headline proclaims, "Prenatal Environment May Dictate Sexual Orientation," just what is it, exactly, that is said to have been dictated?[1] Is it whom someone desires? Whom one has sex with? What a person calls him or herself? Because these various aspects of sexuality don't always point toward the same orientation, it matters very much which one a scientist picks. In this chapter, I'll show that brain organization research on sexual orientation is fraught with internal contradictions. In particular, scientists' extremely different ways of measuring "homosexuality" have contributed to a network of studies that look convincing and mutually supportive on the surface, but in fact are fundamentally at odds with one another.

In interviews, when I asked prominent scientists who study brain organization what they thought about various theories of sexual orientation, most of them were uninterested in these questions. Dr. A, whom I introduced in Chapter 6 by way of his comments on the "commonsense" difference between masculine and feminine sexuality, told me a story to illustrate his conviction that current theoretical work on sexuality is pointless and distracting. Referring to the Klein Sexual Orientation Grid (Klein, Sepekoff, and Wolf 1985), a measure that treats sexual orientation as multidimensional and changeable over time, Dr. A's comments show how brain organization researchers filter out the complications that trouble scientists in other fields:

> Fritz Klein had some complicated grid for determining sexual orientation, and [Dr. N], who was doing the bulk of these interviews, said, "You know, these are all nice scholarly heuristics, but 99 times out of a hundred, listen to the subject and they'll tell you who they are." You don't run into too many straight men who lie and say they're gay. You may have some gay men who will tell you they're heterosexual, but not if they know that the confidentiality of the project is high. And under questioning and interactions and interchange, as well as filling out questionnaires, if they're consistent, we keep them as subjects. If they check off straight and then they start checking off gay and then they tell you straight and they tell you gay, we're not sure what's going on. We're not sure if the person is just having fun with us, we're not sure if the person is themselves not sure of who they are, and when that would happen, we'd exclude them from participating further in the study. (Dr. A. interview, January 21, 1999)

To Dr. A and Dr. N, who are among the most influential scientists in the world studying biological influences on human sexuality, subjects who give equivocal or contradictory answers to questions about sexual orientation are either obstructive or confused. Ironically, Dr. A's admonition to "listen to the subject" ends up being qualified by "*if* they're consistent." Dr. A doesn't consider the possibility that subjects' hedging and ambiguity reflect meaningful complexity—that the phenomenon of sexual orientation *is* complex and sometimes ambiguous. Instead, he thinks of sexual orientation as a simple categorical trait—the objections of modern intellectuals and old cranks like Alfred Kinsey notwithstanding, you *can* sort people into discrete types. In this view, subjects who don't fit the profile are simply lying, or perhaps more charitably, self-deluded.

But in more than two decades of conducting sexuality research, I've noticed that subjects sometimes object to questions when they feel forced into unsatisfying choices to describe their sexuality. More often, they struggle to provide the "right" answers, and this can lead to confusing and contradictory responses. In fact, that's one major reason to have multiple questions that are supposed to assess the same thing—not to screen out "bad" research subjects, but to screen out bad questions. Perhaps Dr. A's subjects hedge and equivocate because his questions or response categories are not adequately nuanced. How much same-sex attraction or even experience is required to invalidate someone's identity as straight, and vice versa? Such questions are literally out of the question in brain organization research.

While other scientists vigorously debate what sexual orientation is and how best to measure it, brain organization researchers almost never address the fundamental questions involved. The majority of studies linking early hormone exposures with human sexuality have focused primarily or exclusively on sexual orientation—yet most brain organization studies that are "about" sexual orientation either have not defined sexual orientation at all, or have used vague and contradictory definitions that often do not agree with the measures scientists have used.[2] In the last decade or so, scientists who study prenatal hormones and sexual orientation have been somewhat more likely to include definitions and even detailed information on measures in their research reports, but most continue to simply declare that their own definition is best, leaving the impression that measures are mundane and predictable rather than plucked from a wide range of alternatives. However, as we shall see, the "quiet" disagreements among brain organization scientists about sexual orientation are profound.

The point is not that some scientists choose the "wrong" measures for sexual orientation. The point, instead, is that the measures they choose often put them at odds with one another. As a result, their studies just don't add up to the consistent findings about brain organization that are commonly claimed. In fact, a good many of the studies supposedly showing that prenatal hormones influence sexual orientation *contradict* other studies that supposedly show the same thing. Between slippery measures and other problems, such as illogical dose–response inferences about hormone effects, the overall evidence that early hormones affect sexual orientation is murky, at best.

This chapter focuses first on problems related to measurement and next on inconsistent patterns of evidence in the studies. I am particularly interested in how scientists define homosexuality in this body of research, and in the way measurement problems amplify some breaches in evidence that appear when comparing the major subsets of studies. That is, some studies start with hormones, and other studies start with endpoints (sexual orientation). Somehow the evidence from all these studies should meet up, but it does not. In part because researchers have very different notions of what counts as homosexual (usually related to what measures will give "better" results in the individual studies), it's impossible to get the findings to align in a way that supports brain organization theory.

I explore measurement of sexual orientation by focusing on three major issues. First, what part of sexuality *is* sexual orientation? Some of the most commonly identified elements include the sex of actual sexual partners;

self-description as heterosexual, homosexual, or bisexual; the sex of the people one falls in love with; and (the one that most scientists prefer) the degree of desire for same- or other-sex partners. Moreover, is there a single, definitive criterion, or do multiple aspects of sexuality constitute orientation? Second, do scientists think of sexual orientation as being toward men versus women, or toward one's own sex versus the other sex? This distinction might seem trivial, but it turns out to have major implications for study design, because it determines which people are considered to have similar versus contrasting sexual orientations. That is, are lesbians like straight men, because both are attracted to women? Or are they like gay men, because both are attracted to people of their own sex? The third question involves issues of quantity: what are the cutpoints, or boundaries, between heterosexuals, homosexuals, and bisexuals? How much interest in or sexual activity with people of the same sex disqualifies someone from a heterosexual orientation, and vice versa?

In the second part of this chapter, I again use symmetry principles to think about how the findings in studies of sexual orientation align with one another. My main strategy is to compare the two major kinds of brain organization studies, cohort studies and case-control studies. As I described in Chapter 3, quasi experiments like those used in human brain organization studies have to be judged together, not individually. With quasi experiments, it is especially important to see if there is triangulation of findings, meaning that evidence from studies with significantly different designs should point in the same direction. Do the cohort studies (where scientists begin with early hormone exposures, then investigate psychosexual characteristics) point to similar conclusions as the case-control studies (where scientists begin with "distinct" psychosexual groups, then look for evidence that these subjects had different prenatal hormone exposures)?[3] As you will see, the very different pattern of measurement in the cohort compared with the case control studies is a key factor in judging how the findings stack up across the two sets of studies.

After considering the overall pattern of findings, I return briefly to the issue of generalizability. In research on sexual orientation and prenatal hormone exposure, no one is really interested in the prenatal hormone histories of specific individuals who are gay or straight: people want to know something about the mechanisms of sexual orientation development in general. So the question of generalizability is a critical issue, and the major question that must be answered is whether "homosexuality" in brain organization research means the same thing as "homosexuality" outside the lab.

Let us begin, then, with the basics: what *is* sexual orientation?

Measurement Issue #1:
Constituents of Sexual Orientation

Sexual attraction? Behavior? Love? Identity? Sexual orientation is a great example of a "commonsense" concept that seems fairly transparent, but turns out to be complicated and slippery when you try to pin it down. And deciding *how* to pin it down generally requires knowing *why* you want to pin it down. In epidemiological work on HIV/AIDS, for instance, it is often more important to know what people *actually do* sexually than what they would *prefer* to do. But that is not always the case. Depending on one's objective, it might be most important to understand how people think of themselves (so that we can devise and target public health campaigns in which people can "recognize" themselves), or to understand what specific sexual practices they enjoy and value (so that we can help them negotiate and maintain safe behaviors) (Young and Meyer 2005). So just what is the best indicator of sexual orientation for brain organization research?

It seems wise to begin by considering the experimental foundation for brain organization studies, that is, to begin with the animal models for sexual orientation. When scientists draw parallels between animal behavior and human homosexuality, they nearly always make a series of assumptions that draw a line between male-typical sexual behaviors and orientation to females, on the one hand, and female-typical sexual behaviors and orientation to males, on the other. Their reasoning goes like this: in animals, *masculinizing* hormones lead to the development of "male-typical" sexual behaviors, such as mounting. An animal who mounts requires a partner who allows itself to be mounted (or, in the case of rodents, facilitates a partner's mounting by performing lordosis, raising the rump). Because allowing a partner's mount or performing lordosis is considered "female-typical" behavior, scientists reason that an animal who exhibits mounting is signaling a preference for female partners. An animal who performs lordosis or allows another animal to mount it is taken to be signaling a preference for male partners. (It's worth knowing the fact that normal animals of both sexes both mount and are mounted, and that mounting behavior can be about other things besides sex, like play or dominance. Certainly, mounting is much more common in males, and lordosis or receiving mounts is more common in females, but even these definitively masculine or feminine behaviors aren't the sole purview of either sex. This does not necessarily discredit the model, but it suggests that characterizing even rodent sexual behavior in dichotomous "masculine" or "feminine" terms is a simplification that might foreclose other ways of thinking about sexual variability.)

Researchers have long interpreted associations between *masculinizing* hormone exposures (which, as you will recall from earlier chapters, can be either androgen or estrogen exposures) and mounting behavior to mean that *masculinizing hormones lead to an orientation toward females, regardless of sex.* By the same reasoning, researchers suggest that *feminizing* hormones (or a lack of masculinizing ones) lead to an orientation toward males.

The chain of reasoning that equates male-typical sexual behaviors (especially mounting) to male-typical sexual orientation has been subject to devastating critique (Byne and Parsons 1993; Fausto-Sterling 1985). Briefly, as an animal model for human homosexuality, it is a logical error to designate just one individual in a same-sex pair as "the homosexual" (that is, a female who mounts, or a male who is mounted). Moreover, according to this model, even male–female pairs can be interpreted as "homosexual" if a female mounts a male (a not too rare occurrence, even among animals not experimentally treated with hormones). Of course, human homosexuality is not typically understood as the preference for particular behaviors or sexual positions, but is understood as a preference for partners of one's own sex. In response to such critiques, animal researchers increasingly conduct studies that assess actual preference for female versus male animals rather than testing for the display of certain "sex-typed" sexual behaviors (Watabe and Endo 1994; Adkins-Regan 2002; Roselli, Larkin, Resko, et al. 2004; Bodo and Rissman 2007).

A Closer Look at Animal Sexual Preferences:
Rams on the Down Low?

Unfortunately, some of the contexts in which animals are tested for "sexual" preferences are still not necessarily good models for human sexuality—for example, scientists might measure whether an animal spends more time investigating bedding that has been used by male versus female animals (Bodo and Rissman 2007), or simply spends more time "visiting" male versus female animals when given the opportunity to approach tethered conspecifics of either sex (Watabe and Endo 1994). Enormous assumptions must be made to interpret these measures as sexual orientation. For example, is time spent with a tethered female an indicator of sexual interest in the female, or an indicator of some aversion to the male? These are not the same thing, because the latter might just be conflict avoidance.

The most stringent sexual orientation test that any team has used was devised by Charles Roselli and his colleagues at Oregon State University
(Resko et al. 1996; Roselli, Resko, and Stormshak 2002; Roselli, Larkin,
Schrunk, and Stormshak 2004). Their three-part test for "male orientation" among rams consists of the following:

> Only rams that would not mount estrous females in the preliminary tests sat-
> isfied the criterion for entrance into the Sexual Preference Paradigm. . . .
> Briefly, in November and December of the second year of life, rams were ex-
> posed simultaneously to two restrained estrous females and two males that
> were chosen at random for use. Rams that courted and mounted males in
> preference to females during a 30-min test that was repeated at least three
> times were classified as male-oriented. Male-oriented rams were given an ad-
> ditional preference test 5 days before they were killed. Using these proce-
> dures, we identified six males that would not mount females in the prelimi-
> nary sexual tests, mounted males in a group setting, and mounted males in
> preference to females in the sexual preference tests. (Resko et al. 1996, 121)

It turns out, though, that most, if not all, of these "male-oriented" rams
are having sex with ewes when the scientists aren't looking. In a study that
was designed to test the relationship between successful reproduction (sir-
ing of lambs) and behavioral measures of "sexual performance," 5 male-
oriented rams impregnated at least 330 ewes over 21 days, resulting in
more than 480 offspring.[4] The number of ewes they impregnated and the
number of lambs sired fell roughly midway between the groups of rams
whose behavioral tests had classified them as "high performing female-
oriented" and "low performing female-oriented" (Stellflug, Cockett, and
Lewis 2006). In behavioral tests after breeding, only 3 of the 5 "male-
oriented" rams exhibited *any* sexual activity, and 2 of the 3 continued to
show a strong (though not exclusive) preference for mounting other rams.
Offering the understated conclusion that "individual sexual partner preference tests did not absolutely reflect the breeding performance of the
male-oriented rams," the scientists speculated that aspects of the testing
environment differed sufficiently from the "competitive breeding environment" to make the preference test an imperfect predictor of rams' mating
behavior (466).

There are two take-home messages from this brief discussion of animal
research on sexual orientation. First, it is not an easy task to come up with
good animal models for partner preferences, and much research focuses on
animals' *sexual positions* rather than on preferred category of partners.
Second, even when sexual preference tests in animals focus on identifying
partner preferences, the preferences that look like a "stable, dispositional
trait" in one context look unstable and fluid in other contexts. (Of course,

it's not really clear which aspects of the context are relevant for inducing different sexual behaviors in these rams. It's tempting to speculate that these rams, like an awful lot of humans, act differently when their aim is procreation than when it is recreation. At any rate, given the number of gay people who sometimes have sex with other-sex partners, and straight people with same-sex partners, maybe the sheep is a better animal model for human sexual behavior than we thought!)

What about humans? Social scientists and public health researchers who study human sexuality spend a great deal of time discussing what counts as sexual orientation, and debating how best to measure it (Klein, Sepe-koff, and Wolf 1985; Laumann et al. 1994; Sell 1997; Solarz 1999; Young and Meyer 2005). Some suggest it is best defined by the sexual attraction or desires one feels, while others point to patterns of falling in love or pair-bonding, the (physical) sex of sexual partners, the *gender* (masculinity or femininity) of sexual partners, or even whether one identifies as straight, gay, lesbian, or bisexual. At first blush, though, there seems to be much greater consensus among researchers conducting brain organization stud-ies. For one thing, they all agree that sexual orientation is a "status," meaning that what people do, or what they call themselves, might change, but the basic character of their desires is fixed from a very early stage of development. John Money, always an imaginative writer, explained the difference between a status and sexual behavior via the elaborate example of a "crazed sex-terrorist":

> The Skyscraper Test exemplifies the difference between act and status. One of the versions of this test applies to a person with a homosexual status who is atop the Empire State Building or other high building and is pushed to the edge of the parapet by a gun-toting, crazed sex terrorist with a heterosexual status. Suppose the homosexual is a man and the terrorist a woman who de-mands that he perform oral sex with her or go over the edge. To save his life, he might do it. If so, he would have performed a heterosexual act, but he would not have changed to have a heterosexual status. (Money 1987)

Of course, the idea that sexual orientation is organized by early hor-mone exposures depends on such a definition, because the notion of orga-nization invokes a point of irrevocable commitment of structure or func-tion in the developing organism. Importantly, the reverse is not true—that is, thinking of sexual orientation as a status that is stable from very early in life does not depend on any particular theory about how sexual orienta-tion develops. Thus, it is not the case that skeptics of brain organization theory necessarily reject the idea that sexual orientation is a status. In any case, let us turn away from this single point of agreement among brain organization researchers to consider their many points of *disagreement* about what sexual orientation is.

If "Kinsey 5" Was the Answer,
What Was the Question?

A common measurement tool that helps maintain the illusion of broad agreement among scientists is the familiar "Kinsey Scale." Kinsey rated men's "psychologic reactions and overt experience" as follows:

0 Exclusively heterosexual with no homosexual
1 Predominantly heterosexual, only incidentally homosexual
2 Predominantly heterosexual, but more than incidentally homosexual
3 Equally heterosexual and homosexual
4 Predominantly homosexual, but more than incidentally heterosexual
5 Predominantly homosexual, but incidentally heterosexual
6 Exclusively homosexual.(Kinsey, Pomeroy, and Martin 1948)

Investigators generally write as though "Kinsey scale" fully explains a method of assessing sexual orientation. Yet a Kinsey scale is a format for *answers;* it does not explain what *questions* were asked. There are in fact many different kinds of questions that one might ask about sexual orientation and answer with a Kinsey scale. Kinsey intended the scale to reflect the heterosexual–homosexual balance in people's sexual *activities and responses*. It was not meant to be a tool for rating people *per se*. In fact, Kinsey was adamantly opposed to the use of the terms *heterosexual, bisexual,* or *homosexual* to refer to people, in part because he thought it implied a biological basis for homosexuality, which he thought was an extremely silly theory. Historian Stephanie Kenen has wryly observed that "Kinsey's rating scale was not a resounding success" in this regard, because "the ratings themselves have become reified categories of identity, and references to 'Kinsey 4s' and 'Kinsey 6s' are not uncommon in popular parlance or scientific literature" (Kenen 1997, 208–209).

Alfred Kinsey has posthumously lost the battle over using his famous scale to rate people, rather than behaviors, as homosexual versus heterosexual, but there are still many questions to resolve among those who use the scale to rate people. Which aspects of sexuality are most interesting and important? Brain organization researchers strenuously disagree about this, and various scientists have chosen many different dimensions to measure with Kinsey scales, such as

Sexual behavior (have sexual contacts been with men, women, or
both?) (e.g., Lindesay 1987)

Identification with a particular sexual orientation term such as *homosexual, bisexual,* or *heterosexual* (e.g., Holtzen 1994)

Sexual fantasies (are they about men, women, or both?) (e.g., Bailey,
Willerman, and Parks 1991)

Composite assessments of night dreams, day dreams, erotica, masturbation fantasies, sexual attractions, and sexual partners (e.g., Meyer-
Bahlburg et al. 2008)

All this variety is often obscured by the simple notation that a study
"employed Kinsey scales to rate subjects as homosexual, heterosexual, or
bisexual." Reading beyond the studies that employed Kinsey scales, the
manifestations of sexuality that count as sexual orientation continue to
expand. For example, John Money saw sexual orientation as comprising
multiple dimensions, but believed that *romantic love* rather than sexual
behaviors, fantasies, or even sexual attraction or arousal was "the definitive criterion of homosexual, heterosexual, and bisexual status" (Money
1987).

Even the most widely accepted definition of sexual orientation as "sexual attraction" to a certain category of partners does not resolve the issue,
because opinions differ about the best indicator of sexual attraction. An
excellent example of disagreement about measuring sexual attraction is
found by contrasting Dr. A, mentioned above, with Michael Bailey, another prominent psychologist who does brain organization research. Recall that Dr. A suggested that people can "tell you who they are," so this
would suggest you could also simply ask people if they are attracted to
men, or women, or both. Bailey, though, believes that measures of physiological arousal are superior. In a study that made the front page of the *New
York Times,* Bailey's team, led by graduate student Gerulf Rieger, compared subjective attraction to a measure of physiological response (Rieger,
Chivers, and Bailey 2005).[5] The team had advertised for "'heterosexual,'
'bisexual,' and 'gay' men for a paid study of sexual arousal," and men who
volunteered "were asked about their sexual attraction toward men and
women" as well as "their sexual identity as straight, bisexual, or gay"
(580). Each man was then tested with a penile plethysmograph, which involves showing the man explicit sexual images while measuring his penis
to detect circumference changes associated with erection. The stimuli in
this study were four short erotic films, two of which showed two men having sex with each other, and two of which showed two women having sex
with each other. (To avoid confusion regarding which person in the film

was arousing to the research subjects, the scientists did not use films of men and women having sex together.) As expected, the men who said they were attracted to men were more aroused by pictures of men, and the men who said they were attracted to women were more aroused by pictures of women. But the men who said they were attracted to both men and women, instead of showing arousal patterns that were intermediate, mostly showed more erectile response to the male images (a few mostly showed response to the female images). Bailey concluded from the study that "in men there's no hint that true bisexual arousal exists, and that for men *arousal is orientation*" (quoted in Carey 2005).

This study got a great deal of attention, but very little of it was critical (for exceptions, see the letter by Barker, Iantaffi, and Gupta [2006] and the comments by Fritz Klein, Gilbert Herdt, and Randall Sell in Carey 2005). It provides a dramatic example of the variety of measures that are available even for apparently obvious aspects of sexual orientation like "erotic attraction." It also shows how subtle inconsistencies between abstract definitions and concrete measures slip into studies. Bailey's comment that "*arousal* is orientation" omits other aspects of orientation that he and his co-authors identified when they wrote: "Sexual orientation refers to the degree of sexual attraction, fantasy, and arousal that one experiences for members of the opposite sex, the same sex, or both" (Rieger, Chivers, and Bailey 2005). What is the justification for dropping out subjective attraction and fantasy? The report also begs the question of choosing between different measures for the same aspect of orientation—in this case, arousal. That is, even if we could agree that arousal is the most important or "core" aspect of orientation, Bailey's team doesn't offer any rationale for judging the *physiological* measure to be the best indicator of arousal. By what criterion is blood flow to the penis a better indicator of desire than someone's subjective sense of arousal? One reason to question the assertion that the physiological measure is superior is that more than a third of the men in the study did not have enough erectile response to the films to be classified as aroused at all (581). Does this mean that one-third of men who believe that they are either heterosexual, gay, or bisexual actually have *no* sexual orientation? That seems to be an absurd conclusion. A test that provides interpretable results only for two-thirds of healthy, normal subjects is problematic, to say the least.

And it's not necessarily the case that a "better" physiological measure is the answer to this problem. Biologically oriented researchers have tended to invest great faith in physiological measures as superior, but this requires justification. In particular, the subjective sense of arousal may be a better predictor of other aspects of sexual orientation (such as actual sexual pairings or love relationships). In fact, there was a second measure of arousal that was overlooked when the study was reported in the popular press. Men were directed to move a lever in response to their own sense of sexual arousal while watching the films. By this measure, bisexual men's arousal actually matched quite well with their self-reported patterns of attraction and identity. Other researchers have found that self-described orientation correlates well with "implicit" measures that, like the physiological measure here, "reflect immediate reactions that may not be available to introspection" (Snowden, Wichter, and Gray 2008).

Finally, think back to the sheep for a moment. It is possible that some aspect of the testing context (such as the specific images chosen, the types of sexual activity portrayed in images that included men versus women, and/ or the fact of being watched by the scientists running the experiment) made attraction to men more salient for the bisexual subjects. Recall that the images of women were of *women having sex with each other.* Perhaps bisexual men generally have less interest in "lesbian" imagery than heterosexual men do. (I have no idea if that's true.) The point is that *context and subjectivity matter to sexuality,* and the attempt to get around subjectivity in order to "drill down" to a more valid substrate is misguided. This is a point I take up in the final chapters. For now, let's return to how sexual orientation is measured in brain organization studies.

Getting Real: Sexual Orientation as Multidimensional

Many brain organization researchers ask about more than one aspect of sexual orientation in their studies. Treating sexual orientation as multidimensional fits more closely with the data on sexuality from social science and public health research, in which the most commonly assessed aspects are the gender of actual sexual partners ("behavior"), the gender of ideal sexual partners ("fantasy"), and self-identification as homosexual, heterosexual, or bisexual ("identity") (Sell and Becker 2001; Laumann et al. 1994).[6]

The reason for assessing multiple dimensions, of course, is that many people's sexual orientation does not neatly line up across all the different possible aspects of this phenomenon. Take a look at the Venn diagrams in Figures 7.1 and 7.2. The data are from the National Health and Social Life Survey (Laumann et al. 1994), one of the largest and most comprehensive sexuality surveys conducted to date in the United States. Each of the diagrams shows how three aspects of sexual orientation (behavior, fantasy,

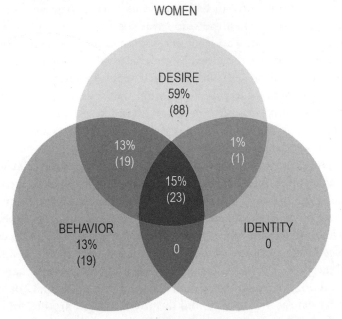

Figure 7.1. Among women, the relationships among same-sex desire, sexual behavior with same-sex partners, and lesbian or bisexual identity. (Laumann et al. 1994, page 298, used with permission of University of Chicago Press.)

and identity) are related in people who reported any kind of same-gender sexuality as adults. The most important thing to notice is the small shaded area at the center of each diagram. This area shows the proportion of people for whom all three aspects of sexual orientation align: just 15 percent of the women and 24 percent of the men indicated sexual desire for same-gender partners, *and* sexual activity with at least one same-gender partner in adulthood, *and* also consider themselves to be either homosexual (including equivalent terms such as gay or lesbian) or bisexual.[7]

These are also the most common dimensions of sexual orientation assessed in brain organization studies, both those that rely on just one dimension and those that ask about two or more dimensions. Note, though, how differently this team of sociologists perceives their finding of discrepancy among the various dimensions of sexual orientation, compared with the researchers (usually psychologists) who conduct brain organization studies. Laumann and colleagues approach the discrepancy itself as the phenomenon of interest, rather than as "noise" to be eliminated. In other words, they measure multiple dimensions in order to answer the question "How are these three aspects of homosexuality related?" (1994, 298).

Further, they use their analysis of continuity and divergence among the three aspects as an opportunity to reflect on various ways that "homosexuality is both organized as a set of behaviors and practices and experienced subjectively" (300). They are also attentive to the way that the subjective experience of sexuality is, at least in certain respects, responsive to the social environment, noting that "the group of people who report behavior and desire but not identity is quite small among the men but fairly sizable among the women, comparable to the women who had sex partners but nothing else and to those who exhibit all three characteristics. This may indicate a slightly lower threshold of homosexual and bisexual identity among men than among women. This would fit with the historically greater visibility of gay men as opposed to lesbians" (300). Finally, they sensibly suggest that at least some of the discrepancies they find are related to their own manner of questioning: "Not surprisingly, since these questions are asked in different places [in our survey] and in different ways (face-to-face vs. self-completion, directly vs. indirectly, etc.), there were some inconsistencies between responses" (298 n. 17). Accordingly, Laumann and colleagues firmly endorse multidimensional assessments:

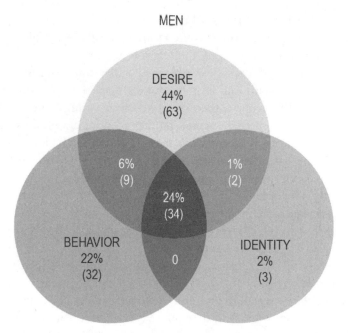

Figure 7.2. Among men, the relationships among same-sex desire, sexual behavior with same-sex partners, and gay or bisexual identity. (Laumann et al. 1994, page 298, used with permission of University of Chicago Press.)

While the measurement of same-gender practices and attitudes is crude at best, with unknown levels of underreporting for each, this preliminary analysis provides unambiguous evidence that no single number can be used to provide an accurate and valid characterization of the incidence and prevalence of homosexuality in the population at large. In sum, homosexuality is fundamentally a multidimensional phenomenon that has manifold meanings and interpretations, depending on context and purpose. (301)

Some brain organization studies do use multidimensional measures (for instance, Ehrhardt et al. 1985; Dittmann, Kappes, and Kappes 1992; McCarty et al. 2006; Meyer-Bahlburg et al. 2008), but the rationale and execution behind such assessments is quite different. Rather than seeking to explore the patterns and meanings of same-sex orientation that multiple measures can reveal, brain organization researchers assess more than one dimension exclusively to increase their ability to detect any same-sex eroticism. I explore problems with this approach further along in this chapter. Meanwhile, it is also fairly common for scientists to agree *in principle* with the idea that sexual orientation is multidimensional, but to omit certain dimensions when they measure the construct, or when they conduct their analyses (as the study of arousal among bisexual men discussed above illustrates). In such cases, the working definition remains one-dimensional (for instance, Ellis et al. 1988; Reite et al. 1995; Bailey, Willerman, and Parks 1991). For a study testing the relation of hormones and sexual orientation via maternal stress, for example, Bailey and colleagues (1991) measured both behavior and fantasy, but dropped the behavior measure. As justification, they cited a theoretical article by John Money (1987) to support their assertion that "sexual fantasy is more closely linked to the concept of sexual orientation than is sexual behavior" (Bailey, Willerman, and Parks 1991). Money actually pointed not to fantasy but to falling in love as the key factor; in any case, why *ask* about sexual behavior if you don't *use* the information?

Just as some criteria are frequently dropped out, other criteria are sometimes slipped into the measure without acknowledgment, because of the ways subjects are recruited into studies. Several highly cited studies by Gunter Dörner (Dörner, Rohde, et al. 1975, 1976, and 1983; Dörner et al. 1980; Dörner, Schenk, et al. 1983) demonstrate this problem. In one study, Dörner *explicitly* asserted that sexual orientation is desire (being "sexually excited by another male") and sexual experience (a "homosexual" has "unambiguous homosexual behavior since puberty"), and *implicitly* suggested that it is role in sexual activity (a homosexual man is one who is "receptive" in intercourse, or exhibits some other human homologue of "female-like" sexual behavior as it is experimentally manipulated in rats)

(Dörner, Rohde, et al. 1975). These components of orientation are not equivalent, and are independent: men can be sexually excited by other men and be sexual "tops" or be sexually versatile (preferring an insertive role in intercourse or having no role preference); and men can be sexually excited by other men but have sexual experiences only with women. In practice, Dörner has never assessed specific position or activity preferences. More-over, his recruitment procedure relied heavily on referrals from clinicians who deal with various "sexual problems" (therapists, sexologists, and venerologists, and so on). He thereby included "having sexual problems" as one of the unstated criteria for being classified as a "homosexual" in his studies.[8]

To be fair, it is difficult to address measurement issues adequately within the page limits of a typical journal article, and in recent years brain organi-zation researchers have been much more likely to publish monographs that do address the definition and appropriate measures for sexual orienta-tion (Wilson and Rahman 2005; Bailey 2003; LeVay 1996). Unfortunately, though, even in these lengthier treatments, there is a tendency for scientists doing brain organization research to make their arguments in a manner that is more like advocacy than like science (see the discussions below about choosing boundaries between homosexual and heterosexual sub-jects, and generalizability of definitions).[9] Meanwhile, in the brain organi-zation studies themselves, investigators tend to sweep measurement issues aside as either unimportant or already settled.

Measurement Issue #2: Framing

There are two basic ways to frame sexual orientation for a study. The first possibility is to conceive of someone's sexual attraction being directed toward women (gynephilic), men (androphilic), or both. The second possi-bility to is conceive of attraction as being toward one's own sex (homo-sexual), the other sex (heterosexual), or both (bisexual). For many pur-poses, it might not make a big difference whether you use a gynephilic/androphilic frame or a homosexual/heterosexual/bisexual frame, but in brain organization research the choice is critical. This is because the frame determines how people are grouped to reflect similar versus different sex-ual orientations in studies. Those groupings, in turn, matter when you are looking at multiple studies to see whether the findings are consistent.

In the gynephile/androphile frame, heterosexual men and lesbians have similar orientations, and gay men and heterosexual women have similar orientations. In the homosexual/heterosexual/bisexual frame, gay men and

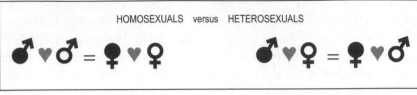

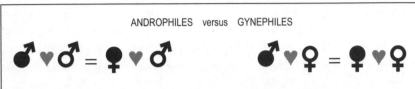

Figure 7.3. Homosexual/heterosexual frame versus androphile/gynephile frame.

lesbians have similar orientations (both are attracted to people of the same sex as themselves), as do heterosexuals of both sexes (attracted to people of the other sex), and bisexuals of both sexes (attracted to both same-sex and other-sex partners). This simple difference is illustrated in Figure 7.3.

The first thing to recognize is that the androphile/gynephile frame is the one that is implicit in the theory of brain organization. It is easiest to understand this by considering how some scientists think of the "representation" of sexual orientation in the brain, and how this line of thought is connected to research in other animals. Not all traits that are thought to be organized by early hormone exposures are credited with corresponding to distinct or "dedicated" brain areas, but sexual orientation often is. Michael Bailey (2003) is among the prominent researchers who have speculated that there is a specific "sexual orientation center of the brain" that differentiates as masculine or feminine under the influence of early hormones (but compare LeVay 1996; see also Morris et al. 2004; Rahman 2005).[10] As Bailey explains it, "If [the sexual orientation center] is masculine, then attraction to women results. If it is feminine, then attraction to men results. Thus, both straight men and lesbians would be expected to have masculine sexual orientation areas, and gay men and straight women to have feminine areas" (Bailey 2003).

As support for this idea, Bailey and others appeal to research in nonhuman animals. Manipulation of early hormone exposures affects various mating and courting behaviors in a range of species. Furthermore, manipulating hormones during development, as well as removing or damaging certain brain areas, indicates that specific neural areas, especially parts of the hypothalamus, are necessary for various components of sexual activity and interest in infrahuman animals. And in sheep, one of the few animals in which natural variations in partner preference have been studied, the size of a cell group in the hypothalamus seems to vary according to both sex and partner preference (in general, the area is larger in rams than in ewes; but larger in "female-oriented" rams than in "male-oriented" rams) (Roselli, Larkin, Resko, et al. 2004). Given the complexity and context-dependence of ram "sexual orientation" that I discussed earlier, it is not entirely clear what this cell group signifies. Moreover, interspecies differences are important, making easy extrapolation from animal experiments to human sexual orientation impossible. Simon LeVay has pointed out an especially useful contrast for understanding the difficulty:

> In rats, for example, lesions in the medial preoptic area interfere with mounting but leave the male rat with some interest in estrous females. They behave as if they still want to do something with the female but have forgotten what it is. In primates, on the other hand, even quite small lesions within the medial preoptic area can cause males to lose all interest in estrous females; they still have a sex drive (as evidenced by their continuing to masturbate), but they seem to lose any notion that females offer a means to satisfy it. (LeVay 1996, 134)

Whether based on behavioral displays or on partner preferences, the animal research on the "organizing" effects of hormones always employs a gynephile/androphile frame for sexual orientation. But psychologists and other scientists often frame human sexual orientation differently. Consider the definition offered by Martin Lalumiere and colleagues (2000) in a review and reanalysis of data on handedness and sexual orientation: "a person's erotic preference for opposite-sex individuals (heterosexuality), same-sex individuals (homosexuality), or both (bisexuality)" (575). Figure 7.3 showed how this "homosexual" versus "heterosexual" framework is different from the androphile/gynephile framework. Figures 7.4 and 7.5 show how each of these frameworks requires a different developmental model for relating early hormones to the development of sexual orientation. Figure 7.4 shows the original model using the gynephile/androphile frame, which does not depend on the sex of the developing individual—the same model applies to males and females. In Figure 7.5, the sexual orientation variable is framed as *homosexuality* versus *heterosexuality*." In this case, to fit with brain organization theory, there must be two different causal models: one for females, and one for males. Exposure to masculinizing hormones supposedly leads to homosexuality for women, but a *lack of exposure* to masculinizing hormones is purported to cause homosexuality in men. Only bisexuality (among those investigators who allow for this category of orientation) has a similar model for males and females:

ORIGINAL MODEL OF BRAIN ORGANIZATION

ATTRACTION TO FEMALES
(GYNEPHILE)

HIGH

H O R M O N E S

ATTRACTION TO MALES
(ANDROPHILE)

**MASCULINIZING HORMONES
IN ORGANIZATION PERIOD**

Figure 7.4. Original model of brain organization, which uses the gynephile/androphile frame.

hormone exposures that are partially, but not completely, masculinizing are thought to underlie bisexuality in both males and females.

And here's where the trouble begins. Remarkable as it seems, some scientists studying brain organization have completely failed to appreciate the implications of alternative ways of framing sexual orientation. Beginning with some of the studies on prenatal stress in the 1980s (Rosenstein and Bigler 1987; Ellis et al. 1988) and continuing with studies of handedness and other markers of brain lateralization (for example, McCormick, Witelson, and Kingstone 1990; Holtzen 1994; McFadden and Champlin 2000), a significant subset of studies have employed the terms *homosexual* and *heterosexual* as if they represented the same phenomenon in women and in men—that is, they switch to the homosexual/heterosexual frame. Yet they do not seem to grasp that this changes how one would model the influence of hormones on sexual orientation, and this leads them to either misinterpret earlier research or design studies that are at odds with the main thread of brain organization. Ellis and colleagues (1988), for example, interpreted animal data on stress to suggest that maternal stress during pregnancy "interfere[s] with fetal production of various sex hormones, especially testosterone. . . . Should this disruption occur when crucial brain parts controlling sexual orientation are being sexually differentiated, per-

BRAIN ORGANIZATION MODEL FOR HOMOSEXUAL / HETEROSEXUAL FRAME

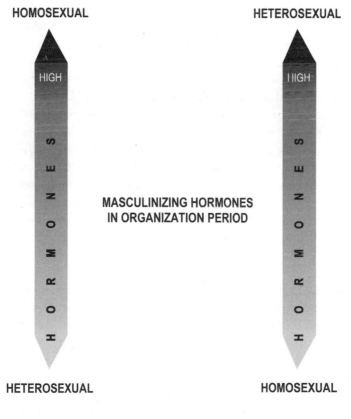

Figure 7.5. Brain organization using the homosexual/heterosexual frame. Note that this frame requires a different model for males versus females.

manent inversions could occur" (153). In the remainder of their discussion of animal experiments, they explicitly noted that the studies concerned *male* rodents, but they structured their own study "to determine if maternal stress during pregnancy was predictive of sexual inversions [i.e., homosexuality and bisexuality]"—*without regard to sex*. They proceeded to construct identical hypotheses regarding an association between prenatal stress and sexual orientation for both men and women.

McCormick, Witelson, and Kingstone (1990) ran into the same problem, but from a different angle. In their literature review, they noted that higher androgens in CAH women are related to "higher ratings of homosexual behavior" and *contrasted* this with CAH men, in whom androgens are also increased but "homosexual behavior was extremely rare" (70). Yet brain organization theory predicts that high androgen exposures in *both* CAH women and CAH men will lead to an orientation toward females (gynephile orientation). Using the "homosexual versus heterosexual" frame, McCormick and colleagues conclude that the hormonal mechanism for development of sexual orientation is *different* in men and women. Of course, if they had used the androphile/gynephile way of framing sexual orientation, they would have interpreted the data to indicate that the mechanism is *identical* in men and women.

Many researchers seem to unwittingly shift theoretical frames as a result of simply linking "atypicality" on two variables—like sexual orientation and handedness, or sexual orientation and prenatal stress. Reasoning that homosexuality is an atypical orientation, they advance the generalized prediction that homosexuals are also likely to be atypical on the second variable. But when researchers fail to differentiate *male* homosexuality from *female* homosexuality, they end up (implicitly or explicitly) hypothesizing that both male and female homosexuality are influenced by the same type of atypical hormone exposures (Rosenstein and Bigler 1987; Ellis et al. 1988; Gladue and Bailey 1995b).[11] Of course, this does not fit either with the animal research on hormones and sex-typed behavior or with the usual predictions and interpretations of the human "cohort" studies of brain organization, all of which argue that exactly the *opposite* type of hormone exposures would lead to male versus female homosexuality. Yet these studies are folded into the brain organization literature as if they concur with studies that use the exact opposite model.

Studies that link sexual orientation with prenatal stress and/or handedness are particularly problematic in this regard. The main tension between these studies and other brain organization research is disagreement about whether homosexuality in men would be related to *higher* or *lower* testosterone during development. The cohort studies (as well as virtually all animal studies on sexual behavior and partner preferences) proceed from the assumption that lower than typical testosterone would lead to male homosexuality. But because left-handedness is slightly more common in men, it is generally thought that *higher* than typical levels of testosterone lead to left-handedness. Thus, studies on various aspects of lateralization introduced a murkier set of predictions—so murky, in fact, that quite a few scientists doing this work have been unable to sort them out.[12]

Linking male homosexuality and left-handedness through prenatal hormone exposures requires a model that involves *lower* than typical testosterone during the critical period when sexual orientation is thought to develop with *higher* than typical testosterone during the period when handedness develops. Norman Geschwind and colleagues (Geschwind and Behan 1984; Geschwind and Galaburda 1985a, 1985b, 1985c) proposed that rodent studies of neuroendocrine response to prenatal stress could resolve this difficulty. As you may recall from earlier chapters, intense stress has been shown to create a brief surge in testosterone, followed by a long-term drop to levels much lower than normal; consequently, male rats subjected to prenatal stress exhibit more-feminine sexual behaviors (Ward 1972, 1984). Geschwind and colleagues suggested that intense stress during the early prenatal period could provide a mechanism whereby gay men (whose sexuality these scientists considered to be "feminized") would nonetheless exhibit the "hypermasculine" characteristic of increased left-handedness. (This solution, of course, requires exquisite timing, and relies upon a developmental sequence in terms of specific critical periods that was purely hypothetical when proposed, and remains so.)

The elaborate wedding of theories proved to be popular, but confusing. Of the first ten studies of lateralization and homosexuality, five included basic errors in their descriptions of the main theories they drew on (Lindesay 1987; McCormick, Witelson, and Kingstone 1990; Marchant-Haycox, McManus, and Wilson 1991; Becker et al. 1992; Holtzen 1994).[13] Three others evaded the issue by predicting a "link" between testosterone, left-handedness, and homosexuality, but without specifying whether higher or lower levels of testosterone would be implicated in this link (Rosenstein and Bigler 1987; Gladue and Bailey 1995b; Reite et al. 1995). In only two of these ten studies did the scientists accurately characterize the various research and theories they built upon (Satz et al. 1991; Götestam, Coates, and Ekstrand 1992).

It is surprising how many articles containing fundamental framing errors have gotten through peer review in well-respected journals, and are repeatedly cited as supporting brain organization theory. In a study that purported to find a link between prenatal stress and homosexuality, Ellis and colleagues (1988) compared homosexual with heterosexual men, and homosexual with heterosexual women, in both cases hypothesizing that homosexuality in offspring is linked to higher maternal stress during the prenatal period. But of course, there is no theoretical reason to link higher maternal stress to homosexuality in females, other than a vague folkloric prejudice that links lesbianism as a "problematic" outcome with various problematic developmental inputs. Still, in spite of methodological problems that go well beyond this issue of framing, this study, as well as a closely related review article (Ellis, Burke, and Ames 1987), are particularly highly cited (44 and 153 citations, respectively, as of July 2008, according to the ISI Web of Science).[14] Further compounding the error, many other scientists cite Ellis and colleagues' (1988) stress research as if it simply supported the theory of brain organization instead of considerably *complicating* it (for example, Holtzen 1994; Hall 2000; Yasuhara, Kempinas, and Pereira 2005; Meek, Schulz, and Keith 2006; Swaab 2007). Similarly, Rosenstein and Bigler (1987) compared male and female heterosexuals (grouped together) to male and female homosexuals (also grouped together) in order to look for an association between homosexuality and handedness. As with Ellis and colleagues, Rosenstein and Bigler assembled their groups in a way that hypothesized a *common* hormonal history for all homosexuals (males and females).

When I began analyzing brain organization research more than ten years ago, I noticed these framing problems and was surprised that not one study in the literature addressed the issue.[15] Recently some scientists have recognized the contradictions in framing and have suggested that their findings should be understood in light of other neurohormonal theories of development rather than classic brain organization theory (Lindesay 1987; McFadden and Champlin 2000; Lippa 2003). For example, in their meta-analysis of sexual orientation and handedness in men and women, Lalumiere, Blanchard, and Zucker (2000) acknowledge:

> The findings that handedness and sexual orientation are associated in both men and women are difficult to explain with current etiological theories of homosexuality, especially the well-known hormonal theory of sexual orientation, which we here call the prenatal androgen exposure theory. The prenatal androgen exposure theory postulates that homosexuality in men is due to undermasculinization and in women to overmasculinization of the brain during a critical period of development. (586)

Nonetheless, even those scientists who have tried to use a broader range of evidence to decide among the various theories that are currently at play for linking sexual orientation with handedness or other "sex-linked" traits have not gone as far as they could. In particular, these discussions still fail to account for the overall pattern of findings among human studies of prenatal stress, lateralization, and the expected versus demonstrated constellation of sex-typed traits among lesbians and gay men under various theories. For example, prenatal stress is often invoked to explain higher rates of left-handedness among gay men, but studies of prenatal stress have themselves repeatedly come up empty-handed: gay men have not experienced higher prenatal stress. Further, as the discussion below will detail, the overall constellation of sex-typed traits might be expected to be more mixed in gay men, but more consistently "masculinized" in lesbians, but this does not fit the data, either.

Measurement Issue #3: Quantifying Orientation

How gay is gay? How straight is straight? In deciding "who counts when you're counting homosexuals" (to quote Stephanie Kenen [1997] again), scientists must consider not only what elements matter, but what are the *cutpoints* for labeling someone as heterosexual, bisexual, or homosexual. Does same-sex orientation count only if it meets certain time-based rules (lifelong; after a certain age; consistent for the past five years)? Do *all* of one's partners have to be same-sex, or a majority, or some, or perhaps only partners with whom there is a certain emotional attachment? Do very small incremental differences in terms of how much attraction is toward persons of the same or the other sex matter? Brain organization researchers have very different answers for these questions, meaning that they select different scores as the cutpoints to divide their subject groups for analysis. This, too, is unacknowledged in the literature.

To illustrate how variable the cutpoints for specific sexual orientation categories are in brain organization research, it's again helpful to refer to Kinsey scales. Each of the studies in Figure 7.6 classified subjects according to scores on some Kinsey-type scale. The variations in shading on the bars show where the scientist set the cutpoint for considering a subject as homosexual, bisexual, or heterosexual. Note that the figure *understates* the amount of variability in the meaning of Kinsey scores, because the aspect(s) of sexuality used to assign a Kinsey score vary from study to study. In some cases the aspects measured were not reported; rarely were actual questions reported. Nonetheless, it is illustrative to see the great variation in how scientists assign Kinsey scores to particular groups (homosexual,

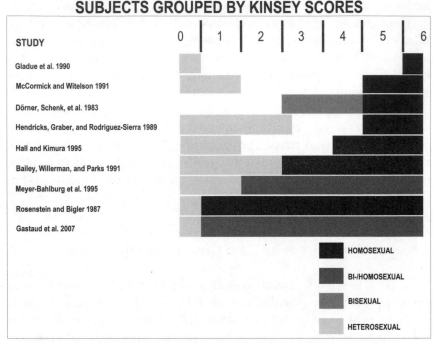

Figure 7.6. Variety in scientists' use of Kinsey scores. Note that in Meyer-Bahlburg et al. 1995 and Gastaud et al. 2007, the labels are implicit rather than explicit (for example, "subjects with homosexual inclinations" rather than "homosexuals").

bisexual, heterosexual). Subject groups were split into homosexuals versus heterosexuals by different scientists at literally every point along the scale between 0 and 6.

As Figure 7.6 shows, one scientist's heterosexuals are another scientist's homosexuals. Compare, for example, Gladue et al. (1990) with Meyer-Bahlburg et al. (1995). Gladue's team excludes from their homosexual category anyone with a lifetime Kinsey score less than 6. In contrast, Meyer-Bahlburg and colleagues' article on "the development of homosexual orientation" among DES-exposed women is based on a sample that includes not a single woman with a Kinsey 6 rating for lifetime "sexual responsiveness" and only one woman with a Kinsey 5 (Meyer-Bahlburg et al. 1995). (Meyer-Bahlburg and colleagues may argue that they do not technically set cutpoints for labeling women as heterosexual, bisexual, and homosexual, but they *functionally* assign the cutpoint for homosexual at a Kinsey 2, because they describe the study in the title and elsewhere in the article as an exploration of the development of "homosexual orientation.")

Decisions about boundaries are related to whether scientists think sexual orientation is categorical or continuous. Those who think of sexual orientation as continuous regard any cutpoints between homosexuals, bisexuals, and heterosexuals as matters of convention, because they imagine that "true" sexual orientations come in many degrees of difference between the poles of homosexuality and heterosexuality. Those who think of sexual orientation as categorical think of cutpoints as reflecting the boundaries between objectively different groups. While both views are widely held among scientists, there is an interesting and somewhat troubling pattern for where these two views show up in the brain organization studies. Briefly, those researchers who group their subjects by prenatal hormone exposure status tend to adopt the continuum approach (Ehrhardt et al. 1985; Dittmann et al. 1990, "Congenital Adrenal Hyperplasia," parts I and II; Meyer-Bahlburg et al. 1995; Zucker et al. 1996; Stikkelbroeck, Beerendonk, et al. 2003; Titus-Ernstoff et al. 2003; Meyer-Bahlburg et al. 2008),[16] while researchers who group people by sexual orientation overwhelmingly adopt the category approach (Dörner, Rohde, et al. 1975, 1983; Swaab and Hofman 1990; Bailey, Willerman, and Parks 1991; Dörner et al. 1991; LeVay 1991; Allen and Gorski 1992; Gladue and Bailey 1995a, 1995b; Ellis and Cole-Harding 2001; Rahman and Wilson 2003; Anders and Hampson 2005; Rahman 2005; Kraemer et al. 2006; Blanchard and Lippa 2007; Witelson et al. 2007).[17] To put this another way, researchers who study people with unusual prenatal hormone exposures (intersex individuals and the offspring of hormone-treated pregnancies) measure adult sexual orientation in very finely graded ways that can pick up the slightest difference between hormone-exposed and unexposed groups. But studies that look for evidence that gay men or lesbians have had different prenatal hormone exposures than their straight counterparts seek to create comparison groups that are as starkly different as possible in terms of sexual orientation, so they choose "extreme" homosexuals and "extreme" heterosexuals.[18] As I will explain below, it is easier in the former group (the cohort studies) to find hormone effects with a very loose criterion for homosexuality, while the latter (case-control) approach will benefit from a very restrictive criterion. It is impossible not to see this as scientific gerrymandering—that is, moving the boundaries of sexual orientation categories around in a way that favors the scientist's own hypothesis.

In an interview, one scientist explained the benefit of looking at "the extremes" in terms of sexual orientation for her studies of handedness and sexual orientation: "I was thinking that, well, if the differences are going to be subtle and small, we need to look at the two most different groups" (Dr. L interview, January 1999). Likewise, Gladue, Green, and Hellman

(1984) explained that their strategy of screening out subjects who did not give absolutely "consistent" responses for each dimension of sexual orientation "was especially critical for the integrity of the homosexual sample," further claiming that "it was essential to distinguish volunteers having exclusively homosexual desires, interests, and experiences lifelong from men having such characteristics intermittently, periodically, or only recently" (1499 n. 10). And indeed, this rationale does make sense on the assumption that subjects with the most "extreme" orientations will be those who had the most extreme cross-sex hormone exposures, *and* that "extreme" homosexuals and heterosexuals will be most likely to differ on other characteristics that are presumably influenced by hormone exposures, such as laterality or cognitive traits. This is the "dose–response" logic I described in Chapter 3. I highlight the expectation here because the evidence runs counter to this expectation, in spite of some scientists' attempts to assemble "pure" sexual orientation groups. I will have more to say about this below, when I review the findings of sexual orientation studies. In any case, it turns out that while these very strict measures make sense if you're considering just one corner of the evidence linking hormones to homosexuality, they create problems when you look at the bigger impact of this research.

THE CHOICE of high-contrast sexual orientation groups for the case-control studies is initially driven by the desire to maximize the chance of finding group differences on hormone-related traits. Yet for some scientists these rigid groupings have literally redefined what it means to be homosexual or heterosexual. Defying social science evidence that there is significant independence among the various aspects of sexual orientation, brain organization scientists treat consistency across behavior, attraction, and identity as "true" or "correct," rather than simply an operational necessity for their particular studies. Ellis and Cole-Harding (2001), for example, go so far as to describe inconsistency across all dimensions of sexual orientation as "erroneous responses."[19]

While this approach is at odds with survey and in-depth qualitative research on how people's desires, behaviors, and identities intersect and diverge (Laumann et al. 1994; Diamond and Savin-Williams 2000; Young 2004), insisting on erotic consistency fits well with some recent theoretical developments in psychology. As Dr. L put it, regarding the late 1980s, "a lot of the literature was questioning whether there was really such a thing as bisexuality" (Dr. L interview, January 1999; see also Bailey 2003). Further, while they differ on the prevalence and stability of bisexuality as a category, many well-known researchers have strongly argued that sexual orientation is basically a categorical phenomenon—at least in men (Hamer and Copeland 1994; LeVay 1996; Wilson and Rahman 2005). In this view, the categories of homosexual, heterosexual, and possibly bisexual are adequate and exhaustive for describing the meaningful patterns of human eroticism. If this is the case, then "weeding out" people who give inconsistent responses on measures of multiple dimensions of sexual orientation is a perfectly good idea. Moreover, some scientists imply—and others state outright—that the "discrete categories" idea of sexual orientation fits better with brain organization theory (LeVay 1996; Bailey 2003; Wilson and Rahman 2005; Kraemer et al. 2006).[20]

There are difficulties, though. It turns out that many researchers have to jettison a high proportion of their research subjects because they don't meet the criterion of "consistency" on sexual orientation. Investigators often allude to, but rarely offer details about, how this affects the composition of the "homosexual" and "heterosexual" groups they compare. For example, Ellis and Cole-Harding's (2001) study of prenatal exposures to stress and alcohol used a method that excluded a "substantial minority" of subjects (with no additional details on the number or proportion involved), because some people's responses weren't totally consistent across the three dimensions of sexual orientation (217). In one study that offered more detail on subjects who were excluded, 20 of 90 (over 22 percent) otherwise eligible men did not meet the criteria for being either *homosexual* or *heterosexual* (Gladue, Green, and Hellman 1984). This is particularly striking given the recruitment strategy for the study, which consisted of advertisements for "homosexual and heterosexual" individuals for a study of endocrine function and sexual orientation. Thus, volunteers likely were already a self-selected group in terms of sexual identity, and more likely than a random group of individuals to identify as either gay or straight, rather than bisexual. Still, over one-fifth of volunteers were not adequately consistent on each of three criteria (sexual fantasies, sexual behaviors, and stability of each since puberty) to be counted as homosexual or heterosexual. Likewise, in one of several recent studies that link brain organization theory to the observation that gay men tend to have more older brothers than heterosexual men have (the "fraternal birth order" effect), Blanchard and Lippa (2007) dropped nearly one-third (3,375 out of 11,654) of all subjects who gave some indication of same-sex orientation but whose answers were deemed "inconsistent" (165).

To place these decisions in context, consider some additional data from the large, population-based study of American adults that I mentioned earlier. In addition to showing how the various aspects of behavior, fantasy, and identity were related, the investigators report the proportion of men and women who have had partners of the same sex, the other sex, and both sexes during various time periods, as well as the proportions who are attracted to persons of the same, the other sex, or both sexes. Interestingly, of all the people who have any same-sex experiences as adults, less than 20 percent of men and just 10 percent of women have been exclusively homosexual in behavior since age 18 (Laumann et al. 1994, 311). Among those who report same-sex attractions and/or behavior, 85 percent of women and 72 percent of men identify as heterosexual (301). Brain organization researchers are not unaware of this issue, at least in terms of their female samples. In the literature on biology and sexual orientation, as well as in the interviews I conducted with prominent scientists, there was general agreement that women are even less likely than men to be totally consistent among the various dimensions of orientation.[21]

One result of filtering out "inconsistent" responses is that many brain organization studies exclude the *majority* of self-described lesbians and many gay men from studies on sexual orientation. This most certainly hampers the utility of brain organization research for elucidating the processes that underlie sexual orientation development. But it is equally important to note that the process of "cleansing" samples to remove apparent sexual inconsistency results in artificially homogeneous "heterosexual" groups, as well. This creates a serious problem of bias. In an artificially homogeneous group, other traits besides sexual orientation will be present to a degree that one would not see in a truly random group of "heterosexuals" or "gay men" or "lesbians." When we analyze the data, there will be irrelevant variables present that can make it seem like there are group differences related to sexual orientation, when in fact there is another hidden variable at play. A historical example will help clarify the problem. Prior to the 1980s, it was common to recruit samples of gay men and lesbians in gay bars, because there were very few other public meeting places for sexual minorities in most communities. Unsurprisingly, these samples showed higher rates of problem drinking than heterosexual people who were recruited from a wide range of community settings.

What about the cohort studies? In one regard these studies look superior, because they assess multiple aspects of sexuality and they do not demand consistency across aspects of sexual orientation. Yet the "homosexuality" found in the studies of hormone-exposed individuals is possibly even further away from homosexuality as understood in the real world. There are two problems with the way even the most careful researchers use multidimensional measures in these studies. The first, content validity, is a technical issue having to do with the way specific dimensions are chosen to be included in these measures, and the second has to do with the way results are eventually labeled in summary statements and titles.

Content validity concerns the key question of meaning in measurement: Does the measure accurately capture the phenomenon of interest? In a classic text on measurement, Carmines and Zeller (1979, 43) define content validity as "the extent to which a specific set of items reflects the construct [i.e., the phenomenon of interest]." Even without settling on a "gold standard" definition of sexual orientation, it is possible to discern content validity problems in some of these comprehensive sexuality assessments. The Sexual Behavior Assessment Schedule–Adult (SEBAS-A) is a good instrument to consider in this regard, both because it was developed by the well-regarded team of Heino Meyer-Bahlburg and Anke Ehrhardt (1983) and because it has been used in several brain organization studies. In studies of DES-exposed women, for example, Meyer-Bahlburg and Ehrhardt combined multiple aspects of sexual orientation assessed via the SEBAS-A to create a global score for sexual orientation (Ehrhardt et al. 1985; Meyer-Bahlburg et al. 1995). When a scale is used for this purpose, the scale items should be a *randomly chosen subset* of the universe of potential items that could reflect sexual orientation, but most of the SEBAS-A items are nonbehavioral aspects of imagery and attraction. In other words, the scale oversamples the nonbehavioral aspects of sexual orientation, thereby exaggerating the importance of imagery items, such as dreams. Further, these questions (especially masturbation fantasies, daydreams, and type of erotica used) may be so closely related that it doesn't make sense to assess them separately, because combining them via simple addition may exaggerate tiny shifts toward same-sex orientation in one's overall profile.

The second issue is how subjects are described in studies that rely on complex assessments of this sort. Sensibly, in the precise passages where scientists discuss subtle differences in sexual orientation, they tend to avoid attaching labels such as *heterosexual* or *homosexual* to their subjects. Yet this careful approach drops away quickly—sometimes even in the article abstracts and titles that scientists themselves choose.

Thus, in the sections where most readers look for the "bottom line," it seems that the exhaustive assessments work on a sort of sexual "one-drop rule": any hint, in any one of the elements, of same-sex attraction or response is glossed as evidence of homosexuality (see, for example, Ehrhardt et al. 1985; Dittmann, Kappes, and Kappes 1992; Meyer-Bahlburg et al. 1995; Dessens et al. 1999). It is little wonder, then, that by the time studies are reported in the popular press, terms like *homosexual* and *lesbian* are used freely.

An interesting example is found in Meyer-Bahlburg and colleagues' (1995) report on women exposed to DES: the sections on methods and results carefully refer to the women in this study by their dichotomized Kinsey scores of 0–1 and 2–6, while the title and introductory section address the development of "homosexual orientation." In a striking understatement, the authors note that "the extent to which bisexuality and homosexuality were increased in DES women was rather modest" (17). In fact, only 1 of 90 women was rated as being "largely homosexual but incidentally heterosexual" and *none* of the 90 was rated as "entirely homosexual" (17). Does it make sense to talk about the "development of homosexual orientation" based on a sample that many people—including most brain organization researchers—would agree *does not include any homosexuals?* If we count the one woman who was rated as "largely homosexual," that's still a rate of lesbianism that's just 1.1 percent, which is *lower* than most estimates for the general population (Laumann et al. 1994; Sell, Wells, and Wypij 1995).

Even if we were to agree that researchers can use whatever definitions they like for "homosexual" or "heterosexual," as long as they are clear and consistent in their own research, the yawning gap between the approaches in the two main sets of studies creates a problem. Although scientists' choices of different measures make sense in terms of studies' internal logic (it gives each kind of study a better chance of finding that there is some kind of hormone effect on sexuality), it is very difficult to reconcile findings across the studies. Bluntly, you can't have "homosexual" groups that are mutually exclusive in two sets of studies, but then claim that the studies complement or build upon each other. Still, as serious as the *measurement* discrepancies are between cohort and case-control studies of brain organization, the *findings* from research on prenatal hormones and human sexual orientation are just as problematic. It's not just that they don't add up "yet" because of incomplete data. Rather, the pattern of findings in the hundreds of studies done to date, and especially between the cohort and the case-control studies, suggests that brain organization theory is not a good explanation for human sexual orientation.

Sexual Disorientation: Lost in a Jungle of Findings

How do these measurement issues figure into the overall assessment of brain organization theory? Recall that all the data from all the studies of sexual orientation combined still form a mere fragment of the evidence that is needed to come all the way from hormones to homosexuality. Because the studies are all quasi experiments, they must work together with each other, as well as with the experimental evidence from nonhuman animals, to form a coherent picture of how early hormones affect human sexual differentiation. The overall network of associations that is predicted by brain organization theory links early hormones with sex-differentiated cognitive skills; gender-related aspects of personality and interests; signs of brain lateralization, including handedness; anthropometry (measurement of any bodily traits that are thought to be influenced by early hormones); and sexual orientation. Yet most studies consist of a search for associations between just two factors at a time, such as sexual orientation and cognitive traits (in the case-control studies), or early hormones and sexual orientation (in the cohort studies). Thus, in addition to the fact that all of the evidence is limited to correlations (which do not entail causation), the expected correlations are tested in pairs, which then are presumed to fit together to complete the overall picture. Figure 7.7 shows the main two-way associations that are predicted by brain organization research on sexual orientation.

Each association is represented either by a solid arrow that runs in a single direction (from early hormones to the domains of personality and behavior that hormones are presumed to influence) or a dotted line that connects two domains, signaling noncausal associations between two traits that are expected to vary together if both are influenced by early hormone exposures. One way to judge whether quasi-experimental evidence is robust is to do something called path analysis. In this case, that means that if brain organization theory is correct, we should be able to trace at least one complete loop on the diagram. For example, some studies might show that early masculinizing hormones are correlated with more "masculine" sexual orientation, and other studies might show that masculinizing hormones are correlated with more "masculine" cognition. If there are additional studies that show masculine cognition to correlate with masculine sexual orientation, then the loop is complete. Conversely, the data are less convincing if some of the loops cannot be closed, and the theory begins to fall apart if the data at hand demonstrate a lack of association where the diagram shows there should be a connection, or negative associations where there should be positive ones (for example, if "masculine" hormone exposures are associated with "feminine" traits).

In part because researchers approach sexual orientation so differently (especially when they switch between the gynephile/androphile and the homosexual/heterosexual frame), it's difficult to lay out the studies together in a way that makes sense. With that limitation in mind, we can still use the evidence at hand to get a preliminary sense of how well the brain organization picture would come into focus if the measurement of sexual orientation weren't an issue. That is, we can plug in the cohort study data to see whether individuals who have had "cross-sex" hormone exposures also show "cross-sex" cognition, personality and interests, bodily traits (other than genital morphology), handedness, and sexual orientation. We can also plug in case-control data to see if "sex atypical" sexual orientation is correlated with sex atypicality on any of these other traits. Note, though, that *only* the cohort studies can give us the information we need to test the hypothetical associations on the left side of the diagram in Figure 7.7 (those that connect early hormones with any individual outcome). And note, too, that *only* the case-control studies give us any information on associations between the various "outcomes" (for instance, between sexual orientation and cognition).[22] That's because the cohort studies generally have small sample sizes, and even fewer subjects with any same-sex orientation, so they do not have sufficient statistical power to look for associations among multiple hypothetical outcomes of brain organization. The solid arrows on Figure 7.7 indicate the associations that cohort studies explore, and the dotted lines indicate the associations explored in case-control studies.

Later I will address how the difference in measurement between researchers conducting cohort studies and those who conduct case-control studies affects the flow of associations. For now, though, let's try to complete some of the loops, first by looking at the evidence that early hormones affect sexual orientation, and next by looking at the evidence connecting each of four other domains either to hormones or to sexual orientation. We can also address the important question of whether the associations are shown for both genetic males and females, only one sex, or neither. This is important because, like other possible breaks in the flow of associations, the evidence does not add up to a convincing picture if some connections can be made only for men, while others can be made only for women.

Let's briefly recap the data reviewed in Chapter 4 to see how well the link between early hormones and sexual orientation is borne out when we have direct information about hormone exposures. The studies that are most relevant concern people who have had higher than typical fetal exposure to "masculinizing" hormones. As I detailed in Chapter 4, these include studies of males and females with congenital adrenal hyperplasia (CAH) and prenatal exposure to the synthetic estrogen DES. (I exclude studies on genetic males with androgen insensitivity from the following discussion because femininity may relate to female rearing rather than a lack of effective androgen exposure. I exclude males with CAH, too, be-

ASSOCIATIONS PREDICTED BY BRAIN ORGANIZATION THEORY

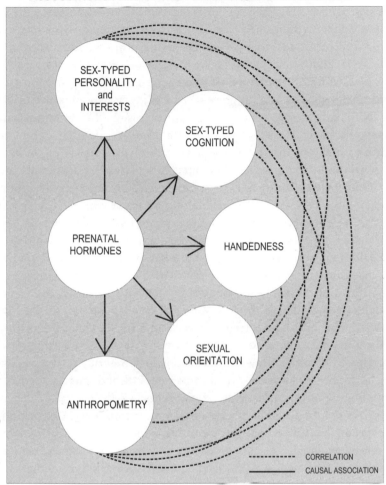

Figure 7.7. Basic two-way associations predicted by brain organization research on sexual orientation. Solid arrows represent causal associations (early hormones are theorized to cause sex differentiation of each trait on the right); dashed lines represent correlations that are predicted if both traits undergo sex differentiation by hormones. Note that the domains are broad, and early hormones hypothetically could affect either a whole domain or limited elements within a given domain.

cause there is no consensus on whether they do in fact have much higher than typical prenatal androgens, or on how androgens above the typical male range would affect them.)

Brain organization theory suggests that women with CAH or prenatal DES exposure should have higher rates of lesbian or bisexual orientation, and that men with CAH or with prenatal DES exposure may have lower rates of gay or bisexual orientation compared with the general population. Let's start with CAH. There is consistent evidence that women with CAH are more likely than their female relatives to have same-sex desire (Ehrhardt, Evers, and Money 1968; Dittmann, Kappes, and Kappes 1992; May, Boyle, and Grant 1996; Zucker et al. 1996; Hines, Brook, and Conway 2004; Morgan et al. 2005; Gastaud et al. 2007; Meyer-Bahlburg et al. 2008). DES exposure is a different story. Contrary to some early evidence that DES exposure may increase lesbian or bisexual orientation among women (Ehrhardt et al. 1985; Meyer-Bahlburg et al. 1995), the largest study by far shows that DES-exposed women are, if anything, *more heterosexual* than their unexposed counterparts (Titus-Ernstoff et al. 2003). (The Titus-Ernstoff study did not assess sexual attractions apart from behavior, but with a sample that includes nearly 5,700 women, any meaningful shift in sexual orientation should certainly have been detected.) Likewise, there is no evidence that sexual orientation is affected among men with DES exposure, though several studies have investigated the question, including the same very large study by Titus-Ernstoff and colleagues (2003), as well as two early studies by Richard Green and colleagues (Yalom, Green, and Fisk 1973; Kester et al. 1980), and unpublished work by Anke Ehrhardt and Heino Meyer-Bahlburg (Anke Ehrhardt, personal communication, 1998). In genetic males there are two situations for which it is difficult to draw any clear conclusion about the link between early hormones and any sex-typed traits in adulthood. First, scientists disagree about whether males with 5-alpha reductase deficiency are generally reared as girls, or whether a substantial proportion are reared within locally specific intersex categories. Second, genetic males who are reassigned and reared as female, because of either congenital anomalies affecting the genitals or traumatic loss of the penis, have a rearing situation that is dramatically unlike any other, because parents, clinicians, and possibly others are aware that the child has been gender-reassigned. From these latter two groups, it is not possible to draw a clear message regarding the effect of early hormones.

Thus, for only one out of the four most relevant groups studied do the data support the prediction that masculinizing prenatal hormones shift sexual orientation toward greater interest in females. Notably, of the four groups, the members of this group, women with CAH, are the only ones with identifiably unusual social experiences related to gender, such as high rates of sex reassignment and genital surgery due to intersex birth, and widespread expectation that they will be gender-atypical. Also, as noted above, when women with CAH are compared with the general population, or when behavioral measures rather than fantasies are the criterion, women with CAH do not seem more likely to be sexually oriented toward women. Now let's examine the specific loops that would hypothetically link sexual orientation with early hormones via another domain, beginning with cognition.

Cognition. Comparing lesbians to heterosexual women, most studies find no differences in terms of cognitive ability (Tuttle and Pillard 1991; Gladue and Bailey 1995b; Wegesin 1998; Neave, Menaged, and Weightman 1999; Rahman, Wilson, and Abrahams 2003; Rahman 2005), and those studies that do find differences have been about as likely to find that lesbians are more feminine (Gladue et al. 1990) as they are to find that lesbians are more masculine (Rahman, Abrahams, and Wilson 2003; Collaer, Reimers, and Manning 2007).

Many studies suggest that gay men are somewhat more "feminine" on selected cognitive traits compared to straight men (for example, Willmott and Brierley 1984; Sanders and Ross-Field 1986; Gladue et al. 1990; Mc-Cormick and Witelson 1991; Sanders and Wright 1997; Wegesin 1998; Neave, Menaged, and Weightman 1999; Rahman and Wilson 2003; Rahman, Wilson, and Abrahams 2004; Collaer, Reimers, and Manning 2007), but most studies find differences in only one or two out of multiple aspects of cognition tested, and a substantial minority of studies find no difference between gay and straight men (Tuttle and Pillard 1991; Gladue and Bailey 1995b; Cohen 2002; Esgate and Flynn 2005).[23]

It could be that more consistent cognitive differences between heterosexual and gay men would be found if studies had larger samples.[24] But small sample sizes cannot fully explain the inconsistent results, because even nonsignificant differences are too often against the predicted direction. Tuttle and Pillard (1991), for example, found that gay men scored in a more masculine direction on three of six cognitive tasks. Similarly, Cohen (2002) found that spatial ability was not broadly different among gay versus straight men, and he also found that among gay men, *higher* scores on a masculinity scale were correlated with *worse* spatial relations. Other studies find a few limited differences, but these are complex. For example, McCormick and Witelson (1991) found cognitive differences between gay and straight men, but only among men who were not consistent right-handers. Likewise, Rahman, Andersson, and Govier (2005) found that gay men were no less likely than straight men to use "male-typical" navigational strategies (using cardinal directions, and identifying distances on a map), though they were more likely than straight men to use the "feminine" strategy of referring to landmarks, which resulted in gay men using more navigational strategies overall—a difference, but not one that neatly supports the idea that gay men are "gender-atypical" or "feminine."

What about evidence of more male-typical cognitive traits in people with higher-than-typical prenatal androgens? Here, the evidence is clear: cognitive traits are not consistently masculine among either males or females with high "masculinizing" fetal hormones, including those with CAH and those with DES exposures. While some studies have shown cognitive differences between CAH-affected and unaffected groups, the differences do not support the theory that androgens specifically enhance masculine, or inhibit feminine, cognitive traits (see Chapter 4 for more details). Instead, there may be subtle, generalized cognitive deficits among people with CAH, especially those with severe forms of the disease (Helleday, Bartfai, et al. 1994; Helleday, Siwers, et al. 1994; Plante et al. 1996; Johannsen et al. 2006). Likewise, there are no consistent cognitive differences among DES-exposed males or females compared with unexposed groups. (See Hines 2004, especially chapter 9, for a good review of cognition among subjects with unusual early hormone exposures.)

Handedness. Lateralized brain function refers to how certain tasks (both cognitive processing and motor tasks like right- versus left-hand dominance) are accomplished by one hemisphere versus the other, or both hemispheres. As Mustanksi, Chivers, and Bailey (2002) have noted, "Despite inconsistent results across studies, narrative reviewers of the literature have suggested that men show a modest increase in functional asymmetry compared to women, meaning that the hemispheres of men are more specialized" (99). Many brain organization studies have been built on this tentative foundation, based on the further supposition that functional asymmetry is related to testosterone exposure. But meta-analyses suggest that the only aspect of functional asymmetry that is clearly sex-differentiated is handedness (Sommer et al. 2008; Vogel, Bowers, and Vogel 2003; see also my discussion in Chapter 3). Because brain organization theory is meant only to explain the development of traits that are sex-differentiated, the following discussion does not include every study that has reported some evidence relating sexual orientation to various indicators of functional asymmetry. Instead, I confine myself to studies of handedness.

In the scenario predicted by classic brain organization theory, gay men would be less left-handed, and lesbians more so, than their same-sex heterosexual counterparts. But as I noted when discussing how sexual orientation is framed, studies of handedness do not consistently build upon the usual brain organization hypothesis that homosexuals of either sex will be sex-atypical. Instead, a great many studies build on the Geschwind-Behan-Galaburda (GBG) hypothesis (Geschwind and Behan 1982; Geschwind and Galaburda 1985b), which suggests a complex web of associations among dyslexia and other learning disabilities, migraine, immune dysfunction, left-handedness, and male homosexuality. Geschwind and his colleagues linked their proposal with classic brain organization theory by way of research showing that under intense stress, pregnant rats first have a large increase in testosterone production, followed by testosterone dropping to atypically low levels. If this same process applied to humans, prenatal stress could account for a link between some traits that are thought to be caused by high levels of testosterone (such as dyslexia and left-handedness) and traits thought to be caused by low levels (as is usual, they slot male homosexuality in here).

But the finer details of the GBG hypothesis often get lost along the way, and the idea has morphed into a fuzzy notion that homosexuality should be correlated with left-handedness and other signs of atypical lateralization. Thus, contrary to one of the central tenets of brain organization theory, some authors have argued that male homosexuality is a sign of "hypermasculine" prenatal hormone exposures (for instance, Lindesay 1987; McFadden and Champlin 2000). Conversely, McCormick, Witelson, and Kingstone (1990) tried to reconcile greater left-handedness among both gay men and lesbians with brain organization theory by suggesting that, contrary to the GBG hypothesis that *higher* prenatal testosterone leads to left-handedness, left-handedness in *men* results from *lower* prenatal testosterone. (They suggested that the development of lateralization, cognitive functions, and sexuality might follow different paths in women and men, but did not suggest how to reconcile their idea with the greater population prevalence of left-handedness in men.)

Given the complexity of the hypotheses involved, and the fact that so many of the associations are hypothetical, it should not be surprising that there is some mutual contradiction among the studies of handedness and sexual orientation. Reviewers differ as to whether the current evidence does or does not point to differences between heterosexuals and homosexuals, especially men (Lalumiere, Blanchard, and Zucker 2000; Mustanski, Chivers, and Bailey 2002). And specific patterns of associations do not hold up. First, and perhaps most importantly, the association between left-handedness and male homosexuality is grafted to brain organization theory by way of a hypothetical association between male homosexuality and prenatal stress—but the studies of prenatal stress on balance don't support the theory (Ellis et al. 1988; Schmidt and Clement 1990; Bailey, Willerman, and Parks 1991; Ellis and Cole-Harding 2001). Second, scientists tend to agree that there should be a stronger pattern of left-handedness among lesbians than among gay men (because male homosexuality would be associated with left-handedness only in cases of prenatal stress early in gestation, whereas female homosexuality would be associated with left-handedness whenever androgen levels were sufficiently high throughout gestation to affect both traits). On its own, the evidence on handedness would seem to support this (Lalumiere, Blanchard, and Zucker 2000). However, this same general constraint on the model should suggest that virtually all aspects of cognition, emotion, or morphology that are affected by early hormones should be linked more consistently in lesbians than in gay men, and this is decidedly not the case. Lesbians do not show consistently "masculinized" traits, nor are there stronger associations in general among sexual orientation and cognition, emotion, or morphology among lesbians than among gay men (quite the converse, in fact). Keep this in mind as we review the other kinds of traits that have been studied to explore a link between sexual orientation and hormones.

The data from studies of handedness among gay men and lesbians are lackluster, at best. But the major difficulty in reconciling these studies with other brain organization research is that the cohort studies (research on individuals with documented prenatal hormone anomalies) do not support the GBG hypothesis: that is, there is no good human evidence that the many traits investigated as indicators of lateralization are actually "perturbed" by early hormone exposures. Just as men and women with CAH are no more likely to have masculine cognitive profiles, they are no more likely to be left-handed. There is suggestive evidence, however, regarding DES exposure and left-handedness. A few small studies have found more left-handedness among DES-exposed women (Schachter 1994; Scheirs and Vingerhoets 1995; Smith and Hines 2000), but the largest study by far, which included 5,686 DES-exposed women, found no evidence of increased left-handedness (Titus-Ernstoff et al. 2003). That same study, though, did find a slight increase in left-handedness among 1,432 DES-exposed men. However, given that DES has a wide range of documented teratogenic effects, subtle shifts in handedness might be seen in DES without specifically implicating the brain organization hypothesis. And again, the full set of associations that might be expected to show up, especially in very large studies with great statistical power, are absolutely contradicted: there is no link between DES and sexual orientation in men or in women (Titus-Ernstoff et al. 2003).

Sheri Berenbaum, a prolific and well-respected researcher of CAH, has offered this sharp rebuttal of the GBG hypothesis:

Anomalies in cerebral dominance, immune functioning, abilities, and neural crest development are hypothesized to correlate with each other because all result from high levels of prenatal testosterone. Studies directly evaluating the effect of testosterone on these traits do not validate the model: sex ratios and animal studies suggest that testosterone has a protective, rather than facilitory, effect on autoimmune diseases; individuals with high levels of early testosterone do not have elevated rates of left-handedness or learning disabilities. (Berenbaum and Denburg 1995, 79)

Gay men and lesbians may be more likely than other people to be left-handed, but the evidence suggests that this cannot be explained by "cross-sex" hormone exposures. Findings of increased left-handedness among gay men and lesbians may be a simple result of noncomparable sampling in the studies done to date (that is, the association is not real, but instead is the result of biased sampling). In particular, heterosexual samples are disproportionately recruited from university students and staff, then compared to lesbian and gay samples recruited from various community sources. Alternatively, even a real association does not necessarily point to a common single factor influencing both handedness and sexual orientation. Instead, various alternative explanations are possible, including the chance that multiple factors operating very early in development make some individuals less "conforming," which in turn could affect a broad range of traits with diverse developmental trajectories.[25]

Anthropometry. As should by now be clear, there is a seemingly endless array of possible traits on which scientists might focus to explore the possibility that early hormone exposures shape sexual orientation. Any trait that is thought to differ, on average, between males and females, or any trait that is also possibly influenced by testosterone early in development, is a potential candidate for investigation. Many studies use an approach that can be broadly described as anthropometry. To date, studies have examined such body parts and proportions as penis size, height and weight and/or body mass index (Perkins 1981; Bogaert and Blanchard 1996; Bogaert 1998; Bogaert, Friesen, and Klentrou 2002; Bogaert 2003; Martin and Nguyen 2004), and the ratio of the second to the fourth digit, or 2D:4D (for instance, Robinson and Manning 2000; Williams et al. 2000; Lippa 2003; Rahman and Wilson 2003; Putz et al. 2004; McFadden et al. 2005; Rahman 2005; Rahman, Korhonen, and Aslam 2005; Kraemer et al. 2006; Collaer, Reimers, and Manning 2007).

A few things should be noted about this line of inquiry. First, the notion that lesbians and gay men have cross-sex body types predates the theory of brain organization (Terry 1999); the notion of homosexuals as "sexually intermediate" was an implicit rationale for applying brain organization theory to the question of human sexuality in the first place (van den Wijngaard 1997). Thus, some studies in this vein are not included in the current review (which is not necessarily a shortcoming, given pervasive biases in subject recruitment for sexual orientation studies well into the 1970s). Second, other than the 2D:4D studies, nearly all such studies in recent years have been conducted by Ray Blanchard and/or Anthony Bogaert; generally data are more convincing when they come from independent research teams. Third, although there are isolated findings of gay versus straight differences on one bodily trait or another, overall these studies do not suggest any consistent pattern regarding cross-sex body types among gay men and lesbians (see the summary in Bogaert, Friesen, and Klentrou 2002). Finally, the strongest evidence for brain organization theory would consist of a triangulation among studies that find some trait to differ between gay men and/or lesbians and their heterosexual counterparts (the case-control studies), on the one hand, and between hormone-exposed and unexposed subjects (the cohort studies, such as CAH versus controls), on the other hand. There are no such patterns.

Prior to the last decade, the default line of inquiry with the anthropometric studies was the old "intermediacy" hypothesis, which fits best with brain organization theory's predictions: gay and bisexual men were expected to be less "masculine" (or more similar to heterosexual women) on any given physical trait measured, while lesbian and bisexual women were expected to be less "feminine" (or more similar to heterosexual men). With the rise of the GBG hypothesis, though, which raised the possibility that gay and bisexual men might have *higher* than typical androgen exposures during development (at least during some periods), a few studies appeared that suggested hypermasculinization of physical traits among gay men. One highly cited (but unreplicated) report, for example, suggested that gay men have larger penises than straight men (Bogaert and Hershberger 1999). Conversely, another report—also unreplicated—suggested that gay men and lesbians both have a cross-sex pattern of long bone development (Martin and Nguyen 2004).[26]

The 2D:4D studies, which constitute the largest subset of anthropometric studies and are currently quite popular, have produced contradictory results that suggest no real effect. The "2D:4D" ratio refers to the relative length of the second (index) finger to the fourth (ring) finger. This ratio is considered sexually dimorphic, because men's 2D:4D ratio tends to be lower, reflecting a relatively longer index finger in women. Of eight studies comparing 2D:4D in gay and straight men, three studies reported higher (feminized) ratios in gay men (Lippa 2003; Collaer, Reimers, and

Manning 2007; McFadden and Shubel 2003); two studies found gay men to have *lower* (hypermasculinized) ratios (Robinson and Manning 2000; Rahman and Wilson 2003); and three found no difference in men's digit ratios by sexual orientation (Williams et al. 2000; Voracek, Manning, and Ponocny 2005; Kraemer et al. 2006).[27] The findings are similarly mixed for women. Three studies report "masculinized" digit ratios to be associated with same-sex orientation in women (Williams et al. 2000; Rahman and Wilson 2003; Kraemer et al. 2006), while one small study (Anders and Hampson 2005) and the two largest studies (Lippa 2003; Collaer, Reimers, and Manning 2007) found no differences between lesbians and heterosexual women.[28]

Thus, the anthropometry studies, even the popular comparison of 2D:4D ratios, do not on balance suggest that gay men or lesbians differ from same-sex comparison groups. Likewise, there are only weak and inconsistent indications that 2D:4D is affected as predicted in people with known unusual prenatal hormone exposures. The strongest evidence to date that 2D:4D ratios are affected by prenatal hormones comes from a recent study by Sheri Berenbaum and colleagues (Berenbaum et al. 2009), who found a statistically significant, but very slight, shift toward the "feminine" pattern of a high 2D:4D ratio among women with complete androgen insensitivity syndrome. Berenbaum and colleagues concluded that "digit ratio is related to effective androgen exposure, but the relation is too small to use digit ratio as a marker for individual differences in androgen exposure" (5123).

Personality and interests. The domain of sex-typed personality and interests is harder to summarize, for a variety of reasons. The specific variables involved range over a wide terrain, including things like aggression, career and hobby interests, nurturing, toy and game preferences, preference for friendships with same-sex or other-sex peers, risk seeking, and interest in objects versus in people, as well as broad composites of "gender role behavior" and masculinity and femininity inventories. Although there are increasing trends toward using more standardized measures of these variables, standardization in this domain lags behind the specificity that has been achieved in tests for aspects of cognitive function. Among the cohort studies, reports seldom focus on those aspects of personality that do not differ, but highlight only those items that are correlated with hormone exposure status. So it is especially difficult to track the precise results in these studies. I defer detailed consideration of these issues to Chapter 8, which is devoted to sex-typed interests, especially as they are assessed in the cohort studies.

I have devoted a section of Chapter 9 to thinking about the relationship between sex-typed personality and interests, on the one hand, and sexual orientation, on the other hand, as a deeply neglected issue of context in brain organization research. Here I just briefly sketch some of the difficulties involved. Unlike the other studies I examine, data on gay men's and lesbians' personality and interests are available from many kinds of research, most of which have nothing to do with brain organization theory. For one thing, sex-atypical personalities and interests might develop much later than most traits that are hypothetically affected by prenatal hormones, so investigators posit a wide range of causal theories for possible differences between lesbians and gay men and their heterosexual counterparts, including entirely social theories. For another thing, quite a bit of data on personality and interests among gay men and lesbians is not generated or interpreted in terms of *any* causal theory of sexual orientation; rather, a great many studies explore how experience as a sexual minority may lead to personality differences, because of such factors as stigma, discrimination, and specific subcultural influences.

With those important caveats in mind, a great many studies suggest that gay men are somewhat atypical (gender-nonconforming) as a group, in terms of interests and other aspects of personality that are considered sex-differentiated; there is less association between gender and sexual orientation in women, but many studies do find more masculine traits among lesbians compared with heterosexual women (for instance, Whitam and Mathy 1991; Lippa 2002; Udry and Chantala 2006; Lippa 2008). The specific variable in this domain that may be of greatest interest is childhood play, especially toy preferences. At least in theory, this is one aspect of personality and interests that is unlikely to be affected by the experience of being gay or lesbian (because play behaviors are usually assessed for early childhood). Retrospective recall bias probably inflates the extent to which lesbians and gay men recall their childhood play behaviors as atypical, but one recent study used home videos from subjects' childhoods to demonstrate that gender nonconformity was in fact evident in children who would later become gay and lesbian adults (Rieger et al. 2008). Below and in the following chapters, I offer some cautions against too readily attributing these associations to the influence of early hormones.

Sex-typed interests, especially childhood play, is the domain in which girls and women with CAH are more consistently masculine than their CAH-unaffected counterparts. However, data do not on balance suggest sex atypicality in terms of personality and interests (often called "gender role behavior") among genetic females with atypical hormone exposures other than CAH, nor for genetic males with sex-atypical early hormone exposures from any cause. Among nonclinical samples where some information on hormones is available (for instance, hormones in amniotic fluid or in maternal blood during pregnancy), there are mostly negative findings. Neither study that examined hormones in amniotic fluid found any relationship between testosterone and sex-typed play (Grimshaw, Sitarenios, and Finegan 1995; Knickmeyer, Baron-Cohen, et al. 2005), in either girls or boys. (Several papers from a longitudinal study by Simon Baron-Cohen's team have reported associations between testosterone in amniotic fluid and some indicators of sex-typed interests, but the findings are both inconsistent across the various reports and unreplicated, as I detail in Chapter 8.) Two studies have tested associations between testosterone and sex-hormone-binding globulin in maternal blood, on the one hand, and gendered behavior, on the other (Udry, Morris, and Kovenock 1995; Udry 2000; Hines, Golombok, et al. 2002). Both studies found associations, but they are both small and contradictory between the two studies. (See Chapter 8 for detailed discussion.)

THIS is a good time to return to the diagrams showing the network of associations that brain organization theory predicts. If the enormous variation in measurement and study design were not at issue, formal meta-analysis would be ideal. Because that is not possible, the diagrams should be understood as heuristic devices rather than precise representations of the evidence. Just as with formal meta-analysis, the kind of synthesis I do here involves a tension between aiming to include as much of the data as possible and aiming to maintain precision in terms of which components of domains show association with other variables, how variables are measured, and the precise nature of the relationships that have been found. For example, an association between sexual orientation and a cognitive trait might be found only in right-handed men, which raises the question of whether and how to include that with associations that are found without regard to handedness. I have resolved the tension in this way: the narrative summaries above draw attention to gaps, discontinuities, and inconsistencies, but the figures that follow err on the side of "lumping" the data to generate a critical-yet-generous summary.

The evidence for virtually every trait studied is different for males versus females, so it is important to map these associations separately by sex. Figure 7.8 shows the pattern of associations for genetic females, and Figure 7.9 shows the pattern for genetic males. Filling in the left side of Figure 7.8 from the cohort studies, early masculinizing hormone exposure is associated with more sexual orientation toward women and more "masculine" toy preferences. But both of these associations are found only in women with CAH, and the sexual orientation difference is found only when CAH women are compared with small and apparently unusual control groups, not when compared with rates of same-sex desire or behavior among women in the general population. The exception might be the subset of women with the most severe form of CAH, among whom a high proportion were initially "announced" and reared as boys, but later sex-reassigned.

Filling in the right side of Figure 7.8 from the case-control studies, lesbians, as a group, tend to be more left-handed, or less consistently right-handed, than heterosexual women, and also seem to have had more "masculine" behavior as children, including toy preferences. Nonetheless, there are two big problems with concluding that these two pairs of associations mutually support brain organization theory. First, you can't trace a causal pathway that allows you to conclude that the association between left-handedness and lesbianism is due to early hormones, because the only association between early hormones and left-handedness is a weak and inconsistent finding among women who experienced prenatal DES exposure—but DES is clearly not associated with shifts in sexual orientation. Second, as demonstrated in the discussion of how sexual orientation is measured, the phenomenon of same-sex orientation is quite different between these two pairs of associations. The "lesbianism" that is linked with prenatal hormone exposures is not at all the same as the "lesbianism" that is linked with left-handedness or childhood toy preferences. In fact, most of the case-control studies use a definition of "lesbianism" that would exclude nearly all the subjects in the cohort studies—even though the latter, as a group, have had the highest exposure to "masculinizing" hormones of any women. In other words, the pattern of data strongly contradicts the dose–response relationship that the theory would predict, a point I elaborate below. Finally, given the broad variability among lesbians on the huge array of variables that have been assessed, the general trend of similarity among lesbians and other women, and the long history of important psychological and somatic "differences" that have been found between lesbians and other women only to be discarded later (Terry 1999), one might do well to recall the advice of Dr. A, and "not get bogged down in dogma . . . thinking of lesbians as sort of like guys" (Dr. A interview, February 5, 1999).

FOR MEN, the pattern looks different, but equally problematic. First, consider the left side of Figure 7.9. From cohort studies, there is no support for the idea that hormonal variation among genetic males who are reared as male is associated with sexual orientation. Likewise, cohort studies don't indicate any association between male-typical cognition and masculinizing hormone exposures. There does seem to be an association between DES exposure (a "masculinizing" condition) and increased left-handedness (also masculine, so in accord with the theory). And finally, there seems to be a very small association between androgen insensitivity syndrome (a "feminizing" condition) and more female-typical digit ratios among genetic males.

Moving to the right side of Figure 7.9, the case-control studies show associations between sexual orientation and both cognition and personality. They also show associations between handedness and sexual orientation—but this association is in the opposite direction to the previous two findings. That is, though data are notably mixed, quite a few studies suggest that gay men as a group have one or more aspects of cognition or personality that are more feminine than heterosexual men, but gay men have a slightly more masculine profile in terms of handedness. (Recall, too, that the hypothesis that is supposed to reconcile this discrepancy—the notion that gay men have experienced greater prenatal stress—is also not supported by evidence.) In the end, even with the generous interpretation that lumps together scattered findings for various aspects of cognition and personality in order to support a link between these domains and sexual orientation, there is no way to trace a loop between sexual orientation and prenatal hormones for genetic males.

What if we loosened the criteria for evidence and included males with 5-alpha reductase deficiency and other males who have anomalous or damaged genitals, and who were therefore sex-reassigned as female and reared more or less as girls? Keeping in mind all the caveats implied by that "more or less," there is a much higher rate of sexual orientation toward women among this group than among genetic females who are reared as girls. And among those who assign themselves back to the male sex in adolescence or adulthood, there is a predictably masculine pattern of personality and interests. If these cases were included, we could close the loops that link sex-typed cognition and personality and interests (but not handedness) to early hormone exposures. In the end, the decision to include or exclude these cases is a subjective matter. Given that there is not even a hint from any of the other cohort study designs that early hormone exposures influence sexual orientation, I believe we should not ask these extremely idiosyncratic cases to bear much weight in the overall network of evidence.

Considered together, data from the studies on cognitive traits, handedness, anthropometry, and personality and interests suggest that, as the particular traits examined in case-control studies of brain organization have

ASSOCIATIONS OBSERVED AMONG (GENETIC) FEMALES

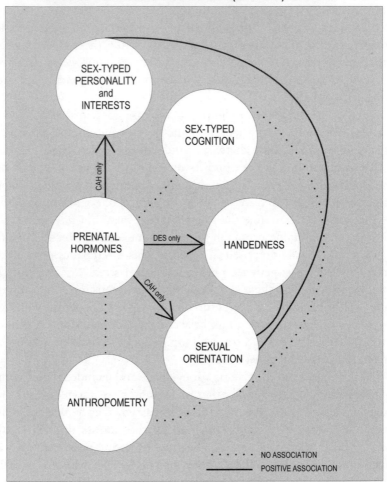

Figure 7.8. Two-way associations in genetic females. Support for the theory requires at a minimum that we could trace a complete loop from prenatal hormones to sexual orientation, from sexual orientation to another domain, and from that second domain back to prenatal hormones. There is a complete loop if we focus only on the case of genetic females with CAH, and on the (weak, but generally positive) association between sexual orientation and sex-typed personality and interests. This chain of associations is better explained by variables of biological and social context than by prenatal hormone exposures (see Chapter 9).

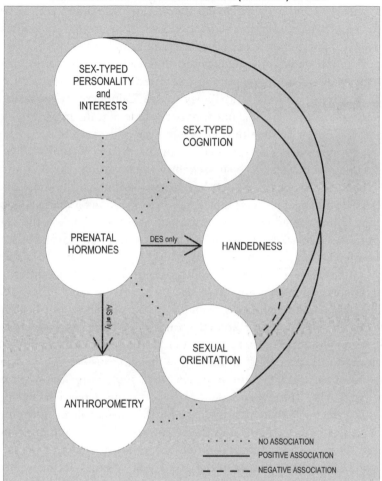

Figure 7.9. Two way associations in genetic males. Support for the theory requires at a minimum that we could trace a complete loop from prenatal hormones to sexual orientation, from sexual orientation to another domain, and from that second domain back to prenatal hormones. Note that the associations between sexual orientation and both sex-typed cognition and sex-typed personality and interests are not entirely consistent. Because there is no complete loop, the theory is not supported for genetic males. Note that these associations do not include genetic males who are reassigned and reared as females, nor males with 5-alpha reductase deficiency; there are too many confounders in rearing conditions and adult gender role for these data to yield a clear message.

proliferated, the evidence gets farther and farther from the theory—yet many scientists, and many laypeople who follow the studies with great interest, cling to the notion that better evidence is just around the corner. It's common for researchers to argue that small sample sizes, which make it difficult to find statistically significant differences between groups, can explain major discrepancies in brain organization research on sexual orientation (Mustanski, Chivers, and Bailey 2002; Lalumiere, Blanchard, and Zucker 2000). However, it is not simply the case that the many failures to support the theory are due to small sample sizes, because the *direction* of differences often defies the predictions of the theory. Virtually all of the studies that find gay men to be more feminine on various traits are at odds with those studies that link left-handedness with male homosexuality. Conversely, the slight increase in left-handedness among lesbians is at odds with the failure to find masculinization on other traits, such as the 2D:4D ratio or spatial relations. Lesbians are not, on balance, more consistently gender-atypical than gay men are, as would be required by the elaborate story that has been spun around homosexuality and left-handedness.

If prenatal hormones really do create a sex-differentiated brain, including directing sexual orientation, then homosexuality should be both more frequent and more extreme among groups with more extreme cross-sex hormone exposures. But that is not, in fact, what brain organization studies show. In fact, the cohort studies show that *extreme cross-sex hormone exposures are associated with only very small shifts in sexual orientation.* How can this be? The evidence defies "dose–response" relationships of the sort that have made such a strong case for the causal role of hormones in the development of genitals, and in certain behaviors among infrahuman animals. Here is where the great difference in measurement of sexual orientation between the cohort and the case-control studies becomes particularly problematic.

Women with CAH have the "highest dose" of prenatal androgens, so it is useful to ask what exactly "same-sex orientation" looks like among women with CAH. Among studies that find more homosexual or bisexual orientation, several find higher rates of *same-sex fantasy or attraction* but not same-sex *behavior* (Dittmann, Kappes, and Kappes 1992; Zucker et al. 1996; Gastaud et al. 2007). An article by Ralf Dittmann and colleagues (1990, "Congenital Adrenal Hyperplasia I") provides a helpful example. In spite of extremely liberal statistics and a vast number of comparisons, no sexual behavior item reached either "significant" or "trend" level, for any age group; the only single item that reached significance—and it does seem to be an interesting item—is that more women with CAH express the wish for a steady or long-term relationship with a female partner (6 of 34 women, or almost 18 percent, versus 0 for the sisters).[29]

As I've noted, while many studies do find increased rates of same-sex behavior among women with CAH compared with their female relatives and other controls, most studies do *not* show same-sex orientation (fantasy *or* behavior) among women with CAH to be any higher than among women in the general population. Several large, population-based studies in the 1990s provided estimates of same-sex orientation and behavior in the general population for the United States, as well as France and Great Britain (Laumann et al. 1994; Sell, Wells, and Wypij 1995; Michael et al. 2001). These studies have estimated that the proportion of women who have same-sex partners sometime in their adult lives is somewhere between 1.8 percent and 3.6 percent, and the proportion who report being attracted to other women runs from 8.6 percent to 18.6 percent. Thus, the differences between women with CAH and control women are not necessarily because women with CAH have unusually high levels of same-sex orientation, *but may be the result of unusually low levels of same-sex orientation among comparison groups*. This might be due to very small sample sizes, especially among comparison groups. Even though CAH itself is rare, the comparison groups are kept artificially small because of a belief that female relatives of girls and women with CAH provide the best control for rearing experiences and social environment variables. In practice, investigators include cousins as well as sisters—and even female relatives of males with CAH, in some cases—so the strategy is of questionable value, particularly because the ideal strategy for studying a rare condition is to have a large unaffected-to-affected ratio.

It's worth pointing out that the largest and most comprehensive study of sexual orientation and CAH done to date confirms the general failure of evidence from CAH to show that increasing levels of prenatal androgens lead to increasing levels of female homosexuality, contrary to the authors' interpretation. Heino Meyer-Bahlburg and colleagues (2008) have conducted the only study of sexual orientation that includes women with the mildest, nonclassical form of CAH. This subset of patients is especially interesting because while a high proportion of women with the classical form of CAH are born with masculinized genitalia, women with the nonclassical form are not. Yet women with nonclassical CAH still have some increase in same-sex orientation compared with their same-sex relatives, leading Meyer-Bahlburg and colleagues to consider their findings as a positive indication of a dose–response relationship between early androgens and male-typical sexual orientation. There are four big problems with this interpretation. First, a dose–response relationship would be best indicated by a progression of increasing same-sex orientation from the lowest level of early androgens, seen in women with nonclassical CAH, through the midrange of early androgens, seen in women with the simple virilizing form of CAH, to the highest range, found in women with the most severe, salt-wasting form. Instead, women with the simple virilizing form showed *less* same-sex eroticism on all aspects of orientation than did the women with nonclassical form (Meyer-Bahlburg et al. 2008, 91). Second, increased same-sex orientation among women with the nonclassical form does not fit well with brain organization theory, because according to most evidence at this point, androgens are not elevated in these women until after the prenatal period when hormones are purported to organize the brain for human psychosexuality (Meyer-Bahlburg et al. 2008, 96). Thus, the increase in same-sex orientation in this group may speak more to other variables—possibly later hormone effects, including indirectly through self-image in response to bodily masculinization, and just as possibly to the social effects of being told during late childhood, adolescence, or even adulthood that androgens are high and that personality and sexuality may be affected (see Chapter 9 for elaboration). Third, only women with the most severe, salt-wasting form had a rate of same-sex eroticism that is clearly higher than general population levels, which for women are estimated to be around 3 percent for same-sex behavior, and up to 19 percent when same-sex attraction is included (Sell, Wells, and Wypij 1995). Fourth, note that the control group in this study is again very small (just 24 unaffected women, compared with a combined CAH sample of 141 women). Given the relatively low prevalence of same-sex orientation in the general population, it is not very surprising to find little indication of same-sex eroticism in this group.

The way that Zucker et al. (1996) characterized their own sample of CAH-affected women applies equally well to the pattern of findings across the multiple studies of sexual orientation in CAH: *"the majority of the CAH women were exclusively heterosexual in their sexual orientation, despite the clear evidence that such women were exposed to prenatal levels of androgen that are in the range of normal males"* (314, emphasis added). Even strong proponents of brain organization theory occasionally notice that this is unexpected, given the predictions of the theory. Wilson and Rahman (2005), for instance, find it "surprising that larger shifts in sexual orientation were not reported" in studies of women with CAH (75).

There is something badly wrong, then, when other researchers summarize the evidence as indicating that in women with CAH, sexual orientation is "much closer to the male pattern" (Witelson, quoted in Weise

2006), or that "lesbianism in these girls is off the charts" (Dr. I interview, August 1998). The popular summaries of data on sexual orientation among women with CAH can be particularly far-fetched. In her pop-science book *Why Men Don't Iron,* Anne Moir asserts that "48 percent of them confess to having homosexual fantasies, and 44 percent are actively lesbian" (Moir 2003, 46). Moir's summary should raise eyebrows, even aside from the gratuitous insinuation that same-sex fantasies are something to "confess" rather than merely "report." This points back to the last question about defining and measuring sexual orientation, to which I promised to return. Given that these scientists don't agree with each other about who counts as a homosexual, how well do they agree with the rest of us? Are the people designated "homosexual" in these studies the same people that are generally thought of as homosexual in the real world?

THROUGHOUT this book, I argue that technical decisions associated with measuring human sexuality are important at a number of levels. It is important to look at what scientists measure when they study sexuality, and it is especially important to see how consistent their definitions are within and between studies. But these technical decisions gain more meaning when they are viewed in larger contexts. Each set of hypotheses and data does not stand on its own, but is delivered by way of a scientific interpretation whereby the researchers guide readers toward the conclusions that they believe are appropriate. As research findings are generalized beyond particular studies or projects, the finer points of operational definitions fade and general terms like *homosexuality* or *masculinization* are left to carry the burden of meaning. Earlier I described measures as "assumption containers," through which certain ideas get carried along through and beyond scientific work, gaining the implicit authority of being "scientific" without ever being tested.

In the dominant schema of brain organization research, sexual orientation is a fairly solid and well-defined variable. People generally are attracted to people of the same sex or of the other sex; many fewer individuals are genuinely attracted to both. While social pressures may restrict the amount of same-sex activity that some people have, and even push some people into opposite-sex activity in order to "conform," people's fantasies and behaviors would generally pretty much go together if their behavior weren't constrained by homophobic convention. According to these researchers, it doesn't really matter that much if you pick one aspect versus another (such as fantasy, behavior, or identity) in order to "capture" sexual orientation in a particular study population. If it is most convenient to recruit subjects by self-identity, then confirm by sexual fantasies, you won't get a different group than if you began by screening a much larger group for behaviors, fantasies, and identity.

These important assertions can be checked against a large body of research on sexual behavior, desires, and identity. From the studies in the public health literature that separately report data on two or more of these dimensions, one can see that it actually makes a great deal of difference which is used as a gauge of orientation (see, for example, Laumann et al. 1994; Sell 1997; Diamond and Savin-Williams 2000; Young et al. 2000). It is also important to note that most brain organization studies do not give enough information to see how their definitions and exclusion criteria affect the composition of their samples, though several investigators indicate in published work and in interviews that they retain only subjects whose sexual orientation perfectly lines up across all aspects of behavior, attractions, and identity. Technically, their selection methods add "consistency" to the operational definition of sexual orientation. When scientists who do this turn around and cite one another to support their definition of sexual orientation as "stable" and "consistent" and a "dichotomous trait" (as Simon LeVay, Dean Hamer, and Michael Bailey all do in recent books), this is misleading. That is because their methods have forced a tight association among fantasies, behavior, and self-declared sexual orientation.

These are teleological methods. Such methods do more than filter out complexity and variation in terms of the subjects of any particular study. And they do more than make brain organization theory seem stronger than it actually is. They add to the overall literature on sexual orientation, creating supposed evidence that sexual orientations come in neat categories, and making disagreement among specific dimensions of orientation look unusual and respondents who "fail" to be consistent look like liars or candidates for therapy.

Moreover, even if one adopts the most generous possible attitude toward the measures used in brain organization studies, the findings across the two main sets of studies cannot be reconciled with one etiological theory. The people with the most extreme cross-sex hormone exposures show little, if any, shift in sexual orientation. They also show no shift from typical populations in most other traits that supposedly signal early hormone effects, and are therefore used to infer different hormone histories between homosexuals and heterosexuals. And among the various studies in which homosexual versus heterosexual subjects are compared for differences that might implicate early hormone exposures, it is not just the case that there are many negative findings (that is, findings of no difference, which might be the result of sample sizes that are too small to find the effect). More importantly, the *direction* of evidence is often broadly inconsistent across the studies. The data on left-handedness, which seems to be increased in both gay men and lesbians, cannot be reconciled with one developmental theory for sexual orientation for both sexes. Yet if sexual orientation develops differently in genetic males and females, then brain organization theory can't explain the association of sexual orientation with cognitive or personality-related traits in men. That's because there is no evidence linking early hormone exposures with sexual orientation in genetic males with typical rearing. In short, the data aren't just weak, they are broadly contradictory.

Some psychologists have expressed awareness that the pattern of evidence doesn't fit the current theory, and suggest that a revision in thinking about the relationship between homosexuality and gender is in order. Even Dr. A, who is quite convinced that sexual orientation is a fairly straightforward variable, thinks it is time to abandon the age-old "theory of intermediacy"—the idea that gay men and lesbians are psychologically sandwiched between the dichotomous poles of masculinity (represented by straight men) and femininity (represented by straight women). He suggests it is better to "not get bogged down in dogma . . . thinking of lesbians as sort of like guys. Think of them as killer women." Further, "maybe heterosexual men are less masculine than gay men in areas that count, like number of sex partners per lifetime" (Dr. A interview, February 5, 1999). Several other researchers, including Michael Bailey (2003), James Lindesay (1987), and Dennis McFadden and Craig Champlin (2000), have suggested that gay men are "hypermasculine" in some important respects.

In my view, this is an interesting development, and promising insofar as it suggests a willingness to leave behind brain organization theory (and the "psychological hermaphroditism" idea in which it is anchored). But as long as researchers try to shoehorn the data on the complexity of sexual desires and the complicated and variable expression of "masculine" and "feminine" aspects of personality and behavior into a tidy and linear model that injects "cross-sex" hormones on one end and spits out homosexuals at the other, such promising insights will go nowhere. What the messy tug-and-clash of research on early hormones and sexual orientation shows most clearly is the inadequacy of the current research paradigm.

CHAPTER EIGHT

Sex-Typed Interests

O F ALL THE HIGH-STAKES claims that are built on brain organization research, none is more important—or more popular—than the idea that early hormones channel our fundamental interests into masculine or feminine directions. Even before the research leaves the pages of scientific journals, hormones are directly linked to career choices and chances, education, the division of labor in families, and the "drive" to be a leader versus a nurturer (for instance, Yalom, Green, and Fisk 1973; Leveroni and Berenbaum 1998; Berenbaum 1999; Udry 2000; Pasterski et al. 2005; Knickmeyer, Baron-Cohen, Raggatt, et al. 2006). The issue shows up with special frequency in the ongoing debate over why relatively few women make it to the senior ranks of physics, engineering, and mathematics careers, especially in academia. Both scientists and laypeople are increasingly likely to attribute sex differences in these fields to a variation in interests, rather than in abilities (not to mention discrimination).

Once again, brain organization theory is right in the middle of the story. In her report "Effects of Early Androgens on Sex-Typed Activities and Interests in Adolescents with Congenital Adrenal Hyperplasia," Sheri Berenbaum (1999, 108) drew a straight line from hormones, to sex-typed interests, to sex disparities in the science professions: "Given associations between interests and occupational choice . . . , it may not be surprising that girls and women are less likely than boys and men to participate in science and engineering. The findings suggest that this sex difference may be influenced by early hormones."

Outside the world of peer-reviewed journals, with their characteristically cautious prose, many psychologists and other scientists make even bolder statements. In an invited address to the American Psychological Association meeting in 2007, Professor Roy Baumeister explained why psychologists are now apt to focus on sex differences in interests instead of abilities: "Several recent works have questioned the whole idea of gender differences in abilities. Even when average differences are found, they tend to be extremely small. In contrast, when you look at what men and women want, what they like, there are genuine differences." Baumeister went on to apply this observation to the vexing question about sex disparities in math and science faculties:

> Maybe women can do math and science perfectly well but they just don't like to. After all, most men don't like math either! Of the small minority of people who do like math, there are probably more men than women. . . . And by the same logic, I suspect most men could learn to change diapers and vacuum under the sofa perfectly well too, and if men don't do those things, it's because they don't want to or don't like to, not because they are constitutionally unable (much as they may occasionally pretend otherwise!). (Baumeister 2007, n.p.).

I think we'd be hard-pressed to find many women who *like* changing diapers or vacuuming under sofas, either. Nonetheless, almost no one disagrees that, on average, there are some sex differences in interests, motivations, or tastes. But what exactly *are* the key differences, and where do they come from?

Many scientists and laypeople alike are convinced that scientific research firmly points to the role of prenatal hormones, especially testosterone (Brizendine 2006; Bazelon 2008; Pinker 2008). But the research in question, like all studies on how early hormone exposures affect later psychology and behavior in humans, are quasi experiments. That means they can't stand on their own, but have to be evaluated together. Symmetry principles, now familiar from the prior chapters where I evaluated studies on masculine and feminine sexuality and sexual orientation, are the only way to get a handle on the massive evidence about hormones and interests—because it isn't enough to know that there is a *lot* of data. We want to know whether the data are *consistent*, and, if they are not, exactly where the weaknesses and contradictions are located.

My point is not to dispute the existence of sex differences in interests

(I'll say it: they do exist), nor to consider whether sex differences are good or what we should do about them ("good" and "should" are never scientific questions, after all, but moral and political ones). My focus, instead, is on what we know about the causes of these differences. Of course, getting more precise about the nature and extent of difference in male versus female interests does play a role in figuring out how such differences arise. So there are two main questions I address in this chapter. First, what *are* masculine and feminine interests—or rather, what do scientists think they are? Here I won't be surveying the vast body of research that tries to establish the nature of sex-typed interests, but instead will map out the ways that scientists doing brain organization research define and measure sex-typed interests. As we have seen before, in various traits related to sexuality, scientists' ideas about the phenomena they study are critical, because shifting definitions and measures mean that studies don't actually fit together in the way that they are typically understood to. Second, I'll build on this mapping of the concepts to ask what we actually know about the origins of sex-typed interests. In the end, I think most readers will agree that scientists have not made a compelling case for the role of early hormones in establishing sex-typed interests. But before saying more about that, let's turn to how scientists think about the nature of masculine and feminine interests.

A Grand Model: Evolution, Sex, and Interests

Currently many scientists see sex-typed interests (which they also sometimes gloss as "motivations" or "values") as both preceding and amplifying other psychological differences, such as those in some specific cognitive skills. To understand the appeal and the importance of the focus on differences in interests, it will help to sketch the whole model of how interests fit into an overall biological model of sex differences. Psychologist Gerianne Alexander (2003) has recently put forward a model of sex differences in interests that draws on evolutionary psychology, experiments on vision in human infants and other animals, models of neural development and function for visual processing, and research that relates prenatal hormone exposures to sex-typed interests in humans and other animals. Alexander's model is useful because she makes explicit the usually implicit assumptions of the scientists who work on sex differences in interests. The general framework of her argument is that sex-typed temperament is "a product of the sexual differentiation of the central nervous system that evolved to predispose an interest in stimuli that promote the acquisition of a gender identity and gender role" (12). In particular she suggests that attention to movement served hunting skills in prehuman and early human males, while attention to particular forms and colors gave selective advantage to females who were naturally attentive to infants.

Alexander argues, in other words, that certain objective "perceptual features"—apart from their cultural valuing as masculine or feminine—make some objects more attractive to females versus males. For example, she draws on an elaborate but selective review of data about visual processing to argue that boys are more attuned to object movement and location, while girls are more attuned to color, especially red or pink. She concludes with the somewhat astonishingly culture-bound assertion that "it may be more than a trivial coincidence that in our current culture we assign blue to boys and pink to girls" (12). One doesn't even have to leave the United States to find this assertion odd. Anyone visiting my own Brooklyn neighborhood filled with South Asian and West Indian immigrants can observe that plenty of men and boys are quite comfortable wearing pink. Another difficulty with Alexander's elaborate story is that the association between pink and femininity is very recent, even in the United States: until the early twentieth century pink was considered a much "stronger" and more masculine color, appropriate for boys, while soft "baby blues" were thought to be the right color for girls (Smith 1989).

Putting aside some of the more questionable details and overreaching of Alexander's particular model, the idea of innate masculine or feminine interests has some elements to recommend it. Alexander and other scientists who work in this area (see, for example, Udry 2000; Baron-Cohen 2003a) suggest that this is a dynamic, "biosocial" or "transactional" process, whereby initial propensities are reinforced through social learning, and that gender-bifurcated reinforcement in turn amplifies the original differences. In short, it is a material and interactive model for gender embodiment—a literal incorporation of social gender into the physical self. This is in principle exactly the kind of model for thinking about the interaction of biological and social factors that feminist biologists and gender theorists have been searching for (for example, Butler 1993; Fausto-Sterling 2000), so the idea potentially has very broad appeal. Further, the model involves a transformation of interests into skills: selective attention to particular stimuli supposedly develops particular perceptual and cognitive strengths, and also literally multiplies sex differences in the physical brain.

From the perspective of development, this is more satisfying than some other models for relating biology to observed differences between the sexes in arenas such as occupation. First, there do seem to be larger sex differences in interests than in cognitive skills. Second, the story doesn't rely on a flat "blueprint" version of neural development, but incorporates in a general way more current information about how experiences physically alter the brain. Therefore, it is more biologically accurate than unidirectional models, and there is a concrete role for gender socialization in this model. Third, on a gut level, interests seem to be more value neutral than skills: it's not that males or females are "better" or "worse" at certain things, it's simply the case that they value different aspects of the world. Finally, as I have demonstrated in earlier chapters, there is very little evidence that steroid hormones have an appreciable, direct influence on sex-typed aspects of human cognition. Perhaps focusing on *interests* is a better way to tie brain organization theory to sex differences in human behavior.

The research itself comprises a large and fairly representative subset of brain organization studies. The most important research for studying how early hormones affect sex-typed interests has been based on the condition of congenital adrenal hyperplasia (CAH). As with sexual orientation and other patterns of sexual behavior, the hypothesis in these studies is that because girls with this condition have very high androgen levels during fetal development, when the brain presumably undergoes sexual differentiation, they will be "masculinized"—that is, they will be interested in the things that boys are interested in. Some other research designs that have been used to investigate the influence of early hormones on interests include studying the offspring of women who were prescribed hormones to prevent miscarriage, and studying girls with male "co-twins" whose testosterone may theoretically have affected the girl's developing brain. (I describe all of the designs in detail in the second half of this chapter.)

The main point for now is to remember that these are observational studies, meaning that scientists can't do experiments on early hormone exposures in humans, but can only observe and try to creatively analyze what is happening without intervening. As with other quasi-experimental research, the first step in putting the evidence together involves considering how traits are defined and measured. There is not currently any explicit definition of "interests" that is broadly shared by scientists in this field, but most researchers at least implicitly link sex-typed interests to evolutionary psychology, suggesting that certain categories of interests were differentially important for males and females in early human history. Much like Gerianne Alexander's proposal, described above, Larry Cahill's version goes like this: males are interested in things that "might relate to the behaviors useful for hunting and for securing a mate" whereas females attend to things that "allow them to hone the skills they will one day need to nurture their young" (Cahill 2005, 45). Simon Baron-Cohen is another psychologist who believes that early hormone exposures "sex" the brain, creating fundamentally divergent interests among men, in general, versus women, in general. His view, in brief, is that women tend to be "empathizers," more interested in people, whereas men tend to be "systemizers," interested in "phenomena that are . . . lawful, finite, and deterministic." Baron-Cohen phrases the sex difference in terms of motivations or "drives": "Empathizing is the drive to identify another person's emotions and thoughts and to respond to these with an appropriate emotion. . . . Systemizing is the drive to analyze the variables in a system to derive the underlying rules that govern its behavior. Systemizing also refers to the drive to construct systems" (2007, 160).

Most other scientists doing this work do not look for perceptual features of stimuli (as Alexander suggests), or to the categorical division of "people" versus "law-governed phenomena" that Baron-Cohen prefers.[1] Instead, they skip the broader theoretical formations and assess subjects' preferences either by drawing parallels to animal research or by reference to the contemporary objects, games, activities, and occupations for which males and females show significantly different preferences.

From Functional Significance to Concrete Traits: Measuring Sex-Typed Interests

Though seldom fully articulated, certain ideas about the functional significance of sex-typed interests have been shared by virtually all scientists exploring the influence of early hormones on humans. At the broadest level, these scientists have presumed that feminine interests revolve around nurturing and self-decoration and that male interests revolve around action, social domination, and mastery of skills.[2]

In the earliest studies, John Money and his colleagues, especially Anke Ehrhardt, simply worked from stereotypes about girls' versus boys' interests and activities, as well as women's versus men's. They conceived sex-typed interests as central to "gender role," which they defined as "everything that a person says and does, to indicate to others or to the self the degree in which one is male or female or ambivalent" (Money and Ehrhardt 1972, 4). Summarizing their first half-decade of research on the topic, and couching their observations as provisional, Money and Ehrhardt explained:

> It is a handicap in the study of sexually dimorphic behavior that, for all the millennia in which men and women have existed, no one yet has an exhaustive list of what to look for. Today's information is not, therefore, final or ab-

solute. With the safeguard of this proviso, one may sum up current findings by saying that genetic females masculinized *in utero* and reared as girls have a high chance of being tomboys in their behavior. (9)

They identified the elements of tomboyism as these: (1) male-typical play, including vigorous activity, outdoor sports ("specifically ball games"), and a preference for "the toys that boys usually play with"; (2) "self-assertiveness in competition for position in the dominance hierarchy of childhood," especially seeking "a position of leadership among younger children"; (3) lack of interest in self-adornment, with a preference instead for "functionalism and utility in clothing, hairstyle, jewelry, and cosmetics"; (4) lack of interest in "rehearsal of maternalism," specifically meaning that "dollplay is negligible" and motherhood "is viewed in a perfunctory way as something to be postponed rather than hastened"; (5) "priority of career over marriage, preferably combining both," but specifically aiming for "achievement" in their careers; and (6) erotic responses that "resembles that of men rather than women" (10).

Erotic interests have for many reasons been the subject of more investigations than other aspects of sex-typed interests, but I put them aside in this chapter. Erotics are usually not a major referent in discussions about sex-typed interests and demographic patterns in labor or education, and erotic interests are not readily assessed in children. Studies of other sex-typed interests in young children have been especially important, because of the assumption that these interests are less shaped by sex-role expectations and stereotyping than the interests and activities or even cognitive patterns of adolescents and adults. (Given the extensive evidence that sex-role stereotyping begins from the moment of birth, and that adults' perceptions of children's activities, emotions, and traits are shaped by gender expectations, this assumption bears close examination. I consider it extensively in Chapter 9.) Erotics aside, then, this leaves five basic components of sex-typed interests, each of which can be dichotomized into feminine and masculine patterns, as Table 8.1 shows.

All of the early brain organization studies that reported on sex-typed interests followed this general pattern, with only slight adjustments in the expectations for older subjects (for instance, for older subjects, interviews assessed masculine versus feminine "hobbies" rather than "toys and games"). Using clinical-style interviews to "sex-type" the interests of their subjects, researchers reasoned that masculine interests include vigorous activity, competition for leadership, and achievement in a career, while feminine interests include self-adornment, nurturing—whether the dolls of childhood or one's own children in adulthood—and romance and mar-

Table 8.1   The division of childhood interests into "masculine" and "feminine,"
based on Money and Ehrhardt 1972.

FEMININE INTERESTS	MASCULINE INTERESTS
SEDENTARY PLAY (INDOORS, LOW PHYSICAL ENERGY EXPENDITURE)	HIGH ENERGY PLAY (OUTDOORS, ESPECIALLY BALL GAMES AND OTHER SPORTS)
COMPETITION AND DOMINANCE LESS IMPORTANT; PERHAPS MORE COOPERATIVE?	LEADERSHIP SEEKING, COMPETITION FOR HIGH POSITION IN THE "DOMINANCE HIERARCHY"
SELF-ADORNMENT: ENJOYMENT OF DRESSING UP, PLAYING WITH MAKEUP, JEWELRY AND HAIR	FUNCTIONALISM AND UTILITY IN CLOTHING AND HAIRSTYLES; LITTLE USE OF COSMETICS OR JEWELRY
REHEARSAL OF MATERNALISM THROUGH DOLL PLAY, BABYSITTING, ANTICIPATION OF ACTUAL MOTHERHOOD AND PLANS FOR MANY CHILDREN	DISINTEREST IN SMALL CHILDREN, NO DOLL PLAY, "PERFUNCTORY" ANTICIPATION OF PARENTHOOD AND PREFERENCE FOR FEW CHILDREN
FANTASIES AND DREAMS OF ROMANCE AND MARRIAGE; MARRIAGE MORE IMPORTANT THAN CAREER	PRIORITY OF CAREER OVER MARRIAGE

riage (Ehrhardt and Money 1967; Money, Ehrhardt, and Masica 1968; Ehrhardt, Epstein, and Money 1968; Ehrhardt, Evers, and Money 1968; Masica, Money, and Ehrhardt 1971; Yalom, Green, and Fisk 1973; Money and Ogunro 1974; McGuire, Ryan, and Omenn 1975; Ehrhardt, Grisanti, and Meyer-Bahlburg 1977; Meyer-Bahlburg, Grisanti, and Ehrhardt 1977; Kester et al. 1980; Ehrhardt et al. 1984).[3] The findings were most consistent for the two main intersex syndromes that were studied, CAH and androgen insensitivity syndrome (AIS): genetic girls with the former syndrome (in which prenatal androgens are very high) were consistently found to have more "masculine" interests than control girls. Conversely, genetically male (but female-reared) girls and women with AIS (whose tissues cannot respond to the androgens their testes produce, so they have a very feminine appearance) were found to have typically feminine interests. In both cases, the "valence" of sex-typed interests seemed to match with whether prenatal hormones were masculinizing or not, but scientists agreed that the CAH evidence is more informative regarding the role of hormones, because both hormones and social experiences are female-typical for AIS girls and women.[4] For that reason, I do not include studies of AIS in the review of findings in this chapter.

The past few decades have seen the introduction of many childhood behavior and interest inventories that highlight gender-differentiated aspects of development; several were created to discriminate between children with and without gender identity disorders (Meyer-Bahlburg, Sandberg, Dolezal, and Yager 1994). Typically the standardized checklists that researchers prefer have been validated by having samples of "normal" males and females endorse interests and activities as masculine or feminine. This may be done either by rating how well some item describes their own interests and activities, or by rating the item as "usual" or "appropriate" for males versus females in general. For example, Melissa Hines (Hines, Golombok, et al. 2002; Hines, Johnston, et al. 2002) as well as other scientists (such as Vreugdenhil et al. 2002) have used the Pre-School Activities Inventory (PSAI) (Golombok and Rust 1993), a measure on which a parent indicates the child's involvement with sex-typical toys, games, and activities. A popular measure for the sex-typed activities and interests of older children, also based on parents' report, is the Child Game Participation Questionnaire (CGPQ; based on Bates and Bentler 1973 as modified by Meyer-Bahlburg, Sandberg, Dolezal, and Yager 1994). Many brain organization studies have used either the CGPQ itself (for instance, Berenbaum and Snyder 1995; Meyer-Bahlburg et al. 2004; Sandberg et al. 1995) or CGPQ-derived measures (Rodgers, Fagot, and Winebarger 1998; Knickmeyer, Wheelwright, et al. 2005; Jurgensen et al. 2007). The Child Behavior and Attitude Questionnaire has also been used, often in conjunction with the CGPQ (for example, Sandberg et al. 1995; Jurgensen al. 2007).

Researchers have many good reasons for using a wide variety of measures, and for thinking broadly about how to represent sex-typed interests. Nonetheless, that makes it a complex process to identify studies that offer data on the relationship between prenatal hormones and "masculine" or "feminine" interests. Because of the strong tendency to focus on "positive findings," evidence about sex-typed interests is usually highlighted only in studies that seem to support a link between hormone exposures and sex-typed interests. Of course, as I explained when introducing symmetry principles in Chapter 3, evaluation of a quasi-experimental network must be comprehensive—and it especially must avoid the hazards of "cherry-picking" data by overlooking findings that don't support the theory. Thus, people familiar with the literature may be surprised to see some studies included here that aren't usually mentioned as bearing on "sex-typed traits."

It is especially important to be alert to this because researchers often derive "sex-typed interests" from measures that have not been developed specifically for assessing gender-related development, but that have items or subscales on which boys and girls typically differ. One example is the Child Behavior Check List (CBCL) (Achenbach and Edelbrock 1979), which was devised to assess behavioral problems. Another is the Perceived Competence Scale for Children (PCSC) (Harter 1982), which was devised to assess a child's own sense of competence in the cognitive, social, and physical domains, as well as her or his general sense of self-worth. Gordon and colleagues (1986) explained that they used the PCSC to assess gender-related interests, because "perceived competence is directly related to a child's *motivation* to engage in tasks and behaviors and his ability to successfully complete them" (131, emphasis added). It is important to recognize that different investigators sometimes use the same scale in different ways, and their interpretations can affect whether or not children with atypical hormone exposures seem to be gender-atypical (for detailed illustration, see the discussion below of the measurement and findings related to "aggression").

Sometimes researchers use different measures because they are researching people in a different age group, or who speak another language. Because their studies involve very young children, Simon Baron-Cohen and his collaborators (especially Rebecca Knickmeyer and Svetlana Lutchmaya) generally have not been able to use the same kinds of questionnaires or even observational measures of play, toy preferences, and so on that other scientists use. Instead, they have devised measures to tap the "people orientation" and "system orientation" that they believe underlie the female and male brain types, respectively, and that can be used with infants and toddlers. These measures have included, for example, the amount of time spent looking at a hanging mobile versus a face at 1 day postbirth (Connellan et al. 2000), or the amount of eye contact a 12-month-old child makes with its parent (Lutchmaya, Baron-Cohen, and Raggatt 2002). As an attempt to directly measure a child's budding "empathy," this group has also done a study in which children were asked to narrate an animated film involving simple geometric figures. The researchers assessed how often the children used "intentional" versus "neutral" propositions to describe the movement of the figures, reasoning that intentional propositions reflect a child's skill at understanding others' perspectives and action strategies (Knickmeyer, Baron-Cohen, Raggatt, et al. 2006).

Brain organization researchers often infer sex-typed interests from "gender role behavior," sometimes simply assuming that people's activities reflect their interests, and sometimes directly inquiring about the degree of ease, comfort, or pleasure derived from engaging in masculine- and feminine-defined activities. For example, although the Recalled Childhood Gender Questionnaire (RCGQ-R) (Meyer-Bahlburg et al. 2006) is not always understood as specifically tapping "interests," the items (including preferred games, toys, playmates, and clothing in childhood) are quite similar to scales that have been specifically constructed for that purpose.[5] Measures of adult gender role reflect the amount of interest or attention a person directs toward home versus career, type of job they hold and/or aspire to, hobbies, relationship patterns and role division with romantic partners or spouses, and personality dimensions (such as "independence" or "tough-mindedness") that typically show sex differences.

To show how researchers' ideas regarding gender role are largely centered on aspects of interests, motivations, and priorities, consider the way sociologist Richard Udry and colleagues (1995) devised a measure of "gendered behavior." They created a single variable that included nineteen measures (some of which are themselves complex measures) split into four conceptual dimensions: "(a) HOME captures the relative emphasis on home versus career and its consequences; (b) FEM captures characteristically feminine interests; (c) JOB captures feminine gender-related job attributes; (d) M-F captures masculinity-femininity dimensions of personality tests" (361). Table 8.2 is reproduced from this study, because it gives a good sense of the range and specific formulation of the questions that are used to capture sex-typed interests in similar studies.

Udry and colleagues' measure is helpful for reflecting on the complex decisions involved in measuring sex-typed interests. The issue of *what* to include is important, as is whether to use multiple items or single items to capture a particular kind of interest. Both *reliability* (the consistency of a measure, meaning that it is less vulnerable to measurement error) and *validity* (the extent to which a measure accurately captures the phenomenon of interest) tend to be increased by combining multiple items into scales or inventories that reflect a single underlying trait. But *how* items are combined and *how* they are analyzed also matter very much (especially whether scores are reported for composites only, or if results are also reported for individual items). Many studies use so many individual items and composite scales that it is easy to lose track of how many possible indicators of sex-typed interests there are in a given study. This poses interpretive problems, such as deciding what constitutes multiple comparisons that should be statistically corrected. (As I have mentioned in earlier chapters, multiple comparisons are like rolling the dice more times than you are allowed in a game—it unfairly increases the chances of coming up with favorable combinations.)

Udry's "gendered behavior" variable is an enormously broad composite, comprising an unspecified number of specific items and scales over all of the four domains that he identified. Because space is limited in a journal article, we don't see all the items or an explanation of how Udry has combined them. Without that kind of detail, we can't judge whether he is tap-

Table 8.2   Example of a composite measure of gendered behavior. Reprinted with permission from Udry et al. 1995.

Factor	Description
HOME	Ever married to a man: yes is feminine. No. of live births; high is feminine. Index of sex role orientation (Dreyer, Woods & Sherman, 1991); traditional is feminine; 16-item scale of gender role attitudes. Importance of career; not important is feminine. Importance of children; importance is feminine. Domestic division of labour scale. Sex-typed activities; 25-item questionnaire listing activities.
FEM	Importance of Marriage; low is important, or feminine. Feminine appearance factor (interviewer ratings of feminine demeanour, facial attractiveness, use of jewelry scale, use of cosmetics scale). Strong vocational interest inventory (20 items that most discriminate males and females); high is feminine (have occupational interest responses more like females) (Hansen & Campbell, 1985). Likes baby care (selected items from maternal attitude questionnaire) (Miller, 1980).
JOB	Proportion female in current or, if not employed at present, in last occupation. Featherman socioeconomic index of current or last occupation; low is feminine (Stevens & Featherman, 1981). Proportion female in work unit on last job.
M–F	Bem Sex Role Inventory, Feminine score (Bem, 1981); 10 female items. Bem Sex Role Inventory, masculine score; 10 mle items. Adjective check list, scored M%; masculine is high=checked adjectives to describe self that a higher proportion of Americans rate as masculine; mean of 300 possible items (Williams & Best, 1990). Personality research form, masculinity score (Berzins, Welling & Wetter, 1978, 1981). Personality research form, femininity score.

ping a single construct (a trait he calls "gendered behavior") or whether he's adding up apples and oranges. That is why scientists prefer to use standard measures that have been validated. The point here is not to say it is a bad measure, just to explain how complicated it is to try to represent a broad abstract construct like "gendered behavior." On the positive side,

Udry incorporates all four conceptual dimensions into a single measure, thereby avoiding the multiple comparisons problem that plagues many other studies.

As we review the specific evidence below, it is important to keep these issues about complex measures and multiple comparisons in mind. Most importantly, recall that cause-and-effect conclusions are always difficult to make in quasi-experimental research, but that results are considered most robust if they hold up across different measures of the same construct.

The Example of Aggression

It can be difficult to find all the studies that bear on sex-typed traits because scientists most often look at multiple traits to see if they can discern any influence from hormones, and research reports tend to emphasize the traits where there is some evidence of such influence. The heterogeneity of measures that I just described further complicates the issue. In the end it is difficult, not only to track the findings, but also to interpret them. To illustrate this problem, I will explore how the trait of aggression is conceived and measured in various studies, how negative findings are especially likely to get lost, and how ultimately, evaluations in the literature distort the evidence. Aggression is not a special trait in this regard—a focus on concern with appearance, nurturing, dominance, and other aspects of sex-typed interests would show the same problems.

To begin with, aggression itself is somewhat contentious as an aspect of sex-typed interests. On a theoretical level, aggression has been linked with a presumed masculine interest in social domination; evolutionary psychologists have also suggested that aggression has adaptive significance for males because it encourages success in competition with other males for female sexual partners. Baron-Cohen (2003a, 124) puts it this way: "In evolutionary terms, the bravest and most skilled fighters in male-male competition would have earned the highest social status, and thus secured the most wives and offspring." (The link between male dominance and reproductive success is almost an article of faith in evolutionary psychology, but it is not consistently supported by empirical research on either humans or other primates [see Haraway 1978; Parish and Waal 2000; Lucas et al. 2004]. While the greater propensity for certain forms of aggression, especially physical violence, among males is not itself especially controversial, questions such as the extent of sex differences, and most importantly the explanation for the sex difference, are all open to dispute [Fischer and Mosquera 2001; Young and Balaban 2003].)

The first analysis of aggression in humans with unusual hormone expo-sures was included in Ehrhardt, Epstein, and Money's (1968) report on girls with the early-treated adrenogenital syndrome. Though the investiga-tors found these girls to be "tomboys" who were more likely than typical girls to be competitive and enjoy high-energy play, the girls were no more aggressive than unaffected girls (selected from a public school sample, and matched for sex, race, age, IQ, and socioeconomic status). (In this case, ag-gression was measured by actual fighting.) Later, in their 1972 book *Man and Woman, Boy and Girl,* Money and Ehrhardt maintained that "aggres-siveness is not a primary trait of tomboyism. In fact, aggressiveness per se is probably not a primary trait of boyishness either, except as a shibboleth of shoddy popular psychology and the news media" (10).

A mere two years later, though, Anke Ehrhardt and her collaborator Su-san Baker (1974) turned away from this view, embracing the idea that ag-gression, along with rough-and-tumble play, was "one of the most consis-tent sex differences found in normal boys and girls" (38). This thinking was in line with their contemporaries. Eleanor Maccoby and Carol Jack-lin, for example, argued in their massive study *The Psychology of Sex Dif-ferences* (1974) that the consistency of sex differences in aggression across human cultures and within nonhuman primates suggested a likely biologi-cal basis for aggression. Ehrhardt and Baker's work anchored a long line of researchers who have sought, and sometimes found hints of, links be-tween early hormones and aggression in humans. Though they found, in agreement with Ehrhardt's earlier study at Johns Hopkins, that "mascu-linized" girls with CAH did not differ from their unaffected sisters or from their mothers in fighting behavior, Ehrhardt and Baker determined to take a second look, because "fighting behavior appears to be one of the most consistent sex differences cross-culturally and in comparisons between various mammalian species" (41). They reasoned that their initial data on fighting were "quite crude" and promised that an analysis based on more-detailed data would come later. But analysis of the more-detailed data on aggression never were published, suggesting that those, too, did not yield any evidence that the girls with CAH were more aggressive.

A few other investigators have asked subjects (or their parents) about actual experiences with fighting, temper displays, and other aggressive be-haviors (for example, Yalom, Green, and Fisk 1973; Lish et al. 1991; Pasterski et al. 2007). It has been most common, though, to use paper-and-pencil tests where subjects are presented either with hypothetical scenarios and asked to choose their most likely response (Reinisch 1981; Berenbaum and Resnick 1997; Alexander and Peterson 2004; Bailey and Hurd 2005; Cohen-Bendahan, Buitelaar, et al. 2005), or with behavior inventories such as the Child Behavior Check List (CBCL) (Achenbach and Edelbrock 1979) used by Gordon et al. (1986) and Berenbaum and colleagues (2004). Also, many brain organization studies employ personality scales, almost all of which have items or subscales that measure aggression, such as the Bem Sex Role Inventory (Bem 1974) used by Hurtig and Rosenthal (1987), Udry, Morris, and Kovenock (1995), and Lish et al. (1991); the NEO Five Factor Inventory (Costa and McCrae 1992) used by Fink, Manning, and Neave (2004); and the Multidimensional Personality Questionnaire (Tellegen 1982), used by Berenbaum and colleagues (Berenbaum and Resnick 1997; Berenbaum et al. 2004).[6]

The way aggression is measured and analyzed in a longitudinal study of males and females with congenital adrenal hyperplasia illustrates a potential hazard of multiple measures of the same construct: investigators and the readers of their studies can lose track of findings. In the one report from that study that focuses exclusively on aggression, Berenbaum and Resnick (1997) assessed aggression in three samples of people with CAH. As others (such as Berenbaum and Snyder 1995; Pasterski et al. 2007) have generally found, males with CAH did not differ from unaffected males in any of the samples. Females with CAH, though, did differ from control females, showing some indication of increased aggression in the two adolescent/adult samples.

Yet the results are more complicated than that brief summary might suggest. For one thing, the results differ depending on which measure of aggression is used. Among the adolescents and adults in Sample 1, the CAH-affected females scored significantly higher on the aggression subscale of the Multidimensional Personality Questionnaire (MPQ). In Sample 2, also adolescents and adults, the difference was not statistically significant, and the trend was in the *opposite* direction, with the CAH group showing *lower* aggression scores on the MPQ. (The MPQ was not used with Sample 3, the children under age 13.) For Sample 2 as well as Sample 3, Berenbaum and Resnick also used another measure, the Reinisch Aggression Inventory (RAI). There were no differences between the children with CAH and unaffected children, but the adolescents and adults in Sample 2 did show increased aggression compared with their unaffected relatives. (Because these same women tended toward *decreased* aggression compared with controls based on the other measure, the results should be viewed as tentative, at best.)

A second complication is that a later report, based on all samples in Berenbaum's longitudinal study of CAH and again using the MPQ, contradicts the suggestion of increased aggression that was found in Sample 1 of the 1997 article. Highlighting the overall satisfactory level of psychological adjustment of children and adults with CAH, Berenbaum and colleagues (2004) noted that there were no significant differences between females with and without CAH on any measure. What they do not highlight, however, is the fact that the measures in this study include two aggression scales: the very same MPQ scale on which a single subsample showed a difference in the earlier study, and the Child Behavior Check List (CBCL), a measure of behavioral problems that includes an aggression subscale.[7] Although Berenbaum herself had used the MPQ to measure aggression, and other investigators (such as Gordon et al. 1986) looking for effects of early androgen on aggression have used the CBCL, Berenbaum and her colleagues were not focusing on aggression in this analysis. Instead, they interpreted the similarity between CAH and unaffected controls on these measures to indicate that there is no *general* disruption of psychological health and development in CAH. Contrasting the null findings of this report with other analyses showing differences on sex-typed traits, they argued that increased androgens in CAH have a *specific* effect on sexual differentiation of the brain (Berenbaum et al. 2004).

I don't mean to single out Dr. Berenbaum, nor Dr. Pasterski, nor studies of aggression in particular, but to point out a general problem with studies of sex-typed interests. Sometimes the same measure is interpreted as indicating a "sex-typed trait," and sometimes it is not. Worse, it seems that the same measure is less likely to be interpreted as indicating a sex-typed trait when the findings don't support brain organization theory, so negative findings are especially likely to drop off the radar screen.[8] That makes it exceedingly difficult to map the full findings of brain organization studies. But it *is* possible to track the findings, and that's what I do in the rest of this chapter.

The Nitty-Gritty: How Do Findings on Sex-Typed Interests Align for Different Research Models?

As long as one stays at a very general level, it seems accurate to say that a great many studies point toward the influence of early hormones on sex-typed interests. But scientific evidence cannot be meaningfully evaluated at a very general level, especially when the evidence comes from quasi experiments. Instead, it's important to look at the data, paying attention to consistency and variation across two dimensions. First, for each specific research model (such as studies of CAH, or of exogenous hormone exposures, or of variation in prenatal hormones among nonclinical populations), what are the specific aspects of sex-typed interests that have been assessed, and for which of these does there seem to be a correlation with early hormone exposures? Second, looking across the different study designs, where are the findings consistent, and where do studies either fail to align, or actually contradict each other?

Once the findings have been evaluated across these dimensions, it's important to consider the totality of evidence presented. In other words, for those data that do look consistent across various study populations and designs, is an interpretation in terms of organizing effects of steroid hormones the most persuasive interpretation of differences between the study groups? Deciding how to interpret the findings requires deciding, in turn, how well scientists have addressed competing explanations besides early hormone exposures. As we have seen, one key issue is the importance of context—both social and biological context—for understanding the relationship between hormone exposures and behavioral outcomes. (I will return to this in detail in Chapter 9.)

The following discussion is organized around five main research models where there are data bearing on early hormone exposures and sex-typed interests: people with CAH compared to unaffected relatives or peers; people with prenatal exposure to other exogenous hormones (these are people whose mothers were prescribed estrogen and/or progestogens because of "problem" pregnancies); studies in nonclinical ("normal") people for whom some markers of prenatal hormone levels are available (for instance, from stored fluid samples from mothers who underwent amniocentesis); studies of twins who have same-sex versus other-sex co-twins; and studies that relate sex-typed interests to a physical trait that may reflect early testosterone exposure, the 2D:4D finger-length ratio.[9] Following the conceptualization of interests among investigators in this field, I abstracted "interests" variables according to the principle that these are distinct from cognitive *skills* but may encourage the development of the latter through selective attention and practice.[10] Let's take a look at the findings.

WE'LL take twin studies as our first model (the hormone-transfer model). Some investigators have tried to relate variation in sex-typed interests to prenatal hormones by studying mixed-sex versus same-sex twin pairs. The idea behind the twin studies is that the presence of a male co-twin will raise the available testosterone during the critical period of sex differentiation of the brain (because testosterone from his testes may circulate in the fetal sac). Thus, girls or boys with male co-twins should have more masculine interests (as well as other traits), compared with those who have a female co-twin. Three published reports have looked at gender-typed play in twins, and two of them found no differences based on the sex of co-twin

(Henderson and Berenbaum 1997; Rodgers, Fagot, and Winebarger 1998). A third, by Resnick, Gottesman, and McGue (1993), reported that girls with male co-twins scored higher on a measure of "sensation-seeking" but that the sex of the co-twin made no difference for males. It's hard to interpret this because the measure itself—sensation-seeking—has rarely been used in other studies. Moreover, as we shall see from many studies following different designs, it is extremely common to find that some marker of hormone exposure relates to interests in one sex but not in the other. The difficulty is that there is no consistent pattern in terms of which sex seems to be affected by normal variations in hormone levels (or even *greater* than normal variations).

The only other twin study to report a difference in sex-typed interests was a report by Celina Cohen-Bendahan and colleagues (Cohen-Bendahan, Buitelaar, et al. 2005), which found that girls with a male co-twin were more aggressive; no other traits related to the presence of a male co-twin. Dr. Cohen-Bendahan herself cautioned me to not place too much faith in the findings of these twin studies. Dr. Cohen-Bendahan has turned away from this research model as unpromising, after learning that a great many investigators have examined their twin data for evidence of "hormone transfer," but have repeatedly found no evidence that either girls or boys with a male co-twin were any different from children with female co-twins (Cohen-Bendahan interview, October 2007).

A MODEL that is currently very popular is the 2D:4D (digit-length ratios) model. This model correlates sex-typed interests with a potential "biomarker" of prenatal androgen exposure known as the 2D:4D ratio. This is the ratio of the length of the second (or index) finger to the fourth (or ring) finger. We saw earlier that in humans the 2D:4D ratio is, on average, lower in males than in females, and there are a number of indications that the ratio may be affected by androgen exposures in early development. For example, Baron-Cohen's team found that masculinized 2D:4D ratios correlated with higher testosterone-to-estrogen ratios in amniotic fluid (Lutchmaya et al. 2004).

Before looking at the specific findings, it is important to briefly revisit some of the reasons to be cautious about this model. These include the fact that there is great variation in 2D:4D by geographic region and ethnicity, while sex differences are smaller (in some populations nonexistent), and there is a great deal of overlap in 2D:4D ratios between males and females in the same population (Loehlin et al. 2006; Manning et al. 2004). Moreover, it is unclear whether 2D:4D is actually masculinized in CAH (which it should be, if it is a marker of prenatal testosterone exposure) (contrast

Brown et al. 2002; Okten, Kalyoncu, and Yaris 2002; Buck et al. 2003). Further, recent evidence also suggests that women's digit length is variable over the menstrual cycle and is also affected by use of hormonal contraceptives (Mayhew et al. 2007). If 2D:4D is not stable but is appreciably affected by circulating hormones, then the finger-length ratio may not be a good biomarker of early hormone exposures, and the research model would be invalid for exploring brain organization theory. (The association with contraception use is an important confounder when 2D:4D is examined in relation to sexual orientation, too.)

Still, more than a dozen published studies have linked 2D:4D to sex-typed interests, and most of these have been published since 2000, which indicates that the model is currently under very active exploration among these scientists. (Note that there are several hundred reports in total that investigate 2D:4D as a putative marker for early androgen exposure [Voracek and Loibl 2009]. Here I focus on those studies that concern (non-erotic) sex-typed interests.) Several reports have linked the digit ratio with aggression, but findings disagree as to whether the effect holds for men only, women only, or both (Bailey 2005; Benderlioglu and Nelson 2004; Coyne et al. 2007; McIntyre et al. 2007; Hampson, Ellis, and Tenk 2008). Voracek and Stieger (2009) suggest that a growing number of "nil findings" suggest that there is no association in either sex. Several studies have looked at other gendered personality traits (seen as related to interests or motivation), and these, too, are mixed. Fink, Manning, and Neave (2004), for instance, showed that "agreeableness" was correlated with 2D:4D in women but not men, and elsewhere reported that "sensation-seeking" was correlated with 2D:4D in men, but not women (Fink et al. 2006). Moore, Quinter, and Freeman (2005) found no link between assertiveness and 2D:4D in women. Two studies have found that digit-length ratios were correlated with occupational preferences in men, but the correlations were in the opposite direction from one another. Weis, Firker, and Hennig (2007), found the predicted correlation of "masculine" preferences with lower 2D:4D, while McIntyre (2003) found the converse association.

Several studies have used more conventional measures of gender role or gender conformity, but these also have not yielded consistent relationships between 2D:4D and other sex-typed traits. Rammsayer and Troche (2007) found mixed evidence, which they interpreted as showing, overall, a positive but weak link between 2D:4D and gender in men (not in women). Csatho et al. (2003), on the other hand, found that a more masculine 2D:4D correlated with "higher, masculinized bias scores" on the Bem Sex Role Inventory among women. Voracek and Dressler (2006) directly addressed Baron-Cohen's typology of "empathizers" and "systemizers," but found no support for the idea that traits linked to empathy or systemizing were linked with digit-length ratios in either men or women.

In the end, studies tend to find only weak links, if any, between 2D:4D and sex-typed interests, and there is contradictory evidence regarding whether the digit ratio is a better marker for early androgen exposure in men versus in women. Even the people who think this model is worthwhile tend to damn it with faint praise, as when Hampson, Ellis, and Tenk (2008) concluded: "The 2D:4D digit ratio may be a valid, though weak, predictor of selective sex-dependent traits that are sensitive to testosterone" (133).[11] Given the uncertainties about the model itself, as well as the inconsistency of the results even from large studies, I am inclined to suggest that this model doesn't substantially add to our knowledge about the origin of sex-typed interests.

A THIRD model can be found in a small cluster of studies that draw on physical evidence related to fetal hormones to explore sex-typed interests in nonclinical populations. These include two studies correlating hormones in amniotic fluid, collected for amniocentesis, with later interests (Grimshaw, Sitarenios, and Finegan 1995; Lutchmaya, Baron-Cohen, and Raggatt 2002; Knickmeyer, Baron-Cohen, et al. 2005; Knickmeyer, Wheelwright, et al. 2005; Knickmeyer, Baron-Cohen, Raggatt, et al. 2006) and two studies for which maternal blood serves as the indicator of prenatal hormones (Udry, Morris, and Kovenock 1995; Udry 2000; Hines, Golombok, et al. 2002).

The two maternal serum studies did find a positive correlation between prenatal hormones and "masculine" scores on gender role behavior in females, but there is an important contradiction between the two. Richard Udry (Udry, Morris, and Kovenock 1995; Udry 2000) has reported that higher levels of sex-hormone-binding globulin (SHBG), but not testosterone, predict more masculine scores on a composite measure of women's adult "gendered" behavior. The second maternal serum study, by Melissa Hines and colleagues (Hines, Golombok, et al. 2002), found a small but statistically significant positive correlation between masculine play behavior (measured via the Pre-School Activities Inventory, or PSAI) and testosterone. The correlation was apparent in girls only, and there was no connection between sex-typed play and SHBG for either sex.

Aside from the contradiction between these two studies of hormones in maternal blood in terms of whether testosterone itself or SHBG is a more appropriate marker of fetal androgen exposure, there are reasons for caution. First, testosterone in maternal serum does not seem to reliably indicate fetal testosterone levels (Cohen-Bendahan, van Goozen et al. 2005), so the model is not a terribly strong test of brain organization theory. Second, Udry's methods are not well described, and his findings are inconsistent between his 1995 and 2000 report (Cohen-Bendahan, van Goozen, et al. 2005).[12] Third, in Hines's study, though the measures used were much more standard and her methods well described, testosterone in maternal blood explained only 2 percent of the variance in gender role—an exceedingly small effect. The correlation could certainly be real, especially given the ample room for measurement error in the variables of interest (which would tend to reduce any observed correlation, hence underestimating the effect size).[13]

Now consider the amniotic fluid studies. There are two different studies that relate hormones in amniotic fluid to the sex-typed interests of offspring, and one of these (a longitudinal study conducted by Simon Baron-Cohen and colleagues, especially Rebecca Knickmeyer and Svetlana Lutchmaya) has produced multiple reports relating to sex-typed interests. (Both studies have also yielded reports on cognitive patterns, but I restrict my focus here to the findings on sex-typed interests.) Neither study that examined hormones in amniotic fluid has found any relationship between testosterone and sex-typed play (Grimshaw, Sitarenios, and Finegan 1995; Knickmeyer, Baron-Cohen, et al. 2005), either in girls or in boys.[14]

The remaining reports from Baron-Cohen's team are somewhat difficult to relate to other studies, because the measures of sex-typed interests are idiosyncratic. (Baron-Cohen's lab has worked with children from the neonatal stage, when none of the standard measures of gendered behavior are applicable.) Recall that Baron-Cohen's theory is that the most fundamental difference in female and male personality and orientation toward the world boils down to an interest in people versus an interest in phenomena that can be "systemized." To measure early tendencies toward such sex-typed interests, Baron-Cohen and his team have devised measures that are meant to reflect interests in people versus interests in objects in very small children, as well as measures that are meant to indicate components of "empathizing." A number of the constructs they have investigated have shown positive correlations with prenatal testosterone levels from amniotic fluid, including "restricted interests" and quality of social relationships (Knickmeyer, Baron-Cohen, et al. 2005); the child's tendency to attribute intentional propositions to cartoon objects (Knickmeyer, Baron-Cohen, Raggatt, et al. 2006), which may reflect a child's ability to imagine other people's thoughts and intentions; and the amount of eye contact an infant makes with its parent (Lutchmaya, Baron-Cohen, and Raggatt 2002).[15] Nonetheless, a number of methodological inconsistencies across the various reports, as well as the fact that all of these reports come from a single team and even a single cohort of children, suggest caution in attributing too much importance to the findings.[16] Examples of important inconsistencies include the way that they treat statistical outliers (girls with very high levels of fetal testosterone) in the various reports, and variations in their statistical modeling for relating both testosterone and other hormones (such as estradiol) to sex-typed traits. Perhaps even more important is the way this team's findings contradict the best-designed study of testosterone in maternal blood, which is the study by Melissa Hines, Golombok, et al. (2002), described earlier. In that report, Hines found that testosterone in maternal serum correlated in a small but significant way to the score obtained by offspring on the Pre-School Activities Inventory, but that the relationship was apparent only for girls, not boys. In the reports by Baron-Cohen's team, several aspects of sex-typed interests have shown a correlation with testosterone among the boys, but none have among the girls (within-sex analyses are more informative for this sort of study than the analyses that look at the combined sample of children, even when they statistically control for sex). Given that there are very small numbers of girls in this study, the failure to find a significant relationship between prenatal testosterone and interests is not surprising, but it is somewhat surprising to find this correlation for boys, given that Hines's much larger study obviously had the statistical power to detect even very small effects, and found none. One can imagine several ways to explain the discrepancy: maybe Hines's result is not meaningful, because maternal serum is not a good reflection of prenatal testosterone (this is a very real possibility, as Hines found no association between testosterone levels in maternal blood and the sex of the fetus). Or maybe Baron-Cohen's result is not meaningful, because of sample limitations and methodological issues (such as the idiosyncratic way that they have treated outliers in the analysis of their very small sample). But the bottom line is that these anomalous results require one to have faith in either Hines's results, or Baron-Cohen's, or possibly neither one—but not both.

A fourth model is found in a large subset of studies focused on sex-typed interests in people exposed in utero to exogenous hormones: synthetic or natural progestogens or estrogens, alone or in combination. The biggest hurdle in interpreting these studies has been the great variety of hormone regimens involved, and the fact that many study groups are composed of subjects exposed not only to quite different amounts of specific hormones, but to different actual substances, and during varying periods of gestation. Another major difficulty is that hypotheses regarding behavioral effects of early exposure to progestogens are especially complex, because subtle structural differences among specific progestogens lead to extremely different biochemical activity, the exact mechanisms of which are still not well understood.

As a group, progestogens induce changes in the endometrium (the lining of the uterus) that are necessary to support pregnancy, but beyond that commonality there are important differences.[17] Most critically, it is not possible to sort progestogens into "purely" androgenic, antiandrogenic, estrogenic, or antiestrogenic groups, because the substances have affinities for a range of receptors and can either augment or counteract the metabolic effects usually associated with testosterone or estrogen (Dubey et al. 2008). Thus it is hazardous to generalize about the effects on sexual development that this entire class of hormones may have.

Nonetheless, because they are cited in literature reviews, studies of prenatal progestin exposure must be included when considering the role of early hormones on sex-typed interests.[18] In 1967 Ehrhardt and Money reported that girls who were born intersex subsequent to prenatal progestin exposure were in some ways masculinized (they did not employ a control group, so I don't detail the study). A few years later, Ehrhardt and her collaborator Heino Meyer-Bahlburg again studied progestin-exposed children, but this time focused on a single progestin, MPA. Unlike the progestins she had studied before, which had masculinized the genitalia of genetic females, MPA is considered a "true progestin"—it *interferes* with androgens. Ehrhardt and Meyer-Bahlburg concluded that MPA feminized girls' behavior, though MPA-exposed and unexposed boys showed no differences. But the group contrasts were quite small: even with extremely generous statistical tests (which the authors justify based on the very small sample), only two items out of a large but unspecified number of comparisons showed a significant difference between the MPA-exposed and unexposed girls: on average, MPA-exposed girls were less likely to be tomboys and preferred more feminine clothing. Notably, exposed girls did not differ from unexposed girls on most of the usual items of sex-typed interest, such as interest in marriage or children, sports interest or ability, or on their interest in boys' versus girls' toys and games. Nor was there a consistent trend: in general, the unexposed girls were almost as likely to have more feminine scores than the MPA-exposed girls as they were to have more masculine scores (see histograms in Ehrhardt, Grisanti, and Meyer-Bahlburg 1977, 394–395). Still, in spite of the very small sample size (13 boys and 15 girls), the narrow findings, and the fact that after thirty years it has never been corroborated, this study is often cited with a confident gloss in reviews (as in Reinisch, Ziemba-Davis, and Sanders 1991; but see Hines, Golombok, et al. 2002 for a more cautious interpretation).

Contrast the long shelf life of Ehrhardt and colleagues' study with a much larger analysis of MPA-exposed children that was published a decade later. In a study that has not been cited by a single research report or review article on the "organizing" effects of hormones, Jaffe and colleagues (Jaffe et al. 1989) compared 74 teenage boys and 98 teenage girls who had been exposed to MPA with 459 boys and 546 girls who had not been exposed. Using standardized measures (including the Bem Sex Role Inventory and the Buss-Durkee aggression scales) as well as interviews, and employing a design with adequate power to detect even very small group differences, Jaffee and colleagues found no indication that MPA exposure was associated with more feminine or less masculine behavior in either boys or girls. In fact, the only significant finding was in the *opposite* direction to their working hypothesis: MPA-exposed boys were somewhat more likely to be identified by teachers as "naughty in school" (Jaffe et al. 1989). How does the article that supports brain organization theory compare with the one that contradicts it, in terms of subsequent attention? The supportive article (Ehrhardt, Grisanti, and Meyer-Bahlburg 1977)—with less than one-fifth as many subjects—has garnered more than 1.3 citations per year since publication (41 total), while the contradictory article (Jaffe et al. 1989) has a mere 0.16 citations per year (3 total).[19] It is not clear to me why this study is overlooked. Perhaps it is due to the general enthusiasm that scientists tend to feel for studies with positive findings, or perhaps it is because it appeared in a journal *(Contraception)* that is not one of the mainstays of brain organization researchers.[20]

While Jaffe's findings have been totally ignored by brain organization researchers, a very small study that was never even published shows up with surprising regularity in reviews of the behavioral effects of early exposure to hormones. Katharina Dalton presented a study of 12 progesterone-exposed girls at a 1975 conference (cited in Ehrhardt, Grisanti, and Meyer-Bahlburg 1977). Though the methodological flaws are serious enough that they are routinely raised, Dalton's unpublished results have been repeated in numerous brain organization studies and reviews (Meyer-Bahlburg, Grisanti, and Ehrhardt 1977; Reinisch 1977; Kester et al. 1980), lending further credence to the idea that MPA feminizes behavior and personality.[21]

The largest studies of exogenous hormone exposures have involved the synthetic estrogen diethylstilbestrol (DES), either alone or in combination with progestins. In Chapters 4 and 6, I described two studies of hormone-exposed males conducted by Richard Green and colleagues (Yalom, Green, and Fisk 1973; Kester et al. 1980). Both reports included findings relevant to sex-typed interests. In the earlier report, "Prenatal Exposure to Female

Hormones," Yalom, Green, and Fisk (1973) found that teenage boys exposed in utero to a regimen of estrogen (DES) plus progesterone were "less aggressive-assertive" than control subjects; they also "[were] less competitive, placed less emphasis on physical prowess, were less success-oriented, and displayed less determination to achieve their goals than the combined contrast groups" (557–558). Twenty 6-year-old boys exposed to the same regimen were not found to differ from controls on measures of toy and game interest, aggressiveness, rough-and-tumble play, or other measures of sex-typed interests, though teachers found these younger boys to be less assertive and athletic. A grain of salt is in order for accepting even these modest findings: the researchers made a staggering number of comparisons, used very generous statistical methods (especially given the lack of clarity regarding hypotheses for estrogens and progestins), and found quite a few significant differences and trends that were in the opposite direction to their hypotheses.[22]

The second study, led by Patricia Kester (a graduate student of Green's), offers a particularly murky picture of the effects of prenatal hormones—again because of the huge number of comparisons, the heterogeneity of the hormone exposures involved, and the lack of clear-cut findings in terms of the direction of effects (Kester et al. 1980). The authors suggested that their data showed the DES-exposed group to be more masculine compared with those exposed to other hormone regimens (especially natural progesterone) and to unexposed controls, but this pattern was limited to a few items (conventionally masculine reading preferences, less cross-dressing, more boys as playmates, and more interest in sports). In contrast, the DES-exposed boys obtained more *feminine* scores on the Bem Sex Role Inventory, and lower scores on the aggression measures; most of their interests and behaviors were indistinguishable from the other groups. The findings regarding progestogen exposures were likewise mixed.[23]

Two other teams conducted analyses of exposure to various estrogens and progestins. In 1984 Ehrhardt, Meyer-Bahlburg, and colleagues repeated their 1977 findings on MPA-exposed girls and added children who had been exposed to other hormones prescribed for "problem pregnancies." They concluded that hormone-exposed boys were not very different from unexposed boys but "hormone treatment seemed to be associated with some degree of increased stereotypic femininity" in girls (Ehrhardt et al. 1984). However, in spite of an overly generous statistical approach, there was only one statistically significant difference between the exposed and unexposed girls: prenatal exposure to a hormone regimen was associated with a preference for "child rearing over career in adulthood" (table V, 470).[24]

June Reinisch and colleagues studied 17 girls and 8 boys who were prenatally exposed to various estrogens and synthetic progestins, and two reports from their study deal with sex-typed interests.[25] Reinisch and Karow (1977) split these children into three treatment groups, reflecting either a high estrogen-to-progestin ratio (the "estrogen" group), a high progestin-to-estrogen ratio (the "progestin" group), and more balanced ratios (a "mixed" group). They reported significant differences between the progestin and estrogen groups, with the former children being rated as more "independent, sensitive, self-assured, individualistic, and self-sufficient" (257)—characteristics that have overall been interpreted as reflecting a masculinization of personality (Jacklin, Maccoby, and Doering 1983). In another report on these same children, Reinisch reported that, in a paper-and-pencil test, progestin-exposed children were more likely to endorse physical aggression as their probable response to a conflict situation, again indicating that synthetic progestins had masculinized behavior (Reinisch 1981).

Yet, like the other data on progestins, these findings require scrutiny. First, a three-way comparison among a total of 25 children is already questionable. And second, only a subset of the children in this already small study were exposed to progestogens that are currently considered "virilizing"; some of them were exposed to the substance, MPA, that Ehrhardt and Meyer-Bahlburg found to be linked to more *feminine* behavior. It is not possible to retrospectively sort this out, as Reinisch did not provide adequate detail to know how many of the subjects were exposed to which hormones, except to note that 5 of 8 boys, and 4 of 16 girls were exposed to the "most virilizing" form of synthetic progestogen, 19-nor-17α-ethinyltestosterone (19-NET) (Reinisch 1981).

As I've described for other behavioral domains in previous chapters, one of the most interesting trends in the literature is the shift in hypotheses— and a troubling parallel shift in the data—regarding synthetic estrogen exposures. Early studies, like the ones described above, mostly suggested that the synthetic estrogen diethylstilbestrol (DES) would be associated with more feminine interests. Kester and colleagues' (1980) finding that men exposed to DES alone were in some ways more masculine prompted them to pay closer attention to developments in animal research on estrogen exposures than had previously been the case in human studies. They noted that their "provocative" findings could be explained by "evidence that estrogen, like testosterone, can exert a masculinizing influence on the developing central nervous system" (283), and cited seven animal studies dating back more than a decade to support the point. As I showed in the review of how ideas about sexuality changed in these studies (Chapter 6), it took a few years for researchers studying humans to pick up this shift in expectation. But once they did, investigators began looking for—and sometimes finding—evidence that DES would masculinize (or defeminize) sex-typed interests.

Like the findings on how DES might affect sexuality, though, the idea that DES might affect sex-typed interests fizzled. One early study suggested that gender-role was "masculinized" among DES-exposed women (Ehrhardt et al. 1989), but a reanalysis of a larger and improved sample reversed that finding (Lish et al. 1991; Lish et al. 1992). Other large-scale studies have consistently found no differences in sex-typed interests among either women or men exposed to DES (Newbold 1993; Titus-Ernstoff et al. 2003).[26]

Interestingly, even given the methods that consistently bias studies toward supporting brain organization theory, there is no clear trend in terms of a suggestion that exogenous hormones—whether alone or in combination—affect sex-typed interests. Disappointingly, one still often sees claims in the scientific literature that exogenous progestogens and/or estrogens affect sex-typed interests (Baron-Cohen 2003a; Berenbaum 1998; Cohen-Bendahan, van de Beek, and Berenbaum 2005), perhaps because some of the larger studies that contradict the earlier and more tentative suggestion of organizing effects are still fairly recent.

THE FIFTH model focuses on congenital adrenal hyperplasia (CAH). By far the most consistent body of evidence related to sex-typed interests comes from the study of children and adults with CAH, so I review these in the greatest detail.[27] As with other domains, relationships between hormones and sex-typed interests have been demonstrated only for (genetic) females, not males. In particular, boys with CAH do not differ from unaffected relatives in terms of toy, game, or playmate preferences (Berenbaum and Hines 1992; Berenbaum and Snyder 1995; Hines, Brook, and Conway 2004) or aggression (Berenbaum and Resnick 1997). Early on it seemed that boys with CAH might have higher energy expenditure and greater interest and/or skill in sports than their unaffected counterparts (Ehrhardt and Baker 1974); later studies not only failed to confirm this, but sometimes even found a reversed pattern, with CAH boys showing less athletic skill or lower levels of rough play (McGuire, Ryan, and Omenn 1975; Hines and Kaufman 1994). Likewise, few studies have found differences between males with CAH and unaffected males on other measures of sex-typed interests and personality. McGuire, Ryan, and Omenn (1975) found (contrary to hypothesis) higher scores on the "Adult Feminine" scale of a masculinity-femininity inventory. The authors interpreted this result as "a consequence of their short stature and traumatizing developmental history," which was "characterized by rapid growth which then subsided so that by age 12 they were frequently the shortest males in their age group, with strong feelings of anger, insecurity, and self-deprecation" (187). Similarly, Berenbaum and colleagues' (2004) analysis of psychological adjustment in CAH found that adolescent and adult males with this syndrome have significantly higher levels of negative emotionality and stress reaction than unaffected males. In general, though, scientists agree that males with CAH are not gender-atypical: they are neither more nor less masculine than other boys and men.

Girls and women with CAH are a different story. According to Sheri Berenbaum, one of the most prolific researchers of this syndrome and the principal investigator of a longitudinal study of CAH that has yielded nearly a dozen original research reports, this model has provided convincing evidence that androgen effects on sex-typed interests are not just discernible, but large (Berenbaum 1998; Berenbaum and Resnick 2007). In a chapter written in response to the question "Why aren't more women in science?" Berenbaum and her collaborator Susan Resnick (2007, 150–151) asserted that CAH is associated with a "masculinized" pattern for "sex-related interests, including childhood toy preferences, adolescent activity interests, and adult hobbies," as well as masculine vocational interests, reduced interest in babies, greater readiness to use physical aggression, and a preference for boys as playmates. My review suggests that "masculinized" interests among girls and women with CAH are much narrower than this.

The most consistent findings, without question, concern play behaviors among genetic females with CAH. Eight independent studies (some of which have yielded multiple publications) have examined toy and game preferences among children with CAH, and seven of the eight found that girls with CAH were more likely than unaffected girls—sometimes *much* more likely—to prefer toys and games that are considered "male-typical," and less likely to prefer toys and games considered "female-typical" (Ehrhardt and Baker 1974; Resnick et al. 1986; Dittmann et al. 1990, "Congenital Adrenal Hyperplasia I"; Berenbaum and Hines 1992; Berenbaum, Duck, and Bryk 2000; Servin et al. 2003; Pasterski et al. 2005; Meyer-Bahlburg et al. 2006). In particular, girls with CAH (or women with CAH who are retrospectively questioned about their childhood behavior, as is the method in several of the largest studies) report much less interest in dolls. Several studies have also asked about playmates. While most boys and girls prefer same-sex playmates, girls with CAH, on average, are more likely to prefer male playmates than are unaffected girls (Meyer-Bahlburg et al. 2006; Berenbaum and Snyder 1995; Ehrhardt and Baker 1974; Hines and Kaufman 1994; Servin et al. 2003).

As noted above, toy and game preferences in childhood also constitute a major portion of childhood "gender role" or "gender-related behavior" instruments. At least eight separate studies have compared CAH-affected and unaffected girls on such composite measures of gendered behavior, and most of them have found that girls with CAH are "more masculine" than control girls (Slijper 1984; Dittmann et al. 1990, "Congenital Adrenal Hyperplasia I"; Berenbaum and Hines 1992; Zucker et al. 1996; Servin et al. 2003; Hines, Brook, and Conway 2004; Meyer-Bahlburg et al. 2006). Only one study found *no* differences (McGuire, Ryan, and Omenn 1975).

While there seems to be no question that girls with CAH are more "masculine" in their choice of toys, games, and playmates than other girls, no other aspects of interests or behavior show such strong findings. Perhaps surprisingly, given that high energy expenditure and "rough-and-tumble" play are central to the tomboy identity that is often conferred on CAH girls, systematic investigations have weakened the early claims that girls with CAH prefer rough or high-energy play (Dittmann et al. 1990, "Congenital Adrenal Hyperplasia I"; Hines and Kaufman 1994; but see Pasterski et al. 2007 for recent support).[28]

The interests and preferences that are assessed in studies of CAH are fairly removed from the aspects of career and family life that are at the heart of debates about biology and sex differences. The exceptions in this regard are specific vocational interests, and the relative priority one places on family and home life versus a career. Most studies have found that, compared with unaffected girls, girls with CAH are less likely to prioritize "marriage and motherhood" over careers (Ehrhardt, Epstein, and Money 1968; Hurtig and Rosenthal 1987; Dittmann et al. 1990, "Congenital Adrenal Hyperplasia I"), though one well-known study (Ehrhardt and Baker 1974) failed to find this difference. The few studies that have presented specific findings on career aspirations or achievements among CAH-affected females offer mixed evidence. In the only study examining actual occupations (rather than attitudes or aspirations), Kuhnle and Bullinger (1997) examined the occupations of 45 women with CAH and found that most of the women were employed in female-typical positions, "and none had reached leading positions" (513). Still, at least two studies have found that girls with CAH show more interest in masculine occupations (Berenbaum 1999; Servin et al. 2003).[29]

What about "real world" measures of interest in marriage and children? What do the family lives of women with CAH look like? Most studies have found much lower rates of marriage and childbearing among women with CAH, but many scientists doing these studies caution against attributing this too much to the effects of early hormones on the brain (Hurtig and Rosenthal 1987; Mulaikal, Migeon, and Rock 1987; Stikkelbroeck, Hermus, et al. 2003). For example, impaired fertility is a direct physical result of the disease in many cases.

Finally, the evidence to date suggests that girls and women with CAH are not more aggressive than unaffected females. Only one study has found the predicted increase in aggression among females with CAH: Pasterski and colleagues (2007) found that parents were significantly more likely to report aggressive behavior among their girls with CAH compared with their unaffected daughters. But five other studies have not found increased aggression in females with CAH (Helleday et al. 1993; Dittmann et al. 1990, "Congenital Adrenal Hyperplasia I"; Gordon et al. 1986; Ehrhardt and Baker 1974; Slijper 1984).[30] As I explained above, another study that is typically interpreted to show increased aggression in females with CAH (Berenbaum and Resnick 1997) bears reconsideration based on a larger follow-up study (Berenbaum et al. 2004). The first study found that two different samples of adolescents and adults each scored higher than controls on one measure of aggression, while the sample of children did not differ from controls. The more comprehensive study included one of the two measures on which one of the samples was earlier found to be more aggressive, and this time no differences were found between people with CAH and controls. Given this pattern of results, even Dr. Berenbaum is currently inclined to view them as "not strong" (Berenbaum interview, June 2008). I would go further, and suggest that the overall evidence points to no increased aggression associated with CAH.

The tendency in the literature is to interpret the findings on aggression quite differently than I have done here, but that is because many studies that bear on this question are overlooked. The literature review in the 2007 study by Pasterski and colleagues is illustrative. That report includes a table that is meant to reflect the totality of data on CAH and aggression (table 1, 369). The table does include a small study (Money and Schwartz 1977) that my own review omits because no control group was used and no statistics were presented. As Pasterski reports, that study found no suggestion of aggressive behavior related to CAH. Yet the table *omits* most studies that employ direct measures of aggression (Slijper 1984; Gordon et al. 1986; Dittmann et al. 1990, "Congenital Adrenal Hyperplasia I"; Helleday et al. 1993; Berenbaum et al. 2004), none of which find that CAH increases aggression. The table includes Berenbaum and Resnick's early (1997) finding of girls with CAH being more aggressive, but not the larger follow-up that found no differences (Berenbaum et al. 2004).[31] It seems plausible that Pasterski overlooked these studies precisely because they find no suggestion of increased aggression among females with CAH; as a consequence, aggression is not highlighted in the findings, nor mentioned in the article abstract or keywords. This is how negative results get lost in the literature: a search for "aggression and CAH" will not turn them up.

Let me pause here to emphasize that I do not mean to suggest scientists are "hiding" negative findings, either their own or those of other scientists. Rather, my interviews suggest that busy scientists, like everyone else, are likely to lose track of or never notice all the interesting details of complex studies, especially if those tidbits do not fit well with the dominant theory. Selective reviews of the data—almost certainly not *intentionally* selective, but selective nonetheless—keep alive certain hypotheses that should be put to rest, like increased aggression among women and girls with CAH.

Putting the Pieces Together: Are Prenatal Hormones the Best Explanation for Sex-Typed Interests?

As I've noted throughout this book, it is critical to remember that brain organization studies in humans are quasi experiments, and that as such, they have to be considered together. It is especially important to see how the findings from different research models and study designs fit together, because different study populations generally will involve different uncontrollable (and frequently unknown) confounders, and different designs have specific strengths and weaknesses. For example, children with hormone-involved clinical conditions like CAH are likely to have had sufficiently unusual hormone exposures during the relevant developmental periods in question that—if the theory is correct—they should be different from their "normal" peers. Yet these children also are often born with ambiguous genitalia and have highly unusual rearing experiences and medical histories. In contrast, studies of children for whom there are measures of hormones in amniotic fluid avoid the confounding effects of being born intersex and raised with a grave illness. Yet this model has other shortcomings, such as uncertainty about the utility of a single measure of hormones from amniotic fluid for estimating the "sex-typical" profile of hormone exposures during gestation.

As it turns out, *there is no specific kind of sex-typed interest that is consistently linked to prenatal hormone exposures by more than one research model.*[32] Moreover, the only model that has yielded strong and consistent data across different samples of research subjects, and with different teams of investigators, is the study of girls and women with CAH.

Females with CAH have the highest levels of androgens, as a group, of any human females, and their androgen levels are high throughout gestation. Thus, any trait that is a product of hormonally driven sex differentiation of the brain should be more masculine in CAH-affected girls. This creates a fundamental disconnect between other research models and CAH, in two ways. First, those traits that are not affected in CAH do not seem like good candidates for being influenced by prenatal androgen exposures. This eliminates everything except toy and game preferences, and possibly playmate preferences and occupational interests, from the list of sex-typed interests that might be "organized" by early hormone exposures. Second, other models offer little suggestion that prenatal steroid hormones affect sex-typed play, which is by far the most masculinized interest in girls with CAH. A partial exception may be found in the small effect of fetal testosterone that Hines and colleagues (Hines, Golombok, et al. 2002) found for play preferences. But given the size of the effect, questions about whether maternal serum even reflects fetal hormone levels, and the lack of corroboration from other studies in nonclinical populations, what does this mean in the real world? Probably not much. It remains possible that androgens are a good explanation for the more masculine activity preferences of girls with CAH, but it seems unlikely that androgens are a good explanation for *within-sex* differences generally. (And without showing that hormones are responsible for *within-sex* differences, we can't conclude that they are responsible for *between-sex* differences. That's because boys and girls are systematically socialized quite differently—even when parents don't mean to, and don't think they are doing so, as I will detail in Chapter 9; so we can't actually attribute male–female differences in interests to sex differences in early androgen exposures.)

While some aspects of interests are indisputably masculinized in girls and women with CAH, perhaps an "organizing" effect of prenatal androgens is not the best explanation. Note especially that few studies have attempted to evaluate the effect of illness itself, or the medical intervention that chronic illness entails. As an exception, Froukje Slijper (1984) compared girls with CAH to girls with diabetes as well as to healthy controls, and found that both groups of girls with chronic illness scored in the more masculine range than controls on a gender scale. Slijper thus noted that "being sick plays a role" in the effect on gender role behavior. Slijper's study is a good example of how scientists could actively explore alternative explanations for the differences they find between people with unusual hormone exposures and those whose exposures have presumably been normal. Chapter 9 deals in depth with how scientists tend to "shoehorn" findings into support of brain organization theory by omitting critical issues of context—both biological and social context.

Constraints on the "Sex-Typed Interests" Story

In the end, whether one believes that differences between girls with CAH and other girls are due to a direct organizing effect of androgens on the brain, or to other factors, will probably not depend as much on the data as on how satisfying one finds the CAH research model in terms of our overall knowledge about steroid hormones, developmental processes, and the social world. However, there are some ways in which data on prenatal hormones and sex-typed interests already offer constraints on the larger model of evolutionarily derived and hormonally mediated sex-typed interests.

Let's go back to the issue of cultural change, which I brought up early in this chapter. Not surprisingly, some researchers (and other critics) have raised objections to the way children's interests are interpreted as "masculine" or "feminine" for these studies. Most scientists have not been concerned that the measures were sexist or old-fashioned (at least that isn't their explicit concern). Instead, they worry that the measures are simply out of date. As McGuire and colleagues put the issue in 1975, when presenting new findings on CAH and reviewing some of Money and Ehrhardt's early work:

> It is likely, with regard to the general content of the interview, that cultural and social changes may have ameliorated many sex-typed differences which were far more pronounced during the mid-1960s, when Money's research was undertaken. The shifting of women's dress, attitudes, and behaviors toward those traditionally designated as masculine has accelerated greatly since that time and necessitates regular updating of tests designed to measure sex-typed behaviors and attitudes. Rosenberg and Sutton-Smith reported as early as 1961 the need for revising their sex-typed norms; all of the changes represented the inclusion of masculine toys and games in girls' repertoires, while boys' toys and game choices remained the same. (McGuire, Ryan, and Omenn 1975, 187)

This may ring a bell. It is similar to the shifts in sex-typed norms for sexuality I documented in Chapter 5: several aspects of sexuality that had been considered "masculine" were increasingly seen as normal for women, but the standard picture of "masculine" sexuality did not appreciably change. Yet, unlike the way that measures of sexuality have been quietly transformed, many of the measures of *non-erotic* sex-typed interests favored by scientists have never been updated.[33] Theoretically, this could mean that there is more continuity in measures of sex-typed interests than there is in sexuality, but as I've shown, the complexity and variety of measures involved in sex-typed interests means that isn't the case. Meanwhile, there is still the real problem of how to deal with actual changes in behavioral norms, and addressing this issue will mean revisiting some of the core assumptions about *why* boys and girls seem to prefer certain toys. In particular, some of the changes undermine the idea that toy preferences are "evolutionarily derived" and reflect certain objective features that have "functional significance" for males versus females.

Children's play has definitely undergone some changes (think of all the girls playing soccer and basketball now compared with the 1970s), but measures of play haven't caught up. Routinely, toy inventories code all building toys (like Legos and Lincoln Logs), as well as vehicles and sports equipment, as "masculine," because overall, boys are more likely than girls to play with these toys. But how much of a sex difference does there have to be to make something a "boys' toy"? At the Barnard College graduation in 2008, a gifted young graduate, Ruth Talansky, had been invited to give "academic reflections" on her experience. Addressing her peers at this all-female college, Talansky structured her talk around the image of Legos, explaining that she chose the image because "we all played with them, all the time, as children."

A very interesting example of how abstract models about boys' and girls' toys' can be misleading is found in a study of girls with CAH in Sweden. Anna Servin and colleagues (Servin et al. 2003) are among the few scientific teams who have used actual observation of play behaviors, rather than interviews or questionnaires, to determine children's play preferences. Following a procedure developed by Sheri Berenbaum and Melissa Hines (Berenbaum and Hines 1992), the scientists arranged toys that had previously been defined as masculine, feminine, or neutral in a semicircle on the floor, and invited children, one at a time, to play with the toys however they wished. Table 8.3 reproduces the results table on toy play from this report. Although girls with CAH spent significantly more time playing with masculine toys, and less time playing with feminine toys, than did the control girls, the most popular toy with *all* of the girls was a toy coded as masculine: the Lincoln Logs. Noting that "twice as much time was spent with that toy as with any other toy," the scientists addressed the discrepancy by speculating that the character of this particular toy, as well as "its novelty to Swedish children," makes it inherently time consuming: "This toy contains many different parts . . . and the means of putting the logs to-

Table 8.3
*Number of Seconds Spent Playing with Masculine and Feminine
Toys as a Function of Group*

Toy	Control girls		CAH girls	
	M	*SD*	*M*	*SD*
Masculine toys				
Garage	123.5	133.4	161.3	181.3
Bus	24.7	79.7	36.3	102.7
Xmen	18.0	44.1	137.3	171.1
Lincoln Logs	253.4	272.5	298.7	266.0
Total	419.7	245.4	633.5	266.4
Feminine toys				
Plastic tea set	36.0	49.0	8.8	16.4
Barbie and Ken dolls	118.0	165.7	58.6	119.1
Doll head	29.0	47.0	20.6	61.9
Baby doll and blanket	39.5	79.6	4.5	15.4
Total	222.4	199.8	92.5	151.5

Note. CAH = congenital adrenal hyperplasia.

gether must be figured out" (447). This fits the nature of tasks that Baron-Cohen has designated "systemizing"—the sine qua non of "boy play." But it was the most popular activity with girls.

Somehow the scientists seemed to overlook the fact that the *second* most popular toy among both sets of girls was also a so-called masculine toy: a garage with four cars (Servin et al. 2003, table 2, 445). (Certainly Swedish children are familiar with garages and cars, so the draw there couldn't have been the sheer novelty of it all.) This time it isn't necessarily the idea of "systemizing" that takes a hit. Instead, it is the idea that "motion" and objects with moving parts are intrinsically interesting to boys rather than girls.

It is even more interesting to compare the amount of time the children spent playing with these "boys' toys" compared to the quintessential "girls' toy" they were offered: a baby doll with a blanket. On average, *the normal control girls spent three times as long playing with the garage and toy cars as they did playing with the baby doll* (the Lincoln Logs–to-doll ratio was more than 6 to 1). The only "girls' toy" that was in the ballpark (if you'll excuse the boyish metaphor) with these boys' toys was a pair of Barbie and Ken dolls. (I suspect that Barbie and Ken were riding around in some of those cars.) A glance at another result further confounds the designation of toys as "masculine" and "feminine." Each child was offered the choice of a masculine toy (car), feminine toy (doll), or neutral toy (ball) to keep as a gift. This alone is somewhat remarkable, because other scientists routinely code ball play as masculine (for instance, Money and Ehrhardt 1972; Alexander and Hines 2002). But the real news here is the sad fate, again, of the baby doll: even among the control girls, this toy was chosen least frequently (control girls were about 36 percent more likely to choose the ball than the doll; see table 3, 445).

How should we interpret these results? Servin and colleagues are perfectly correct in asserting that the girls with CAH did spend more time with the so-called boys' toys, on average (though, as they note, the differences aren't significant for all of the toys). So there is a group difference. But what about the idea that "boys favor construction and transportation toys, whereas girls favor toys such as dolls" (Alexander and Hines 2002, 468)? Obviously, the larger story about sex-typed preferences is damaged by the nitty-gritty details of studies like this.[34]

Returning to the broader picture of sex-typed interests, it's hard to reconcile some of the changes over the past few decades with the idea that hormonally hardwired differences in interests are the key force behind sex distribution in education, occupations, and the division of labor in families. Just as the sexual revolution affected ideas about normal sexuality, as well as actual sexual practices, the second-wave women's movement (among other forces) profoundly altered both ideals about gender roles and the actual work and family lives of average men and women. Changes in the sex ratio of students in higher education, and of new graduates entering the professions, are arguably even more dramatic than the sexual revolution. Here, long-standing patterns of sex difference not only have narrowed, but have reversed. As sociologists Claudia Buchmann and Thomas DiPrete (2006) observed:

> Trend statistics reflect a striking reversal of a gender gap in education that once favored males. In 1960, 65 percent of all bachelor degrees were awarded to men . . . Women continued to lag behind men in college graduation rates during the 1960s and 1970s, until 1982, when they reached parity with men. From 1982 onward, the percentage of bachelor's degrees awarded to women continued to climb such that by 2004 women received 58 percent of all bachelor's degrees. (515–516)

This trend isn't limited to the United States, but is a strikingly global pattern (Buchmann and DiPrete 2006). The changes are even more dramatic when you consider professional degrees. In 1960, just 2 percent of all law degrees in the United States were awarded to women; by 1993 that figure had climbed to 42 percent. In the same period, the proportion of dentistry degrees earned by women rose from 1 percent to 34 percent (Smith 1995);

in medicine, it went from 6 percent in 1961–1962 to over 49 percent in 2007–2008 (AAMC 2009).[35] Younger readers may not realize, and older readers might need reminding, that until very recently it was common opinion that women were generally not temperamentally equipped for these careers.

CODA: Fun with Vervet Barbie

When sex-typed interests are mentioned in popular science coverage, it is now *de rigueur* to include a study that seems to show sex-typed toy preferences in a nonhuman primate, the vervet monkey. Gerianne Alexander and Melissa Hines (2002) devised a study of toy preferences in vervet monkeys as a response to criticisms of brain organization studies, especially studies of girls with CAH. Noting that critics have suggested that "toy preferences in androgenized girls could reflect altered learning histories or altered cognitive development," Alexander and Hines sought a way to demonstrate that "toy preferences may be associated with factors other than human social and cognitive development" (469). In other words, they wanted a model that would help corroborate the interpretation that biological factors, especially hormone exposures, explain the findings of masculinized play in girls with CAH. Their summary of the findings from that study reads like a robust endorsement of biological influence on sex-typed play, and it is worth quoting at length from their summary to understand how much importance they and others have extracted from these results:

> The percent of contact time with toys typically preferred by boys (a car and a ball) was greater in male vervets ($n = 33$) than in female vervets ($n = 30$) ($p < .05$), whereas the percent of contact time with toys typically preferred by girls (a doll and a pot) was greater in female vervets ($p < .01$). In contrast, contact time with toys preferred equally by boys and girls (a picture book and a stuffed dog) was comparable in male and female vervets. The results suggest that sexually differentiated object preferences arose early in human evolution, prior to the emergence of a distinct hominid lineage. This implies that sexually dimorphic preferences for features (e.g., color, shape, movement) may have evolved from differential selection pressures based on the different behavioral roles of males and females, and that evolved object feature preferences may contribute to present day sexually dimorphic toy preferences in children. (467)

On the face of it, the study is a bit difficult to swallow. What, exactly, could be the functional significance of a cooking pot for a female vervet, or a toy police car for a male vervet? But the scientists' reference to "object feature preferences" seems, at first, to offer a way around this. As long as the toys are grouped as "masculine," "feminine," and "neutral," and as long as the attention is strictly on a comparison between the male and female animals, the idea that males and females have different preferences for color, shape, and movement seems plausible. But what happens if we remember that sometimes these toys are coded differently (as when Servin and colleagues coded "ball" as a neutral toy)? And what happens if we look at how "popular" any given toy was *among* the male and female animals (rather than looking only at sex differences)?

Let's start with the male vervets. Overall, as Alexander and Hines note, the male vervets didn't show any significant difference in preferences for "boys'" versus "girls'" toys. Looking at the specific items, male vervets showed the strongest preference for the dog—a "neutral" toy that does not exhibit any of the supposed underlying perceptual features of "boys' toys." The male vervets showed slightly lower but virtually identical preferences for the next three toys in line—the ball, the cooking pot, and the car. How about the female vervets? They showed the greatest preference for the cooking pot, followed closely by the dog, and *then* the doll, each of which got a roughly similar proportion of the overall contact. (Female vervets had much less contact with the other toys, which could mean lots of things. Perhaps the male vervets were more likely to be already "using" those toys, so they weren't available for the females. The point is that certain aspects of the testing situation, such as testing one toy at a time in a group of animals, make it impossible to sort out some important questions.[36])

Now return to Alexander and Hines's explanation for their results. How could these items have differential "functional significance" for male versus female vervets? Certainly vervets aren't practicing future gender roles when they play with cooking pots or police cars. To address this obvious problem, Alexander and Hines argue that "the primate brain has evolved specialized recognition systems for categories with adaptive significance, such as emotional expressions and facial identity" (474). But these features would apply to both the "female-coded" doll and the "neutral" plush dog. Noting that there were no sex differences in preference for "toy categories based on an animate-like (doll, dog) or inanimate-like (car, ball, book, pan) distinction" (474), Alexander and Hines appeal to color to explain why the female vervets spent proportionately more of their time than the males did in contact with the doll and the cooking pot: the doll face is pink, and the pot is red. This is misleading, though, because the female vervets spent slightly *more* time with the (nonpink, nonred) dog than they did with the pink-faced doll.[37]

The same shortcoming is apparent in their explanation of the male vervets' behavior. Reasoning that so-called boys' toys like the ball and the car are "objects with an ability to be used actively," Alexander and Hines speculate that these objects "afford greater opportunities for engaging in rough or active play" (475). The male vervets "favorite" toy—by a wide margin—was the plush dog, a toy that does not apparently lend itself to rough or active play. And let's not forget that the (boyish) police car was roughly tied in contact time with the (girlish) cooking pot. Finally, it would be extremely useful to see what the vervets actually *did* with the various toys. How does a vervet know that the purpose of a cooking pot is not to bang it, throw it, or use it to whack another vervet? Only by knowing the human purposes of these objects can we sort them into categories that *seem* to objectively reflect an "ability to be used actively."[38]

At the end of the day, when we look systematically and concretely at studies of male versus female "interests," confident claims that these interests are significantly directed by prenatal hormone exposures appear to be breathtakingly overblown. This is not to say that we know definitively that hormones *don't* have any such effects (remember basic logic: you can't prove a negative). But it is definitely the case that such effects are not proven and (depending on your interpretive orientation toward the evidence) are either unlikely to be meaningful or likely to be extremely subtle and limited to fairly narrow domains.

So shall I end with the usual line that "more research will be necessary" to sort this all out? Yes and no. Surely we need more research, but it definitely should not be more of the same. For reasons that I will explain in Chapters 9 and 10, I think it would be an extremely poor investment, both scientifically and socially, to continue pouring resources into trying to divide the indivisible—which is what a search for hormone effects, apart from the rest of the developmental context, amounts to.

CHAPTER NINE

Taking Context Seriously

U P TO THIS POINT I have merely hinted at ways in which brain organization research ignores important elements of the social and biological context of development. Scientists who study the influence of prenatal
hormones on sexuality and gender focus only on the early hormone environment and either omit, or seriously minimize, factors that are unusual
for people with intersex conditions. In conducting and evaluating studies
of sexual minorities, brain organization researchers fail to consider how
sexual minority cultures may encourage some of the associations between
sexual orientation and gender that scientists instead attribute to the mutual influence of early hormones. Likewise, scientists' own research practices, especially their sampling schemes and their approach to assessing
sexual orientation, may lead to false associations between sexual orientation and the aspects of psychology and personal history they are assessing.
Finally, almost all scientists who do brain organization work systematically neglect well-established evidence that the brain and the neuroendocrine system (not to mention the rest of the body) are not stable founda-
tions from which behavior and cognition emerge, but develop and change
in a constant dialectic with social and material "inputs," including the individual's own behavior, learning, and mood states. Too often such biases
push these scientists to attribute group differences—between sexual minorities versus heterosexuals, in the latter case, and intersex people versus

healthy comparison groups, in the former—to hormone exposures, without seriously entertaining other explanations that are at least as plausible.

This chapter addresses some key aspects of context that are systematically excluded from brain organization research. This discussion cannot be exhaustive, but I hope to show that we already know enough to understand that current research designs used in human brain organization research are inadequate and effectively "load the dice" toward an interpretation that early hormones are a major determinant of gendered behavior and sexuality. Because studies of CAH are so pivotal to the overall evidence for brain organization, I'll use CAH as the main example in this chapter. I also discuss the social and biological context in other intersex conditions, as well as the very specific case of normal males who are sex-reassigned. Toward the end of this chapter, I put the very question of sex differences into context by shifting the focus to intragroup differences and intergroup similarities as a different way to understand sex, sexuality, and gender.

Under the Knife and under the Gaze: Gender and Sexuality in CAH

Let's begin with a thought experiment to help us reflect anew on the distance between experiments on animals versus the quasi-experimental research that explores hormonal organization of the brain in humans. Experiments tell us a lot about context—that is, after all, the true meaning of the placebo effect. When we want to evaluate a hormone manipulation or a surgery in nonhuman animals, we don't compare mice, birds, or rats who have had injections or surgeries to animals who have not. We conduct sham surgeries or inject inert substances. In part this is done to neutralize the expectations of scientific observers: we need to obscure which animals had the intervention so that when they are monitored for subsequent behavior or physical changes, the observers won't unconsciously see what they expect. Just as importantly, we do these things because we don't know the effect of all the physical manipulations that go along with the intervention: the simple act of cutting tissue, the trauma of recovery, the reactions to anesthesia, and so on. These methods involve a healthy respect for what we don't know. Similarly, in clinical research with humans, scientists always prefer randomization of experimental and control groups, when possible, because a "post hoc" attempt to identify and statistically control for differences between groups requires that the scientist must already know what factors *might* be different between the groups, and *might* be important for the outcomes under investigation. In other words, too much relies on what the scientist already knows or believes about what causes the outcome she is studying. But if the flip of a coin determines who gets an exposure and who doesn't, and there are enough people in the study, chances are good that most other relevant factors besides the exposure that's being studied will be fairly evenly distributed between the treatment group and the group that gets the placebo.

So here is the thought experiment. Imagine that I am a scientist who would like to understand how medical treatment affects the psychosexuality of girls with congenital adrenal hyperplasia. I devise an experiment in which some children are randomly assigned to receive only routine pediatric visits for eighteen years, and other children receive treatments that mirror the typical intervention received by children with CAH [1] This would include subjecting 90 percent or so to "feminizing" genital surgeries to reduce the size of the clitoris and, in most, reshape the vagina. (I omit the treatment with hormones such as glucocorticoids, because this would create serious illness in children without CAH.) Most of these children would get at least one additional surgery during childhood or adolescence. In early adulthood, I would assess them to see if their sexuality or sex-typed interests or cognition were any different from the children who received routine pediatric care. Wait—certainly no ethics board would approve this research, even if I were unethical enough to propose it. And this wouldn't be a perfect experiment anyway, because I would not be comparing "real" surgeries to "sham" surgeries, but girls with surgery to girls without surgery.

What if instead I limit my proposal to an experiment regarding medical *monitoring*—so I would not subject participants to surgeries, but would simply randomly assign half of the subjects to receive medical exams every three months or so, including close genital inspection and testing for size and flexibility; repeated questioning about gender identity and behavior (such as whether the child feels like a boy or girl and why, and what sorts of toys they enjoy); querying of the parents regarding the child's masculinity or femininity; informing parents and the girls themselves that they may or may not be fertile, but that they can certainly adopt children; and so on. Again, no ethics board in the world would allow such an experiment. Why? Because we can easily anticipate that such treatment would have negative effects on the girls involved, as well as their parents. It might make them feel insecure, vulnerable, and possibly even violated by the intense scrutiny and genital manipulation. It might affect their relationships with their bodies, the medical system, their peers, and even their parents, for colluding with the treatment process. So how is it that these experiences—routine in the lives of children and families with CAH—can be utterly discounted in the studies of intersex children and adults?

The Effect of Medical Interventions
on Sexuality in CAH

There have been very few formal evaluations of how girls and women with CAH have experienced the treatment process, but the data that exist present a stunningly negative portrait. Focusing first just on genital surgery, all studies confirm that the idealized version of treatment, a single surgery in early childhood to "normalize" the appearance and function of genitals, is a very rare exception. The norm is a minimum of two surgeries, most commonly both in infancy and in adolescence and/or adulthood, either to finish "normalization" or to correct problems from earlier surgeries, including stenosis, fistulas, and incontinence (eg: Mulaikal, Migeon, and Rock 1987; May, Boyle, and Grant 1996; Crouch et al. 2004; Crouch and Creighton 2007; Stikkelbroeck, Hermus, et al. 2003). The vast majority of women and girls with classic CAH, even in fairly recent studies, have had clitoral surgeries (see, for example, May, Boyle, and Grant 1996; Servin et al. 2003; Pasterski et al. 2005; Meyer-Bahlburg et al. 2006). Of 18 girls studied by May and colleagues in 1996, 15 had total amputation of the clitoris (clitoridectomies) and 2 more had clitoral "recession"—a reduction of the clitoris that involves removal of the erectile tissue (482). Though the intent of this less drastic surgery is to preserve erotic sensation, there are now ample data to conclude that it is virtually impossible to avoid massive nerve damage with any of the clitoral surgeries yet developed (Crouch et al. 2004; Karkazis 2008, esp. chap. 5).

Outcome studies of genital surgery are difficult, in part because constantly evolving techniques mean that any evaluations will necessarily involve "older" surgical approaches (Karkazis 2008). This is of crucial importance to parents and clinicians who must ask questions about the hoped-for benefits of genital surgery (Will it make the child look "normal"? Will it allow "normal" sexual function?) and weigh these answers against the risks (Will clitoral sensation be damaged or destroyed? Will the process require multiple surgeries? Will scarring and growth make the vagina less functional over time?). But the issue of surgical outcome as a moving target is less relevant for the purposes at hand, because the salient point for evidence regarding brain organization is that the women from whom extant data on gender, sexual practices, and sexual orientation have been derived were treated with these same "outmoded" surgical techniques. So it is important to recognize that these surgeries often do not have the intended effect: somewhere between one-third and two-thirds of adult women with CAH have a vaginal opening that is not adequately large or flexible to permit heterosexual intercourse (should they desire this), and the vast majority of women have either impaired clitoral sensation or no sensation at all (Crouch et al. 2004). One of the few long-term outcome studies to report on multiple components of sexual function strongly indicates that, for most women with CAH, vaginal penetration is painful. Gastaud and colleagues (2007) found, first, that "37% (13 of 35) said they never had heterosexual intercourse with vaginal penetration"; moreover, 81 percent of the women who *did* have heterosexual intercourse experienced pain during vaginal penetration (1393). This report also included data on lubrication, orgasm, and overall satisfaction with sex acts. They found severe impairment in all aspects of sexual function in women who had been born with the most severely affected genitalia (Prader IV–V); notably, even though these researchers had expected to find lower sexual functioning in this group, they reported that the "extent of these differences . . . was unanticipated." Given the extent and number of "feminizing" surgeries in this group, which is even substantially higher than for the other women with CAH, it is curious that this finding should come as a surprise. Moreover, the women with CAH who had less severely affected genitalia (Prader stages I–III) also reported significantly higher rates of vaginal pain, and less lubrication and orgasm, than the healthy women to whom they were compared (1394). Another study found similar results, even when CAH women were compared, not to healthy controls, but to women with early-diagnosed diabetes mellitus (which may also impair sexual function): CAH women were found to be less sexually experienced in all areas, and reported higher levels of penetration difficulties, pain, and lubrication problems, than the diabetic women (May, Boyle, and Grant 1996).

While the idea that women with CAH are more likely to be bisexual or lesbian is perhaps the most widely popularized finding from studies of sexuality in this condition, the evidence has long indicated that the most profound difference in sexuality between women with CAH and unaffected women is that the former have very low rates of sexual activity and partnerships across the board, as well as lower levels of sexual arousability, genital sensation, and orgasmic capacity. In a very large study that is now more than twenty years old, Mulaikal and colleagues found that the sexual orientation and activity women reported was more closely related to their vaginal condition than to the degree of prenatal androgen (180). This is not surprising, given that heterosexual eroticism typically centers on penile-vaginal intercourse. Whether this is simply taken for granted as the normal and natural state of affairs (as is the case in virtually all brain organization research) or bemoaned as a male-centered definition of sexuality that women can and should challenge (Koedt 1976; Tiefer 2004; Jackson 1999), there is virtually no one who questions that intercourse is, in this time and place, the very definition of "sex" for heterosexuals. (In spite of the great consternation that followed Bill Clinton's assertion that he "did not have sex with [Monica Lewinsky]," empirical research has repeatedly demonstrated that heterosexuals do, in fact, interpret "having sex" to be synonymous with penile-vaginal intercourse (Bogart et al. 2000; Sanders and Reinisch 1999). Is it so surprising, then, that women who, as a group, have extremely low rates of "typical" vaginal function would also have somewhat increased rates of at least fantasizing about female partners, where sexual intimacy would not necessarily hinge on the depth and flexibility of the vagina?[2] And that is only to consider the physiological effect on genitals that medical treatment of CAH has.

What about the psychological effects? In one follow-up study, women with CAH "used a language of 'rape', 'invasion' and 'violation' when talking about vaginal examinations and other procedures carried out during visits to paediatric and adult clinics" (May, Boyle, and Grant 1996, 491). This is consistent with the stories that intersex adults as well as the parents of intersex children told to medical anthropologist Katrina Karkazis (2008), which she relates in her book in *Fixing Sex: Intersex, Medical Authority, and Lived Experience*. Medical visits every three or four months are often considered necessary to monitor children's hormone levels as well as their response to treatment. As Karkazis documents, girls with CAH, as well as their parents, often experience the genital scrutiny as "intrusive and dehumanizing," as well as of questionable medical value (190). The words of several of Karkazis's informants, all of them parents of girls with intersex conditions, powerfully make this point:

> If she's helping other doctors to learn about it, that's fine, but when she gets older I don't want everyone poking and prodding her and every doctor experience to be, "I've got to look between your legs."

> My daughter knows where the focus [is], and she doesn't like it much. . . . [the doctor] has made a big point about how she has to examine her. She has to. I accepted that the first couple of times, but things have been tense enough between us about the fact that we aren't going to do the surgery.

> I constantly ask [the endocrinologist]: "I know you have to monitor her clitoris, but at what point can this stop? At what point can my daughter say, No, I'm not comfortable with you looking at my bottom like that?" He hasn't

given an answer yet, but I want to raise my daughter to be able to tell someone "No" when someone is doing something that doesn't feel comfortable.
I'm cringing to think about this happening when she's thirteen. This is an awful enough period for a young girl, how does that affect her confidence that
every time she goes in the doctor's office, someone's pulling her pants down
and looking at her? (Karkazis 2008, 190–191)

It is not only heterosexual intercourse, or even sexual partnerships, that
are affected by the physical and psychological interventions meant to "normalize" sexual function for women with CAH. Masturbation, too, is less
frequent and less pleasurable for women with CAH. The report by May,
Boyle, and Grant (1996) offers particularly poignant data on the relationship between masturbation and the medical interventions that women
with CAH have endured. Less than half of the women with CAH reported
ever masturbating, compared with three-quarters of the women with diabetes, but women with CAH "commonly spoke about masturbation as a
necessary medical procedure rather than primarily as a sexual activity"
(484). That is, they associated masturbation with dilation required to keep
the vagina open, or as a way to understand the medical interventions they
had received, as several of the women's comments illustrate:

[masturbation] was necessary to keep it [the vagina] open

I have done so, everybody does, I used to have to, I would explore—what had
[the surgeons] done?

Even now, I've never thought about it for enjoyment.
(May, Boyle, and Grant 1996, 484)

Most women masturbate by stimulating the clitoris, which is the most
efficient way to achieve orgasm (Lloyd 2005), but these comments seem to
suggest that women with CAH may more often think of masturbation as
vaginal penetration, perhaps because this has been so explicitly prioritized
in the way they have been medically "prepared" for sexual interaction.

Thinking about masturbation provides a particularly good opportunity
for reflecting on the fact that, in the early studies of CAH, investigators
both hypothesized and evidently *found* that brain masculinization by androgens had led to *higher levels of masturbation,* as well as virtually all
aspects of sexuality that were at that time considered "masculine" (for
instance, Money 1965c; Ehrhardt, Epstein, and Money 1968; Masica,
Money, and Ehrhardt 1971). Even in studies of other populations, such as
boys exposed to estrogens in utero, or boys and girls exposed to progestogens, there was unequivocal consensus among early brain organization
researchers about which aspects of sexuality were masculine: masturba-

tion, visual arousability, libido, number of partners, initiation of sexual activity, being "adventurous" or "liberal" in sexual attitudes, and versatility in coitus (for instance, Ehrhardt, Evers, and Money 1968; Money and Alexander 1969; Masica, Money, and Ehrhardt 1971; Yalom, Green, and Fisk 1973; Lev-Ran 1974a, 1974b; Money and Ogunro 1974; Money and Daléry 1976; Kester et al. 1980; Hurtig et al. 1983). But as evidence from more girls and women with CAH (and also from girls with masculinized genitalia secondary to prenatal progestogens) showed that they tended to be "late bloomers" and more "conservative" or "reserved" in regard to dating and sexuality (for example, Money and Schwartz 1977; Slijper 1984; Hurtig and Rosenthal 1987; Dittmann, Kappes, and Kappes 1992), those expectations were dropped, and investigators focused single-mindedly, for a time, on sexual orientation. In the 1980s and most of the 1990s, investigators sometimes presented information on sexual activity without explicitly invoking brain organization theory (Kuhnle et al. 1993; May, Boyle, and Grant 1996).

Remarkably, though, by the late 1990s brain organization theory returned, but the interpretation made a 180-degree turn from the original prediction. When lower levels of libido, sexual activity, sexual partnerships, and fertility among women with CAH are now interpreted in the framework of brain organization theory, they are taken as possibly or probably indicating brain *masculinization* (Zucker et al. 1996, 2004; Meyer-Bahlburg 1999). This reversal in the interpretation of evidence regarding what constitutes masculinization not only destabilizes the scientific basis for brain organization theory (as I explored in detail in Chapter 6), it requires pushing aside more observable and proximate factors that impinge upon sexual function, in favor of the presumption of early hormonal influence. The consequences are not just that brain organization theory continues to be given more weight than it merits. These studies, with their fixation on early hormonal effects, have contributed to a systematic disregard for how medical intervention harms women with CAH and other intersex individuals who are subjected to cosmetic, but medically unnecessary, genital surgeries.

Let me be clear: I am not implying that clinicians involved in the medical treatment of intersex individuals are malevolent. Karkazis painstaking documents clinicians' dedication and even passion to improve treatment for their intersex patients—but she also reveals that they are so deeply invested in the idea that genitals can and should be "fixed" that they fail to consider how their treatment does, in spite of their very good intentions, harm their patients. Like Karkazis, I have been struck by both the good intentions and the simultaneous tunnel vision on the part of the researchers (some of whom are also clinicians) who use CAH to explore questions of hormonal brain organization. In most studies, brain organization researchers believe that they have adequately "controlled" for important factors in the social environment by comparing girls with CAH to their same-sex relatives (sisters and/or female cousins). (Their dedication to this idea fosters one of the ongoing methodological shortcomings in this research, namely, the very small comparison groups. It is standard epidemiological practice to oversample healthy control groups to understand the effects of relatively rare conditions, but this is not done in these brain organization studies because of the assumption that only female relatives provide a "good" comparison.) In the relatively few studies in which researchers do consider the effect of treatment on psychosexual outcomes, they routinely assume that *treatment can only limit or counteract the effects of the illness or of the early hormone exposures.* That is, treatment, in their view, can only *help* patients be more like normal, healthy comparison groups. They never consider ways that the treatment itself constitutes a serious interference in the development of sexuality and gender for girls with CAH. In the words of Gastaud and colleagues (2007), "Despite expert medical and surgical care by physicians dedicated to this rare disease, women with CAH still suffer major limitations in their sexual function and reproductive life" (1391). Yet, given that 94 percent of the 35 women in this sample had clitoral surgery and vulvoplasty, and 66 percent also had vaginoplasty, it would seem more reasonable to frame the poor outcomes as *because* of the medical care, rather than *despite* it. A particularly compelling argument for this interpretation comes from the comparison of sexual function among women who have been treated from infancy with the sexual histories of late-treated women reported in the very earliest studies (such as Ehrhardt, Epstein, and Money 1968). Those women reported *more* sexual partners, *higher* libido, a lower threshold to arousal, and more sexual versatility than the investigators anticipated (though it should be noted that the "expected" norms to which women were compared were arguably quite dated and conservative). Given that both late-treated and early-treated women have similar prenatal histories, it surely doesn't make sense to look there to explain the difference. Instead, it seems prudent to consider the factors that *are* different between these groups: early diagnosis, early surgeries, and lifelong suppression of androgens among the early-treated women.

There are a few fairly recent studies of sexual activity and function among women with CAH that do not entertain the brain organization hypothesis, and several of these studies do consider the negative physical and psychological sequelae of the management process itself. (May and colleagues' 1996 report is exemplary in this regard.) But because they do not engage the hypothesis of brain androgenization, those studies do not form a part of the network of brain organization research, and consequently they are not cited in the brain organization literature. Thus it is possible to read the brain organization studies of CAH without ever encountering the sobering data on the overwhelming pain and discomfort during sex; the absence of clitoral sensation; and the feelings of shame, bodily inadequacy, and unattractiveness that are common among women with this condition.

This is just one way—though a very serious one—that context is systematically ignored in brain organization studies, especially the crucial studies on intersex subjects. When I note that "context" is ignored, I mean that scientists doing brain organization studies tend to focus narrowly on the level of steroid hormone exposures alone, to the exclusion of virtually all other variables that affect development. Importantly, this includes many other physiological variables, such as the effects of hormones used to treat CAH or other intersex conditions—which are themselves known to have important effects on mood, energy, and metabolism. Of course, all scientific study involves focusing on some variables rather than others—it is impossible to focus on everything, even everything that the scientist has a hunch might be relevant. But it is not acceptable in science to ignore variables that we already know are relevant, especially if these variables have a good chance of affecting the outcome as much as or more than the variables in which we are interested. This is particularly true when subjects cannot be randomly assigned to receive the exposure we are interested in—in this case, steroid hormones during fetal development.

The Self-Fulfilling Prophecy of Masculinization

What about other sex-typed traits that are shown to be masculinized in CAH, particularly childhood play behaviors and other sex-typed interests? Some of these findings can again be readily connected to the physical condition of genitalia. In particular, researchers often speculate that the low rates of childbearing among women with CAH might be due to decreased "maternalism" following high prenatal androgens, even when they explicitly take into account the difficulties these women have with vaginal penetration and conception (for instance, Meyer-Bahlburg 1999; Stikkelbroeck, Hermus, et al. 2003). But before we call on the black box of brain organization to explain maternalism, we have a number of more proximate, observable variables that should be considered. In addition to difficulties with heterosexual coitus and conception, there are psychological factors related to knowledge of the condition that may affect development. For example, parents are told that their daughters with CAH will have impaired fertility, and the girls themselves may learn this from relatively young ages. It does not take a lot of imagination to understand that having a genetically transmitted chronic medical condition, especially one that is associated with often-traumatic early intervention and lifelong management, might also have an effect on fertility.[3]

Proponents of brain organization theory might remind us, though, that long before they are likely to understand the implications of impaired fertility, girls with CAH are reported to have less feminine play, in particular to play less with dolls than is typical (Ehrhardt, Epstein, and Money 1968; Berenbaum and Hines 1992; Servin et al. 2003; Pasterski et al. 2005; Meyer-Bahlburg et al. 2006), as well as to express less interest in infant care than their unaffected sisters (Leveroni and Berenbaum 1998). This is commonly taken to indicate a fundamental shift toward "masculine" interests in these girls, again secondary to brain androgenization. But attention to variables of context would lead us to recall that we also have data on two relevant issues that are routinely overlooked: (1) Girls with CAH are *expected* to be more masculine, and it is well established that expectations of this sort influence behavior; and (2) anxieties and/or simple expectations about masculinity among girls with CAH may lead to *overreporting masculine behavior* by both parents and the girls (or later, women) themselves. These variables might well work together, such that somewhat more "masculine" behaviors are amplified through a tendency to notice and place importance on these behaviors.

The data on how expectations affect behavior come from a wide variety of disciplines, but many studies are from sociology and social psychology. In their paper "Performance Expectations and Behavior in Small Groups" (1969), Berger and Conner summarized more than half a dozen early studies in social psychology as demonstrating that *expectations* about abilities, which often derive from "status" characteristics, shape perceptions, such that people "thought to be more able were more likely to be perceived as having performed well" (187). This process is a perceptual bias referred to in social psychology as a "status generalization." As Driskell (1982) put it some years later, in describing the perception of abilities in "status-unequal groups," "unequal 'status-based' performance expectations are assigned to actors at the outset of interaction" (229). Early research focused on broad aspects of social status such as social class, gender, or race, but later work, especially studies on the effects of social stigma, has demonstrated that status generalization is a broad phenomenon whereby an individual's perceived attributes—prior to any actual interaction—affect both which behaviors are actually noticed as well as how those behaviors are qualitatively evaluated. Preliminary expectations also directly affect people's subsequent behavior (Driskell and Mullen 1990). Taking up the classic sociological dictum that "things perceived as real are real in their consequences" (Thomas and Thomas 1928, 125), the sociologist Bruce Link investigates how stigma of mental illness affects perceptions *of* the mentally ill, and mentally ill people's *self*-perceptions and behavior (Link et al. 1999; Link and Phelan 2001). Link's model suggests a way to understand how expectations affect behavior and observations of behavior in CAH, itself a highly stigmatized condition.

In the case of CAH, there may be several ways that expectations affect both behavior and the way that behaviors are characterized by parents, as well as girls and women with CAH themselves. Girls and women with CAH are often perceived as "masculine" because of genital development, which leads them to be initially announced as males in a substantial minority of cases. They are also perceived as masculine because of brain organization theory itself. Multiple lines of evidence suggest that brain organization theory has guided the expectations of parents with intersex children since shortly after the theory was proposed. Ever since Money and Ehrhardt's popular book *Man and Woman, Boy and Girl* (1972), information about brain organization theory has been available to a wide audience, beyond people with a scientific or personal interest. Moreover, that book offers a glimpse into what people who *were* directly affected by possibly unusual early hormone exposures were generally told: as described by Money and Ehrhardt, the optimal clinical approach to parents was to inform them that they could expect their daughters to be "tomboys." This expectation is widespread and commonplace; you will encounter it in popular science writings (Kimura 2002; Colapinto 2001), television programs (*NOVA*, "Sex: Unknown" 2001), and on websites devoted to CAH (for example, CARES Foundation 2008; Berenbaum 2008). Again, Katrina Karkazis offers helpful insights from her interviews with intersex women and the parents of intersex girls. She notes that "parents of girls with CAH often focus on behavior that is understood as overly masculine, which they attribute to prenatal androgen exposure" (Karkazis 2008). And the influence of writings by Money and others is unmistakable. One couple relayed that they "had a lot of questions about androgen exposure and the brain. We got John Money's books and speculated about how it might affect her personality and abilities" (195).

Because of their heightened awareness, parents may read a surprisingly wide range of behaviors as "cross-sex," in effect creating a rigid behavioral gender binary and attributing these differences to the child's "mixed" biology. Interestingly, when parents have more than one child of the same gender, but only one is born with an intersex condition, they can interpret the same behavior differently in the child with the intersex condition. (194)

The connections among perceived status, perceived attributes, actual behavior, and assessments of behavior constitute a looping process. Gender is a particularly powerful cognitive schema. Although research on how gender as a status triggers this sort of perception-behavior-evaluation loop generally focuses on male–female differences, we can easily apply the observation to girls with CAH, who are seen as having a "diminished" femininity, or rather, are *suspected of being masculine*—a status that is decidedly stigmatized in girls. Much empirical research on gender has suggested that girls are given more leeway than boys in terms of gender conformity, at least in contemporary U.S. society, but girls with CAH may be a special case. Observably masculinized genitals, as well as a strong perception that the brain has been masculinized, confer a slightly ambiguous gender status, such that nonconformity with gender stereotypes may be much more readily noticed in these girls, and may become a persistent element in the child's biography, including her overarching self-narrative. That is, CAH can be thought of as a "frame" that orders the potentially infinite, complex, and sometimes contradictory information about the child's development into a story that can be told and understood.

The concept of framing is often used to capture semideliberate strategies of information presentation in political psychology. The concept also is helpful for thinking about the following process. How do broad gender schemas—which are by nature more flat and dichotomous than anyone expects real people to be—get reconciled with people's actual and persistent failure to be perfectly gender-conforming? And how, in turn, do persistent (small) failures to conform usually fail to undermine the confidence people have in the gender schema itself, as well as the certainty that individual people are "really" men or "really" women? In the words of Gamson and Modigliani (1987), political psychologists who study the rhetorical strategies, emphasis, and rules of presentation that turn potentially incomprehensible and disconnected "events" into "news," frames supply "a central organizing idea or story line that provides meaning to an unfolding strip of events, weaving a connection among them" (143).

A large body of research has documented that attributions of gender profoundly shape the way observers perceive emotions and behavior in children. Often referred to as "baby X studies," such research involves creatively manipulating the perceived sex of identical infants or very small children, and analyzing how observers describe or react to the children when they are labeled as female versus when they are labeled as male. One of the earliest and best-known of these studies involves a short film of a baby in an infant seat who is presented with four different stimuli—a teddy bear, a jack-in-the-box, a doll, and a buzzer—each of which is presented several times in succession. More than 200 observers were shown the film and were randomly assigned the information that the child was a boy versus a girl, without making an obvious point of the infant's sex. There were remarkable differences in the way observers saw the same infant when they thought they were watching "David" versus when they thought they were watching "Dana," and these differences emerged most strongly where the child's response was somewhat ambiguous. That is, in situations where the infant's response was unambiguous, as in the clear pleasure s/he showed with the bear and the doll, or the clear distress s/he showed with the buzzer, the ratings did not differ much, except that the "boy" was seen as displaying more pleasure and less fear across all situations than the "girl" was (Condry and Condry 1976, 817). But one stimulus, the jack-in-the-box, elicited a much more complex response: "At first the infant stares at the box and shows a slight startled reaction when it is first opened. Upon successive presentations the infant becomes more and more agitated and after the third presentation the infant cries when the box is pushed forward (even before it is opened) and screams when the jack-in-the-box jumps up" (816). Given this sequence of responses, those observers who thought they were watching "David" were more likely to describe the child as "angry," while those who thought they were watching "Dana" described the child as "afraid." The Condrys pose an interesting question about how this differential attribution process might affect real-world responses to ambiguous emotional displays: "If you think a child is angry do you treat 'him' differently than if you think 'she' is afraid?" (817).

Many subsequent studies have extended these observations, including by controlling for actual variations in infant characteristics (for example, Zeavey, Katz, and Zalk 1975; Sidorowicz and Lunney 1980; Culp, Cook, and Housley 1983; Lyons and Serbin 1986; Donovan, Taylor, and Leavitt 2007). There is even evidence that modern mothers, who typically know their child's gender before birth, begin this labeling process to interpret the child's "behavior" or personality while still in the womb (Rothman 1986).

Most interesting for the subject at hand is a "baby X" study that used not just male and female labels, but the label "hermaphrodite" for the same child. John Delk and colleagues videotaped the activities of a 22-month old infant who was neutrally dressed and easily perceived as either male or female. As with the vast majority of labeling studies, observers (in this case, 464 medical, nursing, and psychology students at various U.S. universities) were far more likely to label activities as masculine when the ascribed gender was male, and feminine when the ascribed gender was female. When observers were told that the child was a "hermaphrodite," an interesting finding emerged. Women observers rated 32 percent of the activities of the "hermaphrodite" as masculine, but only 19 percent of the "girl's" activities as masculine (532). In other words, the same child, when labeled "hermaphrodite" instead of "girl," was significantly more likely to be perceived as engaging in "masculine" activities when she was in fact doing the exact same thing—in fact, roughly 70 percent more likely.

Lest readers be tempted to speculate that gender labeling is a thing of the past, it is worth noting that recent studies generally indicate labeling effects as strong as those found in the earliest studies, more than thirty years ago. The reliance on cognitive schemas about gender to label behavior not only affects how we perceive people and their specific actions, it loops back to reinforce belief of fundamental gender differences. Importantly, this process leads people to discount actual evidence of intrasex variability and intersex similarity, and operates even in the face of commitments to similarity or a "wish" for similarity. It may be disappointing and even hard to swallow for many parents, especially those who consider themselves feminist, but being egalitarian doesn't seem to affect gender labeling very much. This is critical for understanding how parental expectations may affect gendered behavior in girls with CAH.

Another factor that is likely to add to the labeling effects is that the data on sexuality and gendered behavior are always gathered in a context where CAH is the salient factor; an extensive literature on "priming" effects suggests that both facts and interpretive frames that are made salient in a testing situation exert important effects on people's reports of their own opinions, behavior, and characteristics as well as on their actual behavior after this priming (Payne 2001; Fazio and Olson 2003). Because the "narrative" of CAH involves expectations (and/or fears) that behavior will be masculinized, especially in the realm of sexuality, this is likely to affect the way that girls and women with CAH, and/or their parents, characterize their behaviors. Here, the fact that the increased same-sex eroticism among women with CAH is primarily in the realm of fantasy rather than behavior may be important: reports of fantasies and dreams might be even more subject to priming effects than are reports of behaviors. Moreover, very few studies use direct observation, and recall of prior events or behavior "patterns" are especially susceptible to priming effects (for example, McFarlane, Martin, and Williams 1988; DeCoster and Claypool 2004).

Labeling of girls with CAH is especially complex in those cases where the children have initially been assigned as male, then reassigned female upon diagnosis. As with the "tunnel vision" that researchers show in contemplating only positive effects of treatment on sexuality, they seem to miss the pattern that initial assignment as male and even earlier age at diagnosis are more highly correlated with gendered behavior than are objective measures of virilization (which should relate to actual degree of androgen effects). Slijper (1984), for instance, found that gender score was correlated with whether the child was originally registered as a boy, whereas gender score was *not* correlated with the "objective degree of virilization" based on the Prader score at birth. Others, too, have found that gender behavior is related to such factors as early age at diagnosis or early initiation of treatment (Dittmann et al. 1990, "Congenital Adrenal Hyperplasia II"; Berenbaum, Duck, and Bryk 2000) and with initial assignment as male (Meyer-Bahlburg 2002).

In addition to labeling and priming, there is an enormous research literature documenting parental sex-typing, meaning that parents differentially promote, discourage, reward, or punish behaviors in their children according to gender (for instance, Lytton and Romney 1991). Scientists conducting brain organization studies have tended to neglect evidence of labeling effects (evidenced by the fact that they routinely rely on parental reports of behavior, rather than direct observation, and have made few attempts to rate children's behavior while "blinded" to their CAH status), but they have shown some interest in parental sex-typing. Unfortunately, researchers typically have simply asked parents about their habits regarding reward or encouragement of masculine versus feminine behavior (Ehrhardt and Baker 1974; Berenbaum and Hines 1992), even though extensive research indicates that parents are unaware of their own sex-typing (e.g., Fagot 1978) and that sex-typing effects are larger when parents are observed rather than queried (Lytton and Romney 1991). One team, though, observed parents with their CAH-affected and unaffected daughters, and they found that while parents encouraged sex-typical play in all their children, the daughters with CAH were given *more* positive responses for playing with "girls' toys" than were the unaffected daughters. The investigators interpret the findings as follows:

> This exaggerated encouragement of female-typical play in daughters with CAH, in addition to suggesting that parents do not cause their male-typical toy choices by reinforcing them, also may be a reaction to the increased male-typical play behavior shown by these girls. Girls with CAH may engender more parental encouragement of female typical toy choices because their *parents are trying to normalize their behavior.* (Pasterski et al. 2005, 275, emphasis added)

This study makes an important contribution, but I would suggest that they impose a too-narrow framework in interpreting their findings. It seems to me that the issue of socialization is not quite so deliberate or straightforward as parents treating the girls more like boys, or *encouraging* boy-typical play. One critical fact is likely to be that parents are *anxious* about masculine behavior in their daughters with CAH and have been primed to expect it. Parents are trying to normalize girls' behavior, which they probably fear and perceive to be abnormal. Further, while parents might show greater reward for "girlish" behavior (at least in this research setting, where CAH has been made salient), we should not suppose that would necessarily translate directly into more girlish behavior on the part of girls. Expectations and sex-typed interactions do have effects, but they don't always have the *intended* effects. Bernice Hausman's thoughtful piece on the David Reimer/John-Joan narrative may shed some light here. Hausman (2000) suggests an alternative to the usual idea that "innate male nature"—usually attributed to early testosterone—caused David Reimer to reject his late-imposed female assignment to return to his original male status. Instead Hausman points to the extremely rigid gender socialization that was prescribed to facilitate the child's adjustment, along with an intrusive yearly clinical regimen that bears more than a little resemblance to the treatment for CAH (except that girls with CAH go roughly four times as often): "The child was made to display her body to the physicians at the hospital, submit to psychological testing, and agree to the desirability of vaginal construction." Hausman suggests that an alternative but plausible reading of the child's rejection of femininity is that "Joan resisted the heavy-handed plot to make her into a girl" (123).

I think there's more to the case than that (which I will elaborate shortly), but I do think that Hausman is on to something. Socialization is real and important, but complex. There are undeniable patterns of encouragement and reward by gender, but most of these do not seem to be deliberate. Moreover, putting the question of gender aside and thinking about socialization more generally, parents' intentions do not neatly translate into children's compliance, even in far less complicated situations than a child whose genitals have been maimed.

There is yet another category of variables that might be relevant to the development of gender and sexuality in the context of CAH: other *physical factors*, aside from genitals, that are affected by CAH. First, consider aspects of physical morphology that are easily observable by others and are a basis on which the growing child can compare herself. Height and weight are both unusual in this syndrome. Girls with CAH are progressively more likely to be overweight, and as adults are significantly shorter and heavier than other women. Mulaikal, Migeon, and Rock (1987) found that women with the salt-wasting form had a mean height at the 25th percentile for normal women, and those with the simple virilizing form were at just the 10th percentile (179). Advances in hormonal treatment have improved the height outcomes for CAH patients in recent decades, but studies continue to show significant differences: people with CAH are substantially shorter and more likely to be overweight or obese (Cornean, Hindmarsh, and Brook 1998; Eugster et al. 2001; Merke and Bornstein 2005; Volkl et al. 2006; Goncalves et al. 2009). A recent meta-analysis found that adults with CAH were 10 centimeters (about 4 inches) shorter than average (Eugster et al. 2001).

Even if medical management has been excellent since infancy, there is a highly increased chance of being hairier than other girls and women; this is especially true for women with the nonclassical form. Among these women with the "less extreme" form of CAH, Maria New, a clinician who runs a large CAH clinic, has noted "a history of discomfort and social stress related to their pretreatment experiences with androgen-dependent signs such as acne, hirsutism, and conception difficulties. Unlike classical patients, [nonclassical CAH] patients are not treated from birth and often lack diagnosis and hormonal control for many years" (New 2006, 4208). These findings are particularly useful for reflecting on the recent studies in which Heino Meyer-Bahlburg and colleagues (Meyer-Bahlburg et al. 2006; Meyer-Bahlburg et al. 2008) have evaluated gender and sexuality in women with both classical and nonclassical CAH. On one hand, these studies provide convincing evidence that atypical genitals do not totally explain masculinization in CAH, because women with nonclassical CAH (whose genitals are not affected by the condition) reported both more masculine gendered behavior in childhood, and more same-sex fantasies than unaffected controls. On the other hand, when we broaden the view of important variables to consider in CAH beyond the simple pair of early hormone effects on the brain and effects (including socialization and medical treatment) that flow from having atypical genitals, the interpretation gets murkier. Certainly issues of priming, labeling, and retrospective recall bias should not be discounted. Neither should the factors of body image from high rates of overweight and hirsutism, as well as short stature, nor possible mood effects of steroid hormone disruptions for years prior to diagnosis. While there isn't a direct narrative to connect these factors to higher rates of behaviors considered "masculine," we should be very cautious before discounting the possibility. This is especially true given that women with nonclassical CAH are also directly told, from the point of diagnosis, by clinicians and self-help sites, as well as by clinical research literature, that they have pathologically "masculine" features of physiology, and possibly psychology.

Even with early diagnosis, hormone control is difficult in this condition, which is the primary reason that children with the classical forms are followed so closely. Mood problems, especially depression, anxiety, and affective distress, have been reported in boys and girls with CAH (Kuhnle, Bullinger, and Schwarz 1995; Johannsen, Ripa et al. 2006); in addition to the stigma of the condition and potential trauma related to treatment, mood problems may be traced to imbalances with steroid hormones, including cortisol. Mood-regulating hormones may also play a role in the generally decreased bodily satisfaction and libido in women with CAH— and oversuppression of androgen might be of particular concern here.

A Blind Eye: CAH and the
Epistemology of Ignorance

In sum, girls with CAH are in some small respects "masculinized," and this is likely to be at least in part because they are *expected and believed to be masculinized*. It may also be because potential masculinization in these girls is a source of fear and dread both for their parents, and for many of the girls and women themselves. Another factor might be a function of how "masculinization" is conceptualized: if it is a lower degree of certain behaviors and traits that are considered "feminine"—such as fertility, heterosexual activity and desire, or libido—then it is important to recognize that genital interventions and long-term outcomes may be all the explanation that is necessary. Finally, broader physiological differences—including higher weight–height ratios, shorter stature, hirsutism, disruptions in mood-regulating hormones, and more—may affect behavior through a variety of pathways, none of which are well understood. And last, but not least, it is possible that early androgen effects on the brain could explain some of these effects. But all of these possible effects must be seen in context. Importantly, the effect sizes for behavior differences between girls with CAH and other girls are quite small in most domains—so *whatever* influences are driving the slight difference in psychosexual behavior relative to unaffected women, *the cumulative effects of all influences are relatively weak*. It is important to recognize that even if girls with CAH are to some degree masculinized (from whatever factors), *most reports likely overestimate this masculinization* because they are (a) retrospective rather than concurrent; (b) based on general assessments of behavior patterns rather than on specific observations; and (c) collected under circumstances that make the condition of CAH salient, thus "priming" research subjects to assign more masculinity to subjects with CAH.

Given the extensive factors that are demonstrably and significantly different in the development of girls with CAH compared with unaffected girls, the fact that researchers interested in brain organization focus single-mindedly on the potential effects of early hormone exposures on the brain is troubling, to say the least. In fact, I think that the persistent ignorance of broad physical and psychological health for women with CAH is downright stunning. Historians and philosophers of science are giving increased attention to the way that gaps in knowledge, as well as knowledge itself, are actively produced and maintained. The study of this phenomenon, what Tuana calls "the epistemology of ignorance" and Proctor (2008) calls "agnotology," reveals that specific ideologies, cultural schema, and political interests systematically block certain forms of information and cause people to "forget" or fail to incorporate certain facts into the overall thinking on a subject. Together with other patterns I've documented in this book (specifically: the pervasive tendency to ignore findings that contradict brain organization theory; failing to notice when critical definitional frames are shifted; failing to "connect the dots" regarding dose–response contradictions), the systematic disregard for the context in which girls and women with CAH develop constitutes more than an oversight. The attentive reader may have noticed the relative recency of the outcome studies that begin to systematically document pain, lack of sensation, and other difficulties with sexual function. These studies have been done in part because of brilliant and energetic organizing on the part of some intersex adults, who have begun to insist on having a voice in evaluating and reshaping the medical process that many believe constitutes serious harm (for example, Chase and ISNA 1998). But it is important to recognize that clinicians—including those who research brain organization and those who don't—must surely have heard reports of pain, discomfort, and lack of sensation for decades, since the first children who underwent early surgery reached adulthood in the 1980s. This is a classic case study for agnotology: the "not-known" regarding CAH is not merely information that is not *yet* known, or not *yet* investigated. Forty years into the serious study of this disorder, these lacunae persist because they help to keep brain organization theory alive.

STILL—and here is the amazing thing—at the end of the day, in their overall outcome, women with CAH are remarkably similar to unaffected women, aside from their lower levels of sexual relationships and childbearing. In spite of prenatal androgen levels that are typically in the high-male range, they end up very female-typical in cognition, personality, and sexuality. Most studies show that mental health and functioning are overall relatively high, especially when considering the amount of medical management this group of women must endure. Gender is relatively typical in adulthood, too. While they express somewhat more masculine career aspirations in adolescence, they do not have significantly different occupations or achieve different levels within occupational sectors. They tend to report more same-sex fantasies and attractions, but only that subgroup who were initially announced as male and then reassigned seem to have any higher incidence of same-sex behavior—and this is generally quite a small increase.

The interesting thing, from my perspective, is that women with CAH end up being so very close to typical. As a feminist, I do not assume that "typical" outcomes in terms of gender and sexuality are necessarily optimal. If heterosexuality is not the "goal" of development, for example, then same-sex desires and relationships look like a perfectly good, and possibly better, outcome for many women with CAH. Because even if penile-vaginal intercourse can be "decentered" in a heterosexual relationship, both partners are likely to be frustrated if they cannot, at least sometimes, participate in this activity. It is conceptually important to the heterosexual script and is also likely to be an activity that male partners have found satisfying in other sexual relationships. So contrary to the usual presumption, from the perspective of maximizing opportunity for mutually satisfying sexual partnerships, it might be better if women with CAH were less, rather than more, typical. But typical they are, and this points to one of the most important, and most often overlooked, facts of human development: there are very strong pressures toward convergence. I will return to that point in Chapter 10. For now, let's see how a consideration of context provides plausible alternatives to brain organization for understanding how gender identity develops in three situations that are usually interpreted in terms of prenatal hormones.

Denaturalized Gender: Certainty, Ambivalence, and Change

In *Gender: An Ethnomethodological Approach,* one of the earliest extensive studies of how gender is socially produced, Suzanne Kessler and Wendy McKenna (1978) built upon Harold Garfinkel's revolutionary insights regarding the "natural attitude" toward gender. The "natural attitude" consists of "the 'facts' of gender in terms of Western reality": those things that we "know" to be true (Kessler and McKenna 1978, 114). The component "facts" of the natural attitude include that there are two and only two genders; gender is invariant (once a male, always a male, and so on); genitals are the essential sign of gender; there are no gender transfers "except ceremonial ones (masquerades)"; everyone must be classified by gender (no exceptions); the male/female dichotomy is natural; and "being male or female is natural and not dependent on what anyone decides that you are" (114–115). The concept of the natural attitude toward gender has been an important tool for an entire generation of scholars interested in "unhinging" gender from sex—that is, in destabilizing the idea that gender *must* be a naturally existing category that emanates from original biological differences. By searching for how the "natural attitude" operates, and actively questioning it, it becomes possible to explore how gender may result from social structures and institutional practices. In short, gender can be reconceptualized as an "effect" rather than a mere fact, something that *requires explanation* rather than something that *explains* the social world.

Suzanne Kessler's later work on the management of intersex infants has been of central importance in helping feminists rethink the relationship between the body and gender. In *Lessons from the Intersexed,* Kessler (1998) shows in meticulous detail how the natural attitude toward gender shapes medical perceptions and decisions regarding management of intersex infants. In particular, while the official medical view is that clinicians guide parents to *choose a "best" or "optimal" sex* for intersex children, the operative narrative for both clinicians and parents is that they *uncover the true sex* of the infant, correcting those aspects of anatomy and biochemistry that might not align with this status. Thus, even though intersex children are born with "mixed" or "ambiguous" characteristics of sex, they nonetheless are seen as truly and properly belonging to one sex or the other, and this decision is not regarded as a matter of choice so much as a matter of finding "the *right* answer" to the question of sex.

Against this backdrop, we can consider three scenarios that are generally taken in the brain organization literature to indicate the role of early hormones in the formation of gender identity—which is the conceptual core of gender, believed to form the fundamental basis for all other aspects of gender.

Complete androgen insensitivity syndrome (CAIS): Genetic males with a total inability to respond to testosterone are born appearing to be normal females. They are typically not diagnosed until midchildhood to early adolescence. The significance of possessing XY chromosomes as a phenotypic female is culturally profound. Yet in CAIS the gender is not "denaturalized" until relatively late in development. At birth and through middle childhood, at least, the child and her parents, family, and peers all have had no reason to question that she is a girl. There are no external signs available to interrupt the "natural attitude" to gender, even after a diagnosis of XY with testes and androgen insensitivity. It is assumed that brain tissue is likewise insensitive to androgen, so the medical theory about "invisible signs" of gender agrees perfectly with visible ones.

Congenital adrenal hyperplasia (CAH): Genetic females with CAH are typically born with masculinized genitalia, which may vary from a slightly enlarged clitoris, to fused labia, to a fully formed penis and (empty) scrotum. Some children are initially announced as male, but diagnosis typically happens in infancy and children are overwhelming reared as female. In CAH, the initial confusion regarding sex is managed by a process that looks very much like uncovering the "true" sex of the child. That is, even though the decision is formally framed in terms of choosing the optimal sex, most doctors and parents do not, in these cases, consider that the child is possibly "really" male— particularly given the cultural weight of sex chromosomes and the potential fertility for children raised as female (see Kessler 1998; Karkazis 2008). Along with certainty regarding the child's "true" sex, though, there is some ambivalence regarding the child's femininity. The parents and child are routinely told that the child "may be more masculine" or "may be a tomboy."

Normal males reassigned due to congenital anomaly or trauma: In genetic males with normal responsiveness to androgens, sex is assigned as female only in extremely rare cases. These include children who were born healthy but whose penises were irreparably damaged, and those whose penis was absent or severely damaged at birth. In these cases, there seems to be no possibility of constructing an "adequate" functioning penis, where function is primarily measured by the ability to urinate standing up, and to penetrate a vagina in heterosexual intercourse (two factors that have been, until recently, considered as central to forming a male identity). When such children are assigned or reassigned as female, this is perceived by everyone involved as at best, a compromise, and at worst, a tragedy. These children are not believed to be "really" female by either clinicians or parents. The attempt to rear these children as female is, therefore, an entirely different process, one characterized more by conscious effort and engagement and an attempt to habitually interact with and perceive the child as female while always living with the cognitive dissonance of her/his contrary "true" male sex. This situation puts gender assignment in conflict with the natural attitude.

In terms of outcomes, the simplest are considered to be genetic males with complete androgen insensitivity, who are, according to virtually all published evaluations, unremarkable in all respects from healthy genetic females—except that they do not have female internal reproductive structures and therefore do not menstruate or become pregnant. Compared with "typical" women, they do not have higher rates of masculine gender identity, gender-atypical behavior or interests, or same-sex erotic interests—that is, nothing that brain organization researchers interpret as "masculinization" of psychosexual development. Genetic females with CAH, on the other hand, overwhelmingly identify as female, but consistently show slightly more "male-typical" behavior, as a group, than do most genetic females. Most notably this includes childhood play behaviors and adult eroticism, in that a slightly higher proportion of women with CAH compared with unaffected women report sexual attraction to women. Genetic males who have normal androgen sensitivity but have been assigned (or reassigned) as female are the group with the highest known rate of gender change (Cohen-Kettenis 2005; Meyer-Bahlburg 2005). Meyer-Bahlburg's review of such cases indicates that, by adulthood, up to 50 percent of such people either reassign themselves to the male gender or have serious gender identity problems (Meyer-Bahlburg 2005, 432). From the small number of detailed case reports, it appears that a high proportion of those who retain a female gender identity report sexual attraction to other women (Reiner and Kropp 2004; Reiner 2004). Recently, due to the tragedy and scandal that followed when David Reimer's story was made public, the dominant protocol for sex assignment in cases of ambiguous or damaged genitalia has been challenged. Money and colleagues at Johns Hopkins built that protocol on the understanding that children without "functional" penises should be reassigned as female, as long as the reassignment could happen by around 2 years of age. Lately the idea that children cannot develop a male identity without a penis has been replaced with the idea that children cannot develop a female identity if they have been exposed to a threshold level of androgens in utero (Diamond and Sigmundson 1997; Diamond 1998).

This array of outcomes is overwhelmingly interpreted as evidence that early androgens are responsible for directing gender identity and sexual orientation. As John Colapinto's best-selling book about David Reimer described this most famous case of (failed) sex-reassignment in a previously normal male, rejecting the female assignment was a return to his self "as nature made him" (Colapinto 2001). Michael Bailey has suggested, in his famously blunt way, "If you can't make a male attracted to other males by cutting off his penis, how strong could any psychosocial effect be?"

(quoted in Wade 2007). This comment is a particularly helpful one for pointing out the "natural attitude" at work: What is it about cutting off a penis that would be expected to direct sexual orientation toward men? Gay men, after all, have standard-issue penises. But recall that genitals are the "essential signs" of gender, and—as a great many empirical and theoretical analyses of gender have demonstrated—the natural attitude of gender hinges critically on the presumption of heterosexuality, an aspect of the phenomenon that Garfinkel did not identify. Heterosexuality is assumed to be the ultimate "goal" of gender, as well as its source. Hormones, too, have become an axiomatic component. The notion, now a cliché as well as a theory, is that male identity follows when the brain is "bathed in testosterone."

But the evidence is just as compatible with another explanation that does not require recourse to hormonally organized brains: In the context of a cultural system in which gender is seen as proceeding "naturally" from an individual's single, true sex, gender socialization and development cannot be *willfully* produced; gender development will generally follow what is *perceived as* the "true sex." This latter explanation locates the error of the standard Johns Hopkins protocol not in its disregard of the "hormonal status" or "brain gender" of some children for whom clinicians and parents must actively manage gender, but in its disregard (or ignorance) of gender as a *naturalized,* rather than a *natural* system. In other words, gender rests upon a belief system about what is "natural." It appears "real" only when the process used to produce it is invisible.

Here is the promised return to Hausman's reflections on the David Reimer story. Though Hausman raises the issue of the natural attitude about gender, she didn't quite draw the conclusion that the reason David Reimer's sex reassignment didn't "take" was that it was recognized as "not the real thing." Yet all accounts point to a myriad of ways that the natural attitude was ruptured in this case: no one actually believed that the child "was" a girl—they believed (or rather, hoped) that he *might be made into* a girl. In fact, they believed that, as unlikely as it might be, it was his only chance of survival. In Colapinto's fuller account (2001), it is obvious that parents, the broader family, clinicians, and teachers—among whom the child's original male sex was an open secret—colluded in the heavy-handed enforcement of femininity. The fact that Money recruited transsexual women to try to convince the child to have vaginal construction may well have underscored for Reimer that what was under way was, in fact, a *reconstruction* of gender, a replacement rather than "the original."[4]

One of the core insights gained from using the concept of the natural attitude is that, for it to work seamlessly, the management of gender must be concealed or obscured. *If it cannot be concealed, then a more consistent outcome will follow if everyone involved believes that they are managing problematic "outward signs" rather than actively producing a sex that contradicts the "natural," "true," or "original" sex.* Here, the notion of concealment is not one of human action, but of the power of cultural schemas and institutionalized processes to obscure their own functioning. Thus, the point that a "better" or "more consistent" outcome will follow concealment is meant to indicate that it is precisely where the humans involved (clinicians, but probably most especially parents) understand that they are reshaping or managing gender that the process of gender socialization becomes radically unlike the usual course of events. This should not be confused with a point that Karkazis (2008) raises regarding clinicians' failure to help parents of intersex children connect with each other, namely, that clinicians sometimes seem to believe that "secrecy about the condition would foster a better outcome" (183). "Naturalized gender" requires that there is no one to keep the secret of "true sex" that is believed to be in tension with gender.

In the case of girls with CAH, there are occasionally misgivings about whether or not the female assignment was the best choice, but both patients and their parents overwhelmingly accept the assignment, even in the most extreme cases of virilization (at least in the United States and other Western contexts; on cultural variability in this regard, see Kuhnle and Krahl 2002). On the other hand, when genetic males are reared female because they have either been born with an "inadequate" penis or have experienced traumatic loss of the penis, this cannot fail to be perceived as a profound contradiction. It is a testament to human resilience that people manage this situation, that families do raise such children who have a strong sense of self and belonging. But it seems to me no mystery at all that individuals raised under these circumstances often experience gender in a radically different way than is typical.

There is no simple way to decide, based on the evidence, between the idea I've just advanced about "naturalized gender" and the theory of brain organization by hormones. It depends on how much weight one gives to cognitive and emotional schema, how one evaluates the totality of evidence for human brain organization, and one's general attitude toward the relationship between sex and gender.[5]

Understanding gender as naturalized and as a multilevel system, rather than as merely an individual or interpersonal-level variable has consequences for how we evaluate some of the common narratives about the emergence of gender in typically developing individuals, too. While working on this book, especially while conducting interviews with scientists, but also while discussing the project with friends, colleagues, and strangers, I repeatedly encountered remarkably similar versions of the following story: *I used to believe that gender was a social construct, until I had a child. My spouse and I were dedicated to challenging sex stereotypes, so we did our best to provide gender-neutral toys and encouragement. But my children resisted our efforts—our little boy turned his sex-neutral toys into weapons. And our little girl turned everything we gave her into a doll, giving it a name and a 'life story' and so on"* (for example, interviews with Drs. A, B, I, and J; see also Blum 1997; Summers 2005; Brizendine 2006). I've come to see the story as both a "canned" narrative (a broadly circulated, stylized, easily recognizable story that is available "off the shelf" as a shorthand way for people to express their deep convictions about the naturalness of gender) and a genuine attempt to reconcile people's experiences that their intentions as parents do not map neatly onto their children's behavior.

In this sense I agree with people like Steven Pinker who insist that people are not "blank slates" on whom the intentions of teachers, parents, and other caregivers can simply be written. If this were the case, there would be no rebellion and no social "deviance"—there would be no innovation, either. Where I part with Pinker is that I think he overemphasizes the "groupness" of human nature, and he overemphasizes the intentional aspects of socialization. In other words, I believe that there is plenty of evidence to suggest that people have predispositions of many sorts, and that who we end up becoming is not a simple reflection of what was poured into our empty heads by our parents and our cultures. But my hunch is that, contrary to the seemingly "obvious" evidence that we get from observing the world around us, gender is just not a very useful way to group people's predispositions.

"Not useful," though, shouldn't be confused with "not real." Gender is one of many possible ways to slice of the pie of human variation, and the natural attitude encourages us to see it as the most obvious and fundamental way to divide things. This, in turn, brings about more meaningful difference along the dimension of gender than would otherwise occur— including stimulating physical differences that possibly include average differences in the brain. To repeat Thomas's dictum, "Things perceived as real are real in their consequences" (Thomas and Thomas 1928, 125). In my view, *naturalized gender* becomes *embodied gender* (a self-fulfilling, empirically demonstrable representation of gender on/in the physical body, including the brain) through the following process.[6] Everyone begins with subtle predispositions in the form of perceptual biases and sensitivities, variations in motor skill, and so on. Predispositions are simply initial states from which all feedback and interaction proceeds. These predispositions may be (but aren't necessarily) randomly distributed, but they are not, at least initially, firm personality "traits," nor do they point inevitably or properly to one developmental pathway and one outcome (in other words, atypical gendered behavior or eroticism, insofar as it is present in any individual or group, most certainly is *not* either pathological or a diversion from some "developmental template"). Regardless of the initial distribution of predispositions, each infant is met from the moment of birth and possibly before that with a gender label that engages gender schemas and shapes every single subsequent interaction, as well as the process of self-recognition. Self-recognition as male or as female involves learning and incorporating aspects of the gender schema that operate in one's culture—evolutionary ecologist Joan Roughgarden has beautifully referred to this as a person's "reaching out to cultural norms" (2004, 27).

Meanwhile, "society [is] imposing its expectations on the individual" (27). While the process is almost never deliberate or even conscious, the gender of a person with whom we are interacting is salient. Recall that evidence suggests it is especially salient in ambiguous situations, where the intentions or mental state of another person are somewhat unclear—which is especially true of much interaction with infants. Even if we *are* aware of the gender schemas when interacting, and are consciously attempting to ignore or undermine them, the schema is still a cognitive resource that we cannot entirely escape using, especially in face-to-face interaction. Gender schemas mean that the feedback received by individual children is systematically bifurcated—it is not perfectly stereotyped by gender, but it is demonstrably and pervasively patterned, such that girls get more of certain types of interaction and boys get more of another. Recall the evidence that even parents with explicit commitments to disrupt sex-typing were found, in actuality, to sex-type their children. Thus, even if initial individual predispositions were randomly distributed among all newborn infants, the functional predispositions would still very quickly develop along lines that are gendered. Here, as with the CAH-affected girls, the amazing thing is not difference—this is utterly predictable by applying the variables we can already observe to some known developmental principles. The amazing thing is the outcome: an extraordinary degree of similarity across the sexes, and diversity within them.

Yet we have strong cultural and scientific habits that push us toward focusing on, and finding, differences. Let's revisit the issue of sex-typed interests. Unlike the other physical and psychological domains that I have considered in this book, the fact that these characteristics differ, on average, between the sexes is not a "finding" but instead is a core defining feature of the category. That is, only those items that can reliably distinguish between males and females are selected for inventories of sex-typed interest or gender role behavior, or kept in the inventories when they undergo revision. These characteristics constitute the contemporary definition of masculinity and femininity (that is, gender) in the time and place when the measures are created. Expectations and stereotypes play a particularly large role in this domain, affecting how people perceive the traits, behavior, and emotions of other people, as well as the way people describe themselves (Plant et al. 2000; Pallier 2003).

There is a special kind of difficulty involved when relating traits of gender to sexual orientation. Like sexual orientation, but unlike somatic characteristics or cognitive features, assessment of traits in this domain overwhelmingly involves self-reflection and self-labeling. Questions might ask someone to speculate, for example, on how she would respond in a hypothetical situation of conflict, or ask her to rank her life priorities. There is a deeper undercurrent here that engages some of the most profound questions of the self: What sort of person am I? What is my life about? Differences in culture and subculture obviously shape responses to such questions, and for this reason it is important to remember that lesbian, gay, bisexual, and transgender (LGBT) people are very likely to be part of distinctive sexual subcultures—especially those who are included in brain organization studies, given their sampling schemes.

Lesbian, gay, and bisexual people (those with minority sexual orientations) are mostly not transgender (people who don't identify with their "birth sex" or the gender in which they were reared), but LGBT communities overlap and have entangled histories. It is tempting to think that the distinction between gender and sexuality has influenced the history and politics of LGBT people, but the influence has been at least as strong in the other direction: political aspirations and intragroup struggles have shaped the very contours of *gender* and *sexuality* as these categories became conceptually distinct in the twentieth century. Anthropologist David Valentine (2007) has documented how gay and lesbian activists actively distinguished themselves from "gender deviants" (transsexuals, transvestites, and even such nondiagnosed but socially disdained groups as "fairies" and "bulldaggers") in order to win social respectability and escape from the stigma of mental illness. This was not simply a cynical political strategy, of course, but fit well with some gay men's and lesbians' fundamental sense of being typical men or women in terms of gender.

Regardless of whether individual gay and lesbian people are gender-conforming or not, there is a strong folk notion that lesbians and gay men are sexual "inverts," that is, that they have cross-sex traits (Kite and

Deaux 1987). This predates brain organization theory and was an important impetus for pursuing behavioral endocrinology research on humans in the first place. Gay men and lesbians have long used gender atypicality to signal to potential sexual partners and peers, and various forms of formal and informal gender crossing have played an important role in gay and lesbian social worlds for more than a century, perhaps much longer.[7]

Recall the three-ply-yarn metaphor that I introduced in Chapter 1 as a way to understand the relationships among sex, gender, and sexuality. Our cultural narrative about sex in the broad sense is both dichotomous (either male or female) and neatly aligned across these three realms: sex–gender–sexuality is a package deal. The entanglement of gender atypicality and erotic atypicality may be important in the ongoing process of personality development; for instance, those who perceive themselves as gender-atypical may also be more attuned or open to signs of erotic atypicality, and vice versa. This is relevant for daily life, and also for scientific research. Constructing a lesbian or gay identity in the first place unavoidably involves engaging with the notion of gender inversion, and reporting on gender in the context of a study where sexual orientation has been made salient also involves priming the gender stereotypes associated with one's own orientation. This is not just relevant for gay men and lesbians: to the extent that the structure of recruitment or questions within studies makes sex, gender, or sexuality salient for participants, heterosexual respondents may be inclined to overestimate their gender conformity.

As a final way of thinking about how the relationship between sexuality and gender is itself context-dependent, it is instructive to consider two examples in which same-sex desires do not align with gender atypicality. Over the past few years, much attention has been given to the "Down Low" (DL) phenomenon, men who have sex with other men but do not identify as "gay." The "DL functions not as a nonidentity but as an alternative sexual identity and community denoting same-gender interest, masculine gender roles . . . Black racial/ethnic identity, and a dissociation from both White and Black middle-class gay cultures" (Young and Meyer 2005, 1146). One of the reasons for all the attention to the DL is anxiety over the fact that men on the DL can't be readily identified as such, including by the women with whom some men on the DL have sex. Another example is femme lesbians, a subset of lesbians who have long puzzled sexological researchers and confounded popular stereotypes. Sexologists have often excluded femme lesbians from the category of "true homosexuals" based on their gender typicality, attributing their relationships with other women to weakness or "feminine" traits such as passivity, sentimentality, or tendency to monogamous love (Terry 1999). But lively butch-femme subcultures throughout the world (Wieringa, Blackwood, and Bhaiya 2007; Hollibaugh and Moraga 1983) attest to the existence and persistence of simultaneous femininity and lesbianism in many women. People's willingness and ability to "pass" as heterosexual because of gender conformity affects sexual identification, visibility, and disclosure of oneself as gay or lesbian, and these also affect participation in research on sexual orientation. Thus, both men on the DL and femme lesbians are probably underrepresented in studies of sexual orientation, which are but two examples of why the association of gender conformity and heterosexuality, or nonconformity and homosexuality, should not be taken at face value.

Context matters. In our ongoing development, context matters from the earliest stages of gestation, through infancy, childhood, adolescence, and on into adulthood, especially when we are talking about domains of personality and behavior that are as complex and irreducibly social as gender and sexuality. It also matters in science, affecting the way we recruit our research subjects and interpret the information they offer us, the way subjects perceive the goals of the research and unconsciously emphasize or filter information about themselves and important others (especially children) on whom we ask them to report. But there has been next to no attention to context in brain organization research. And context is demonstrably different for the very groups that scientists compare when they do brain organization studies, so the effect of contextual variables is hidden within the differences they may observe.

Can we conclude, then, that there are no meaningful differences in the initial predispositions of male and female infants, at the group level? No, because we can't remove children from the socialization process in order to test this—that would require intervening to an unacceptable degree with children's initial contact with parents and others, which is critical for healthy emotional attachment. But we also can't conclude that there *are* such differences, and the evidence from brain organization research adds very little reason to suspect that differences in initial predispositions make a meaningful contribution to gender.

Why do I place persistent emphasis on similarity? Is it because I think (as quite a few critics have accused feminists in general of thinking) "difference is bad" and everyone should be the same? No. I think difference and variety are good, and my problem with the current story of "sex in the brain" is chiefly that it *underplays and mislocates the creative sources of human difference.* Gender schemas are powerful, permanent shapers of both our materially based, lived selves and the way we perceive ourselves and others. In this way, gender is like a polarized lens, channeling the "unruly" vibrations of sunlight into a single direction. It's not that light only "looks" less glaring after passing through the lens: it is, in fact, transformed, and more homogeneous. If you view sunlight only after it's been polarized, you'll have a much narrower view of how light can move. Of course, people do not become "gender automatons"—there is variety and flexibility, and I am of the opinion that these are good things.

One of the fascinating things about gender is that the natural attitude can persist even in the face of rapid change. Ambivalent, uneven, and contested social commitments to changing the structure of education, work, and family obligations have dramatically reshaped these variables in slightly more than one generation. Gender differences in cognition and in sexual expression have narrowed significantly during this same time. As with CAH women compared with unaffected women, the interesting question to me is not "What makes men and women different?" Instead, it is this: Given that we do begin with slightly different average inputs in important biological factors for development (levels of steroid hormones that affect growth, metabolism, mood, and so on), which *could* be followed by very slight differences in initial predispositions, and given that this *is* followed with pervasive, multilevel "tracking" of expectations and experiences by gender, *how is it that we end up so similar?*[8] Why is there so much intrasex variability and intersex overlap, even on traits that we take as being fundamentally "proper" to one gender versus the other? That is an interesting puzzle.

So what kinds of research might we do that would foreground interesting, tractable questions about these processes? What kinds of studies would not involve putting blinders on so that we could focus on a singular input like prenatal hormone levels, thereby distorting understanding of how even one factor, let alone many others, do affect development? These are open, exciting questions, and they are the stuff of careers, not final chapters. But in Chapter 10 I'll sketch out a few promising ways that we might think of these things differently.

CHAPTER TEN

Trading Essence for Potential

I N THE MIDDLE OF THE seventeenth century, a self-taught Dutch naturalist who was well regarded both for the microscopic lenses he ground
and for the wonderful array of "little animals" he had observed with this
exciting tool, turned his attention to human semen. In her marvelously entertaining history of the doctrine of preformationism, Clara Pinto-Correia
describes how this naturalist, Antoni van Leeuwenhoek, receives "some intriguing drawings" made by a French aristocrat, himself an accomplished
and enthusiastic amateur scientist. Leeuwenhoek "has already seen little
animals in the semen of several species, and has wondered aloud about
their real nature. Now he has before him Dalenpatius's letter, and with it
the ultimate temptation: the drawings of the Frenchman portray little
men, complete with hats and beards, enclosed inside the heads of the little
animals" (Pinto-Correia 1997, 65). Leeuwenhoek's term for these wildly
abundant little animals—the spermatozoa—was "animalcules." Sometime
not too long after he received this letter, Leeuwenhoek saw with his own
eyes the rudimentary seed of man, encased inside the sperm. Each one
swam about, a bit basic perhaps, but with its own definite personhood already implicit in its compact little being. These seeds in semen, the sperm,
were thought to carry the critical essence of humanity from one generation
to the next. Implanted in the womb, the "little man" (Leeuwenhoek's
term) inside the animalcule would meet the matter that was essential for
growing into a fully realized human being. But the significant essence of its

being was present from the start, as unaltered and unalterable as a tiny Russian doll, giving shape from the inside out to the bigger self that it would come to inhabit.

We are the descendants of these "little men" (in Latin: *homunculus*)—though not in quite the material way the preformationists like Leeuwenhoek would have predicted.[1] Pinto-Correia (1997) suggests that humans, in general, have an enduring fascination with "the idea of boxes inside boxes inside boxes" (6). She defines preformation as "the assumption that the primordial organism already contains inside itself all other organisms of the same species, perfectly preformed, minuscule though they might be" and distinguishes this from preexistence, "the more sophisticated version of the model, in which the primordial organism contains only the basic blueprints of all the related organisms to come" (xxi). The theory of brain organization is a variation on the broad theme of preexistence, where the entity that is transmitted across time, from one generation to the next, is masculinity or femininity itself. Rather than being curled inside the sperm, as the "spermists" believed, or tucked inside the ovum, as "ovists" did, our imaginary timeless "little men" and "little women" are encased in testosterone and estrogen.

Toward a More Biological View of Development

In this chapter, I draw heavily on theoretical and empirical work in developmental and evolutionary biology, especially as articulated by Evan Balaban, Anne Fausto-Sterling, Evelyn Fox Keller, and Richard Lewontin. Balaban, a cognitive and developmental neuroscientist who has made profound contributions to knowledge about biological processes in the development of behavior, expresses strong dissatisfaction with the *way* biology is currently invoked as an explanation for human behavior:

> Talented thinkers from many domains of cognitive science (with collectively little primary training as biologists) have transformed debates about cognitive development into debates about biology, by claiming that evolutionary, genetic, and/or cellular facts dictate the acceptance of (often opposing) opinions about the origin, development or function of the human mind or human behavior. (Balaban 2006, 299)

As Balaban sees it, the current arguments about what makes us who we are miss the mark, because the issue to resolve in understanding development is not the relative contribution of genetic variables (including specific gene products, such as steroid hormones) versus environmental variables (including but not limited to experience). He is clear that the problem is not simply that some thinkers underemphasize either "experience" or "biology"—the problem is that formulating this contrast already reveals a fundamental misunderstanding of biology:

> Biological contributions to cognitive development continue to be conceived predominantly along deterministic lines, with proponents of different positions arguing about the preponderance of gene-based versus experience-based influences that organize brain circuits irreversibly during prenatal or early postnatal life, and evolutionary influences acting through selection on small numbers of genes. (298)

In contrast to this deterministic approach, a growing cadre of biologists advance a way of understanding development that recognizes constraints imposed by our genetic inheritance but moves firmly away from a notion of genes as static, concrete building blocks. In her magnificently compact treatise on how we might better understand the relationship between genes and organisms, Evelyn Fox Keller draws attention to how the notion of genes as "elemental particles," biological analogues to the atoms of physics and molecules of chemistry, long ago outlived its usefulness, if not its influence (Keller 2000, 18). Metaphorically speaking, whatever is "written in our genes" must be a very open-ended story, because gene expression is a dynamic, contingent process that is responsive both to specific conditions during development and to random events. In Balaban's words, "Developmental processes *inseparably fuse* experience-dependent and experience-independent components, have important *stochastic contributions,* and exhibit a greater degree of *mechanistic continuity* between developing and adult nervous systems than previously thought" (Balaban 2006, 298, emphasis added).

This is unfamiliar territory to most nonbiologists, so what exactly does that mean? The three key concepts are the inseparability of experience and heredity, the importance of random events, and the fact that development is a lifelong process. Outcomes in the cognitive domain, in particular, are always contingent, rather than ultimate.

Norms of Reaction

The fusing of experience and heredity is the idea of interaction. But the "interactionism" of brain organization models is almost always *addition* rather than the inseparable fusing that defines the relationship of factors that interact. For me, the easiest way to understand how interaction works, and therefore what is wrong with a deterministic or additive model of development, is through the concept of norms of reaction. In *Triple Helix: Gene, Organism, and Environment,* the distinguished evolutionary biologist Richard Lewontin notes that "a genotype does not specify a unique outcome of development; rather, it specifies a norm of reaction, a pattern of different developmental outcomes in different environments" (Lewontin 2000, 23). To understand norms of reaction, we need to define three concepts: genotype, phenotype, and environment. *Genotype* refers to the actual "genes," the basic units of hereditary material in DNA, while *phenotype* refers to developmental outcomes, the individual's "traits." So phenotype is a somewhat broad concept, referring to anatomical, physiological, and behavioral characteristics that might be defined at the very gross level (such as height, intellectual function), the specific level (femur length, mental rotation skill), or even more precisely, by specifying aspects of the individual's developmental process (age when final femur length is achieved, amount of practice required to acquire *x* amount of improvement in mental rotation, and so on).

Environment is the broadest and most difficult part of this trio to specify, because it is literally the remainder of the developmental "input" beyond the specific hereditary information influencing a trait. Components of the developmental environment that typically come to mind include aspects of the material world external to the individual that impinge upon development (such as sunlight, nutrition, available oxygen, environmental toxins) as well as experiences that are difficult to conceptualize at a material level (nurturing, separation, trauma, exposure to specific cognitive schemas, and the like). Less familiar is the fact that additional characteristics of the individual organism, beyond the genome, constitute part of the developmental environment. This includes everything from the entire remainder of the individual's genome (sometimes bracketed as "genetic background" to indicate those aspects of inheritance that aren't directly relevant for development of a trait), to nongenetic material of the egg, to (in the human example) the individual's developing self-consciousness and responses to internal and external inputs. Hormones that affect development, whether they are an individual's own hormones (endogenous) or hormones from the mother or a co-twin (exogenous), are part of the environment.

A norm of reaction (NOR) is a sort of map that shows the relationship between the genotype and the phenotype across different environments. One of my favorite experiments of all time, because of its elegance and the extreme patience of the researchers involved, is the NOR mapping carried out by the botanists Clausen, Keck, and Hiesey over several decades on native plant species in California (Hiesey, Clausen, and Keck 1942; Clausen and Hiesey 1960; Clausen, Keck, and Hiesey 1947). Their work "aimed directly at clarifying two sets of relationships: the first is the relation of plants to each other, and the second is their relationship to the environment" (Hiesey, Clausen, and Keck 1942, 5). NORs are a way of representing this latter relationship. Conceptually, the method they followed was relatively simple. They began by gathering samples of naturally occurring variants of a number of plant species, referred to as an "ecotype," from climatically distinct regions in California. Individual plants from each of these ecotypes were then cloned by propagating cuttings from the wild-gathered parent plant. The researchers tested the effect of the growing environment by following this procedure: "Clone-members of different ecotypes of perennial species were brought into gardens at contrasting altitudes in California, one located in the mild coastal climate near sea-level at Stanford University, another half-way up the western slope of the Sierra Nevada at 4,600 feet elevation, . . . and a third at timberline near the crest of the Sierras at an altitude of 10,000 feet" (7). Notably, "the 'environment' of the plants of different genetic/geographic origin within each experimental field was made as similar as possible" (Balaban 2006).

It may seem odd to shift to plants at this point in a book about people, but there is a method to my madness. Clausen's team followed a painstaking process of charting the growth of successive generations of each of these transplants, measuring and plotting the characteristics of tens of thousands of individual plants over the years. Obviously this sort of experiment could not be done for people, nor for most other animals. There is a second reason that plants provide a good example: it removes the issue of socialization and learning. Plants are simpler. Yet these experiments dramatically disrupt any simple sense of genetics as a "blueprint" for development, because the appearance of genetically identical clones was strikingly different in the distinct climates. On the one hand, among "clone-members" from one region transplanted to another, "details of structure and general habit of growth are essentially unaltered by transplanting" (Hiesey, Clausen, and Keck 1942, 9). In other words, basic features common to the species remained relatively constant. On the other hand, "striking modifications in size of vegetative parts, especially of leaves and stems, and in the extent of branching of flowering stems" (9) were observed among genetically identical plants grown at different altitudes.

This point is so important, and so antithetical to the popular notion of genetic "blueprints," that it bears driving home with an illustration.

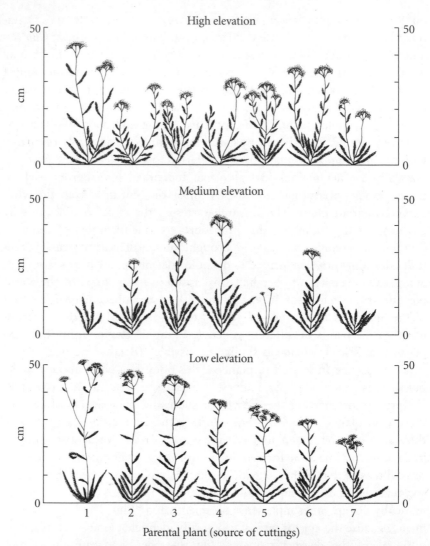

Figure 10.1. Norm of reaction depicting the growth pattern of seven *Achillea* geno-types. (J. Clausen, D. D. Keck, and W. M. Heisey, "Experimental Studies on the Nature of Species. III: Environmental Responses of Climatic Races of Achillea," *Carnegie Institution of Washington Publications* 581 [1958]: 1–129.) Used with permission from Carnegie Institution of Science.

Figure 10.1, from a paper by Clausen and colleagues, has been used by many authors to teach the NOR concept.

This figure shows the relationship between altitude and plant height for seven different genotypes of the single species *Achillea millefolium*. Each column represents a genotype, and each row is a different environment.

Note two interesting things about these plants. There is no way to arrange the plants by height that is "environment-neutral": the plant that is tallest at high elevation and low elevation is one of the shortest plants at medium elevation. And there is no single environment that is uniformly favorable for producing taller plants. Plant number 4, for example, grows tallest at medium elevation, while plant number 3 grows highest at low elevation, and plant number 6 grows highest at high elevation.

The environmental dependence of the relationship between the genotype and the phenotype is not anything special in *Achillea,* by the way, nor is it restricted to plants. Instead, it is a general feature of development that is well recognized by developmental and evolutionary biologists (Nunez-Farfan and Schlichting 2001).

Figure 10.2 provides another way of understanding the variation among *Achillea* plants, and this will help connect the observations in *Achillea* with our knowledge of variation in humans.

In Figure 10.2, each plant represents the "average" growth habit, again focusing primarily on height, for different populations of *Achillea* descended from plants that were initially gathered in distinct climates (ecotypes). To the right of each plant there is a rotated frequency chart that shows the distribution of heights for that ecotype, when plants from that

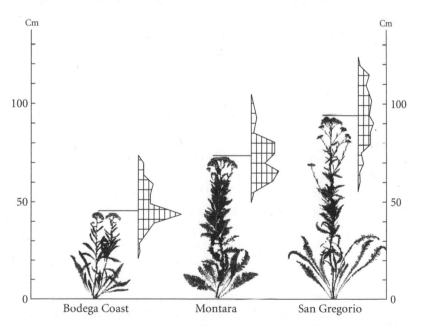

Figure 10.2. Form and variation among three ecotypes of *Achillea.* (Hiesey, Clausen, and Keck 1942, page 14.) Used with permission from University of Chicago Press.

ecotype are grown in a uniform garden at the Stanford experimental field (Hiesey, Clausen, and Keck 1942, 14). Note that the "range of variation within each population overlaps that of the others, but the mean heights are significantly different" (14). This is an interesting example because it is similar to the shape of variation we see when large populations of men and women are compared on certain aspects of cognition and temperament, such as spatial abilities or "nurturing." On its own, the chart seems to tell a story about ecotypes among plants that is similar to the familiar narrative of sex differences: even though there is overlap in the "scores" for individuals in the different groups, the mean scores are higher in some groups and lower in others, and some groups are apparently much more variable than others (look at the broad spread with the relatively long tails for the "San Gregorio" group compared with the sharp central peak and blunt tails of the "Bodega Coast" group). Recall that the greater variability of males is an important element of the overall narrative of how biological factors like hormones, on the front end, explain social factors like the sex composition of math faculties, at the back end (Summers 2005; Baumeister 2007).

There is an additional reason the analogy between ecotypes and the sexes is attractive. The differences among various ecotypes of the same species are the result of evolution: plants adapt within specific local environments. This is similar to the explanation generally offered for observed differences between males and females: insofar as such differences reliably exist, they can be understood to reflect the different demands that the environment has historically made on males versus females. Most narrators of the sex-difference story simply stop here, assuming that (inferred) history leaves not only an *indelible* imprint on future development but a *predictable* one. But here is where Figure 10.1 comes in. What if we were to look at particular traits for ecotypes—or sexes—when aspects of the developmental environment are substantially different?

To answer this, we need to know about the norm of reaction for the group in question. This "group level" norm of reaction would represent the range of phenotypic variation observed for members of the group, across different environments. Because a genotype is a unique property of an individual organism, defined as "the DNA sequences contained within all of an individual's chromosomes" (Balaban 2006, 304), let's think about the NOR at this group level as specifying "ecotype–phenotype" relationships, rather than "genotype–phenotype" relationships. (Although we might well disagree about the nature of the different environmental demands made on males and females in human evolutionary history, let's agree that, in some theoretical sense, males and females are different "ecotypes" of the same species.) Next, it's important to realize that NORs are not *theoretical* maps, they are *empirical* ones: How do actual phenotypes vary in actual environments? Here it is particularly important to make observations of variation for the traits we're interested in, rather than make inferences about how males and females have differed across history, regions, or cultures. Balaban explains, "If the NOR for a particular genotype could be specified over all possible environments, we could regard the phenotypic distance it covers as indicating, in part, the developmental potential of that genotype. We say 'in part' because unless we knew the mechanisms through which the NOR acquires its particular shape, we could not predict phenotypes in a new set of environments" (Balaban 2006, 304).

The norm of reaction for a trait is always incomplete in the sense that we can only map the known environments. Moreover, even if we knew all environments to date, history itself limits the norm of reaction map: What will happen in a new situation? The map can't actually predict what the phenotype will be under new circumstances. Moreover, because of the stochastic components of development, the prediction of developmental outcomes would still not be perfect.[2]

So even this idea of the "developmental potential" of an organism (let's say a human!) can only give us *an idea of how responsive to environmental influences particular phenotypic traits have been in known environments* —not predict what the trait will be in unknown environments. One thing we can know, however, is whether the NOR is flat—meaning that the trait stays the same across different environments. If it is flat, then this means one of two things: either there is very little developmental flexibility in the organism for that trait (the "preexistence model" of one-to-one mapping between the "blueprint" and the developed state); or there could be a lot of "developmental flexibility" in the sense that, in order to produce the same, stable outcome under widely different circumstances, the developmental system has to flexibly change what it does. However, virtually all the traits that are of interest in behavioral cognitive science have nonflat NORs (Balaban 1998), so we would expect a nonflat NOR for any putative outcome of the brain organization process. If it is not flat, this means that the blueprint model cannot be correct; the environment is important in *how* heredity shapes the trait.

In the case of sex differences in behavior, we might want to know not just whether the trait differs across environments (that is, whether the NOR is flat), but whether the magnitude or direction of sex differences also varies across environments. In this case, the NOR curves of the phenotypic trait for each sex would be nonparallel to each other, as well as nonflat. This is important, because as soon as the NORs are nonparallel, then it is readily conceivable that some environments exist for which group differences would be eliminated or even reversed. In what follows, I will argue that most, if not all, of the aspects of human behavior that are presumed to be sex-typed by early hormone exposure have nonflat, non-parallel NORs, meaning that the existence of trait differences, and their sizes as well as their directions, are context-dependent and malleable.

Gender Norms (of Reaction)

The norm of reaction is not a new or a controversial concept. But it is routinely ignored in favor of the popular, simpler, and wrong model of deterministic, additive factors that influence development. There is much more to say about how this concept might help us to reshape our notion of sex, emotion, and cognition. A good way to proceed is to think about what is already known regarding the relationship between the ecotype of sex and the development of various characteristics. How does the developmental environment matter?

There are a number of examples of differences in the way that specific environmental conditions affect sex-typed behavior in males versus females in other species, such that there is no environment-neutral way to specify the sex difference. For example, territorial exploration is a trait that is considered male-typical in rhesus monkeys, but in this species males seem to be more vulnerable to "aberrant" rearing conditions such as isolation. Thus, "isolation-reared males [show] much more frequent and intense fear, and less exploration, than isolation-reared females" (Sacket, Holm, and Landesman-Dwyer, quoted in Jacklin, Maccoby, and Doering 1983, 164). Likewise, both male and female rats have a capacity for "maternal" behavior (which raises the question of why grooming and caring for rat pups isn't just called "parental" behavior). Leboucher has shown that previous exposure to pups has a more dramatic effect in stimulating "maternal" behavior among male versus female rats (Leboucher 1989). Another interesting example concerns the way that males versus females navigate through space. The use of "spatial location" (such as orientation in relation to compass points) versus "local markers" is considered to be a reliable male–female difference in a number of species, with males preferring (or excelling at) the former strategy and females preferring (or excelling at) the latter one. But recent data from Rebecca Herman and Kim Wallen (2007) complicate this perspective. Herman and Wallen studied spatial navigation among rhesus monkeys, manipulating both hormones and the information that was available to the animals for using the two spatial strategies. For the purpose of this discussion, the interesting finding was among the animals that were not hormonally manipulated. Here, in a novel environment in which local markers were not available, the researchers found that the *females,* not the males, were able to make better use of the "male typical" navigational strategy. They also found that blocking the males' prenatal androgen exposures *improved* the use of local markers (not that blocking androgen prevented development of the "male-typical" trait, which in their model would have meant impairing the ability to use "spatial consistency").[3]

These examples suggest that although these traits—maternalism, exploration, spatial relations—are staples of the brain organization literature, it is not accurate to simply talk about these traits as representing "stable differences" between the sexes, even though the differences may in fact be quite real and even substantial in any given time and place. Given what researchers themselves have shown about how *sex differences in these traits can and do change in different environments,* it is teleological to pronounce such environment-dependent states as "sex-typed." That is, because the nature of the sex difference depends on specific environments, the environments that produce more pronounced sex differences are defined in a circular manner as "ideal" or "natural." But—especially when the variation in outcomes produced by environmental change is not pathological—one must ask on what basis is that environment considered ideal? Ideal for what? For emphasizing a particular pattern of difference.

The practice of considering environments that emphasize sex differences as "ideal" becomes even more problematic when one takes into account the degree to which normal males exhibit so-called female-typical behavior, and normal females exhibit so-called male-typical behavior. This blurring of "male" and "female" behavior is substantial even in species where behavior is quite heavily linked to hormones, and even for behaviors that are directly tied to reproduction or rearing of the young (for example, Clemens, Hiroi, and Gorski 1969; Noble 1979; Leboucher 1989; for an excellent review of this issue, see van den Wijngaard 1997). Note that in the popular coverage of sex differences, the concept of substantial overlap is almost completely absent and the impression of a strict dichotomy of sex-typed behavior is even more pronounced (Lancaster 2003; Barnett and Rivers 2004; Cameron 2007).

The example of bones is especially profitable for rethinking the development of sex-typed characteristics in humans, both because gonadal hormones play a central role in bone growth and maintenance and because the most theoretically and empirically rich application of concepts in developmental biology to problems of sex difference has focused on bones

(Fausto-Sterling 2005). Begin with the aspect of bones that is one of the central phenotypic variations routinely cited as a "sex" difference: males are taller, on average, reflecting more length in the long bones. According to psychologist Richard Lippa, a brain organization researcher who has used data from the large BBC Internet survey to analyze sex differences in height along with a variety of characteristics that he relates to brain organization theory, "evolved biological factors are the primary cause of sex differences" in height as well as in libido (what he calls "sex drive") (Lippa 2009). Lippa specifically contrasts the "biological model" for height and libido from a "hybrid model"—where both biological and social structural influences are seen as contributing to sex differences—which is what passes for interactionism in most brain organization research, though it is what developmental biologists would call an additive or deterministic model. The current basic theory behind sex differences in height is this:

> The growth plates or epiphyseal plates are generally more sensitive to the effects of estrogen than to those of testosterone. During puberty in the female, the rising levels of estrogen seal the epiphyseal plate earlier than testosterone does in males. The effects of the male hormone, testosterone, are felt at a later stage. Thus, females stop growing earlier than males do. (skeletalsystem.net)

The scientific actors and relationships described here—the narrative—should be very familiar by now. An observable sex difference in phenotype (height) and a physiological process that involves the so-called sex hormones estrogen and testosterone. The story implies that estrogen would be responsible for hardening of epiphyseal plates only in females, while the actor in males would be testosterone (in fact, *both* hormones are important for stopping bone growth in *both* sexes). The story doesn't help account for a number of interesting problems we could observe about men, women, and height across populations and time. For example, if "sex hormones" are the primary explanatory factors, how do we explain differences in groups that aren't defined by sex? For instance, why are Dutch people so tall, even compared with other Western Europeans with similar health and nutrition metrics (Carvel 2002)? How can we account for the fact that the average height among Dutch women is now somewhat higher than the average height among Spanish men was just about 60 years ago (Cavelaars et al. 2000)? And why has height in Dutch people (and the rest of the industrialized world) been significantly *rising* during the precise period when age at menarche (the time when estrogen levels dramatically rise in girls) has been significantly *dropping* (Fredriks et al. 2000; Ong, Ahmed, and Dunger 2006)?

The problematic habit of classifying hormones as fundamentally related to, and even *for*, sex (rather than about broader functions, such as growth regulation) and as "sex-specific" (that is, properly belonging to one sex or the other, rather than being related to normal function in healthy individuals of both sexes) is part of the difficulty here. But it's only one part. The other part is that when height is conceptualized as a sex (read: biological) difference, height is then theorized as fundamentally flowing from sex in a deterministic way. This model has the unhappy effect of suggesting that the general shape of sex differences is stagnant over time and across populations. Notice, for example, that Lippa (2009) pronounces that sex differences in height are fully biological, even though he has gathered information about only one single time point! But that is a conclusion one cannot draw without data over time, as well as across locations. In fact, we have enough data to show that the NOR map between sex and height is not flat, because (as noted above) there have been highly significant changes in height for both men and women in just the past hundred years. (Recall that while an NOR is, strictly speaking, a reflection of how genetically identical individuals develop in different environments, I'm using the "expanded" concept of NOR as the relation between male and female "ecotypes," respectively, and the trait of height, across different environments.) Consider the fact that average height among men has been increasing faster than average height among women in most European countries over the last seventy years (Cavelaars et al. 2000). This shows us that the NOR is not only flat but nonparallel: different environments affect height differently for women versus men. Once we know this, all bets are off for predicting how sex differences might look in a future, novel environment (unless and until we know the actual developmental mechanisms that mediate the interaction of the genotype and the phenotype, which we are very far from knowing).

Does this mean that there is not really an average difference in height for men versus women? Of course not. In the environments mapped to date, heights are consistently different at the group level, by sex. But that is not a particularly interesting observation. The "fundamental sex difference" story makes it difficult to notice how sex differences, themselves, change shape in different environments. It also makes it hard to absorb information on important influences from social structures. Here is a fascinating example. A recent study in the *Journal of Demography,* based on growth data from two large longitudinal studies, reported that sex differences in children's height are significantly related to "son preference." Song and Burgard (2008) show that, in addition to long-term nutritional status and exposure to infectious diseases, a measure of son preference at the population level helped explain a significant and substantial height advantage among Chinese boys, especially in rural areas. In other words, by a variety of measures, the relative preference for boys seems to be higher in China than in the Philippines, and there is also a larger sex difference in height in China, even when all other major factors related to height are controlled. A plausible explanation for this effect might be that where sons matter more to parents than daughters do, sons are given a larger share of families' nutritional and health-care resources.

Let's pause to consider how we might think of "gender NORs" in the psychosexual domain. To take but one example, sex differences in educational attainment, which are sometimes taken to reflect underlying sex-typed traits such as achievement orientation or competitiveness, have recently undergone several shifts and even reversals. Prior to the late 1970s, males were consistently more likely to proceed to college following high school, but in the following decade the sex difference evened out. More remarkable, since the early 1990s, young women have been significantly *more* likely to enroll in and earn degrees from four-year colleges than young men (Smith 1995). In the nation's largest university system, the California state universities, women were 57 percent of new first-year students in 2000, and 66 percent of graduates in 2005 ("Fewer Men on Campus" 2006).[4] Obviously there has been no major biological change between the sexes during these decades, so the answer must lie in some aspect or aspects of the social environment. Changing gender norms clearly mediate the relationship between sex and education, but there are other important elements, too. Note that educational participation and attainment are also dependent on class and on race (Huang, Taddese, and Walter 2000). The shape of sex differences is *specific* to class-race groups and has *changed differently* in specific groups over the past few decades. The fact that sex differences in educational attainment cannot be accurately generalized, but must be described for particular historical periods, and specific, culturally defined groups, is critical, because it shows that the "reaction norm" linking gender to education is not flat. Note that for bachelor's degrees earned in the United States in 2003, "women earned about two-thirds (67 percent) of degrees granted to Black, non-Hispanics, 63 percent of degrees granted to American Indians/Alaska Natives, 61 percent of degrees granted to Hispanics, 58 percent of degrees granted to White, non-Hispanics, and 55 percent of degrees granted to Asians/Pacific Islanders" (National Center for Education Statistics 2009).

In virtually all fields related to science, technology, engineering, and mathematics, the gender gap has narrowed significantly over the past few decades, though it has narrowed less in some fields than in others (Huang,

Taddese, and Walter 2000). As I have discussed in earlier chapters, proponents of brain organization theory make much of the role of "sex-typed interests" in explaining the disparities that persist. So let us consider what is known specifically about interests. "A review of national surveys of college students . . . reported that, while the career interests of men and women have been becoming increasingly similar during the last 3 decades (1966 to 1996), one of the largest remaining gender gaps was still engineering, a field in which few women intend to study and work" (Huang, Taddese, and Walter 2000, 5). The figures for engineering bachelor's, master's, and doctoral degrees awarded, though, show a spectacular amount of change in this field over a very short period. In 1966, less than 0.5 percent of bachelor's or doctoral degrees in engineering went to women, but by 1996 the figures had jumped to 17.9 percent and 12.3 percent, respectively (5). Looking at other quintessentially male-dominated fields, by 1996 women were earning one-third of mathematics and computer science degrees.

We cannot assume that we have basically reached the limit of change in the shape of gender differences in these or in any fields unless we assert that there cannot (or perhaps should not) be any more change in social structures related to gender. And what, after all, has changed in terms of gender as a multilevel social system? Certainly there have been important moves toward formal equality, as barriers to equal participation in educational programs have been removed. But core gender schemas don't seem to have changed much in this time (Gherardi and Poggio 2001; Fischer 1993). People still hold the same basic stereotypes about the attributes that men versus women should have (Prentice and Carranza 2002) and still tend to attribute "masculine" characteristics like rationality and competence more readily to males, independent of the evidence available regarding the actual attributes of individuals being evaluated (Valian 1999). In fact, there is good evidence of "backlash" effects (social and economic reprisals) for women who try to break the mold of traditional femininity (Rudman and Phelan 2008)—which they must do if they are to enter traditionally male fields or rise to leadership positions. But the issue goes beyond the conscious attempt to break glass ceilings. While brain organization researchers are fond of emphasizing individual "interests" and "choices," other scholars point to factors that shape and constrain these interests, such as the "flow" of work and support that men versus women can count on in their families (Williams 2000), and the gender schemas that shape the perception and actual development of people's behaviors and characteristics (Valian 1999; Harris 2004). Given the irrefutable evidence of change in actual outcomes after relatively short-term and limited alteration of gender structures, doesn't it seem premature to attribute disparities to biological factors in sex-typed interests?

Relating these changes in education and career interests again to the concept of NORs, we can see by looking at education in specific fields not only that the "gender NOR" is nonflat, but also that it is *not parallel* for males and females, because the sex difference has reversed in some fields. Think of accounting, which is a field that seems to be a great example of a domain that would interest "systemizers" (to use Simon Baron-Cohen's notion that fundamental interest in "law-like systems" is triggered by early testosterone). Wootton and Kemmerer (2000) document that women constituted 10 percent of accountants in 1930 and more than 51 percent in 1990, with 53 percent of new accounting degrees awarded to women in 1990. They note, too, that "increases in the aggregate workforce were not accompanied by subsequent proportional increases in participation at the upper-management levels of accounting firms" (169). The gender composition of these professions at different points in time, the rapid change in a matter of a few decades, and the continued predominance of males in upper management can all be readily explained by social, cultural, educational, and economic factors (as Wootton and Kemmerer do). These factors cannot, of course, explain why any individual woman or man in either 1930 or 1990 might have become an accountant, a biologist, a stay-at-home parent, or a lawyer. But because the pattern of sex differences in all of these occupations, except stay-at-home parent, has changed so much in recent history, appealing to sex as the fundamental explanation for occupational choice is not terribly convincing.[5]

But perhaps the arguments about sex differences in math and science or occupations more generally is too familiar. It might be easier to see the norm-of-reaction concept with more physical examples. So let's return to the issue of bones, and move to other aspects of bone growth that look more like the kinds of phenotypes psychologists are interested in. Although early development has a major impact on later characteristics, bone width and density continue to change and develop over time, following the same basic processes that guide early development (this is what Balaban means by "mechanistic continuity" between juvenile and adult systems). Anne Fausto-Sterling has recently used bone density and osteoporosis to push against the sex versus gender distinction itself, asking "what it might mean to claim that our bodies physically imbibe culture" (Fausto-Sterling 2005, 1495). Because bone development is "an area often accepted as an irrefutable site of sex difference" (1498), showing how sex differences in bone development are responsive to culture not only improves our understanding of bone health and diseases, it destabilizes biological sex. This work is part of a larger feminist project that involves documenting the ways in which apparently "natural" properties of sexed bodies are in fact dependent on gendered activities that actively shape bodies, as well as our perceptions of bodies (for example, Oudshoorn 1994; Kessler 1998; Fausto-Sterling 2000). In the case of bone density and osteoporosis, some of the factors that underlie putatively innate biological sex differences include physical activity levels, especially weight-bearing work and exercise; diet, especially calcium intake; exposure to sunlight for vitamin D; and hormones—of these, only hormones have a plausible "biological" basis that is easy to conceptualize apart from gender, socioeconomic position, and culture. And while biological factors like hormones are involved in translating specific inputs like weight-bearing exercise into a sex difference in bone density, we can immediately see that this will be affected by systematic gender differences in weight-bearing exercise. It is harder to remember, though, that steroid hormones are also affected by the social organization of gender. Steroid hormones, including gonadal hormones, are responsive to a wide range of behaviors, mood states, and pharmacologic interventions (such as hormonal birth control) that are profoundly affected by gender relations. Gender, in interaction with socioeconomic position and culture, heavily influences the kinds of work and play that people engage in, the food they eat (including both quantity and types), the likelihood that they will diet, and so on.

One brief example that Fausto-Sterling offers to demonstrate that "culture shapes bones" is the fact of very low bone mineral density (BMD) in the lower back among ultra-Orthodox Jewish adolescents, who also "have lowered physical activity, less exposure to sunlight, and drink less milk than their secular counterparts" (1491). We can elaborate this example to show how gender and culture interact, too. While bone density is typically higher among males than females in this age group, "boys had profoundly lower spinal BMD than did girls" in this community (Taha et al. 2001). Because the sex difference was found even after they controlled for hours per week of weight-bearing exercise and walking, the authors conclude that sex-related biology is an important determinant of bone mineral density. But the ultra-Orthodox communities are highly gender segregated in terms of both activities and spatial organization (males and females are literally required to be in different physical spaces much of the time, and boys spend most of their hours inside, studying the Talmud). It is likely, then, that the study has under-assessed relevant gender differences, such as the amount of sunlight exposure, or other physical movement that does not register for the girls as "weight-bearing exercise" but that nonetheless distinguishes their activities from the extreme stillness that the boys experience. Imagine, though, that they have in fact perfectly measured all factors that are systematically different by gender. That would mean that in this cultural context, the sex difference seems to work differently than in other contexts.

This demonstrates the true meaning of interaction. It's not just that there is some constant amount of bone density that is attributable to sex and some other variable amount that you add on for gender. Nor is it accurate to think of sex as controlling some specific aspects of the *processes* related to bone density, while gender controls other processes. If we think of this interaction in Balaban's terms as the "inseparable fusing" of "experience-dependent" and "experience-independent" aspects of development, it suggests an alternative working definition for sex. If *sex* is the experience-independent part of maleness or femaleness, and gender is the experience-dependent part, what would that mean? For one thing, it would mean that the question of what counts as sex would become a matter for empirical investigation instead of being a matter of definition or philosophy. For another thing, it would mean the immediate constriction of the term *sex* to exclude all those aspects of biology that we already know *are* dependent on experience (where experience includes exposures that depend on personal biography as well as culture and social position in structures like gender, class, and race). Third, even once the term is restricted in this way, the meaning of this term for developmental endpoints must also change from current usage. That is because the *way* in which "sex" works still depends on gender, as well as other aspects of culture. This is most especially true in the realm of behavior, particularly that portion of behavior that we understand as gender and sexuality.

VERY FEW developmental endpoints are truly "final"; instead, they are interim states, with the possibility of growth and change until death. This is the meaning of plasticity. In this ongoing process, the interaction of physiological and experiential variables is iterative, meaning that the current state of the organism *interacts* with each subsequent input—whether that input is experiential or physical (including biochemical). In brain organization studies, both the animal researchers and the scientists studying humans follow two practices that obscure this fact, causing them to misrepresent scientific knowledge on the effects of hormone exposures during the "critical" periods for development of the brain. In literature reviews for new studies, as well as in review articles and theoretical pieces on brain organization, brain organization researchers have systematically omitted data on the modifiability of the supposedly "permanent" behavioral patterns that follow early hormone exposures. As early as 1969 it was known that many of the "organizing" effects of hormones are not permanent, but are easily modifiable by experience. In a little-cited study by researchers at UCLA, for example, scientists found that allowing an androgenized female rat to have just two hours to adapt to a stud male *completely eliminated* the behavioral effects of prenatal testosterone injections (Clemens, Hiroi, and Gorski 1969). Money and Ehrhardt (1972, 85) suggested that "neonatal androgen may have rendered the females more sensitive to the copulatory environment, possibly to olfactory cues, in the manner that is usually typical of males. Once adapted to the environment, they became disinhibited. The *behavior that was then released was not masculine in type, but the feminine response of lordosis*" (emphasis added). Subsequent experiments have shown that a great many of the sex-typed behaviors that are supposedly permanently organized by prenatal hormones can be dramatically modified or even reversed by simple and relatively short-term behavioral interventions such as neonatal handling (Wakshlak and Weinstock 1990), early exposure to pups (in rats) (Leboucher 1989), and sexual experience (Hendricks, Lehman, and Oswalt 1982), to cite just a few examples.

A recent very exciting example of plasticity in humans concerns dyslexia, a cognitive trait that has been theoretically linked to early testosterone and has even been examined in some brain organization studies as a marker of "masculinization" (for example, Götestam, Coates, and Ekstrand 1992). Simos and colleagues (2002) studied children with dyslexia before and after eighty hours of intensive remedial reading instruction. At the beginning of the study, magnetic source imaging showed that the children with dyslexia had a different pattern of brain activation compared with normal children with no reading problems. In particular, they showed very low signals associated with an area that is normally involved in phonological processing. Remarkably, after the intensive intervention the children not only made substantial improvement in their reading skills but also showed much larger signals associated with the phonological processing area that formerly showed low signals.

Another powerful example concerns spatial cognition—one of the hallmarks of psychosexual differences. Feng, Spence, and Pratt (2007) identified a basic information-processing capacity that underlies spatial cognition and showed that differences in this capacity (the distribution of spatial attention) are related to differences in the higher-level process of mental rotation ability. They then showed that a remarkably brief intervention—just ten hours of practice with an action video game—caused "substantial gains in both spatial attention and mental rotation, with women benefiting more than men" (850). The ten-hour training did not completely eliminate the sex difference, but it came extraordinarily close—the mean scores after training were no longer statistically distinguishable between males and females.[6]

Thus, even though early hormones affect neural development, the language of "hardwiring," "blueprints," "latency," "permanent organization," and so on clearly conveys an inaccurate picture of the *nature* of early hormone effects on behavior. As Doell and Longino (1988) noted two decades ago, these metaphors fail to accurately capture how development really works. Even in rats, early hormone exposures do not create a solid foundation on which behavior must forever stand. At first glance the true process might seem to be captured by the notion of developmental "cascades," which several organization theory researchers raised in their interviews with me. The notion of developmental cascades suggests that hormones don't directly *determine* behavior, but create a small push in one direction, which is then amplified by experiences and other inputs that in turn trigger additional inputs, such that a tiny push at the front end can end up in a sizable difference in outcome. But this is only half the story—one in which the small initial differences almost inevitably grow larger as additional effects accumulate. But an early push in a certain direction can be *either enhanced or entirely eliminated* by subsequent experience, such that development from that point forward would proceed as though the early hormone exposure had never happened.

Again, the point is not that hormone effects are not "real." Hormones are important growth mediators, and they do figure into development, including neural development, in a variety of important ways. Nor is the point that males and females aren't "really" different. There are demonstrable differences, on average, between males and females in a variety of characteristics, including some limited cognitive abilities, personality traits, and interests, including sexual interests. The problem is the way that brain organization theory brings together ideas about hormones with observations of male–female differences. The story attributes an unrealistic specificity and permanence to early hormone effects, as well as a demonstrably false inevitability and uniformity to sex differences, which are inaccurate even for those animals whose sexual and other behaviors may turn out to be mechanistically less complicated than ours.

Norms of reaction give us an alternative way to understand the experiments that show how early hormones shape reproductive physiology—variations in gonadal hormones during critical periods of development affect the relationship between genotype (XX or XY) and phenotype for genitals. This is the idea behind brain organization theory, too. The prob-

lem is that the data have never fit the model so well in the case of brains as
the case of genitals. And theoretically, there are many reasons to predict
that the model would not fit. As I described in Chapter 3, one major problem is that brains, unlike genitals, are plastic—so the ultimate phenotype
isn't apparent at birth or even at some point in childhood or adolescence.
In fact, brain development is not completely "finished" at any point prior
to death. But quite a few aspects of our personalities, skills, and sexual desires do seem to be pretty stable through at least our adult lives. So perhaps
the issue of plasticity could be set aside, as long as we specify that we are
interested in the phenotype at a particular point in time. Of course, this is
what brain organization researchers do when they look specifically at play
in children, or gender identity *in adults,* and so on. The problem is that in
between the initial environmental input of early hormones and the point at
which the phenotype is observed, there are many other aspects of the environment that impinge on development.

Let's return to the question of experiments and consider how animal
experiments on brain organization could better incorporate the NOR
principle, while remaining attentive to plasticity. Such experiments would
require testing hormones-during critical-periods as the *first* important
variation in environment, and the phenotype that results from early hormones as an *interim state* in the organism's development. Next, it would
require systematically testing how early hormone exposures affect the behavior of animals who are reared in different ways, sexually socialized in
different ways, exposed to different subsequent hormone environments,
have different diets and exercise regimens, and so on. A fundamental error
in most research to date is that researchers have *assumed* that the "organizing effects" of hormones are permanent, without testing similarly
treated animals across a range of developmental conditions and genomes
to test the stability of those effects. This is to proceed as though all important aspects of development are over, or as though development of the
brain and behavior has a fundamentally different character after the "organization" phase. But it does not.

Thus, the notion of a blueprint or an unchangeable behavioral pattern,
set early on and then lying "latent" and waiting to be "released" does not
really work. Add to the model the understanding of gender as a multilevel
environment that impinges on development in ways that are both pervasive and only dimly recognized at present. What is the bottom line? It is
reasonable to expect that in the context of extremely gender-dichotomous
socialization around sex, within-sex variability can be dramatically atten-
uated and cross-sex difference will be enhanced. This suggestion fits the
empirical evidence regarding domains of development that were reviewed
above (bones, spatial relations, educational interests and attainment). It

has already been demonstrated in these domains that different environments can change the shape of sex differences, reducing their size or even reversing the direction of difference. The question is not whether that is possible. The question, instead, is what kinds of differences should we—socially and politically—accept or even embrace, and conversely, what kinds of skills and traits would we prefer to encourage (or discourage) in everyone. As I have argued, this is not a scientific question, but a political one. Punting this important political issue back to science by suggesting that some things are simply the "state of nature" clearly will not do.

Conclusion: Toward the Open Door

The influence of biological variables that we *think* of as "sex-linked" almost certainly plays a role in the overall iterative-and-looping process of development. But in terms of a scientific program for understanding development, we've reached the end of that road—in fact, we've gone way off the road into the woods and are now stuck in the deep mud of "innate sex differences." As I have demonstrated throughout this book, the data are not compelling when placed together in the "network of associations," and models of development that are current in biology give us little reason to continue pursuing this line of reasoning. So here are a few closing observations that might help us climb out of the mud:

Steroid hormones are important, but they aren't best conceptualized as "sex hormones." They do lots of things; "sex hormones" was the original conceptualization that drove the research and classification on hormones, but it doesn't fit the data on what hormones do any better than other possible schemes. And the "sex hormone" framework demonstrably blocks recognition of complex and accurate information (Oudshoorn 1994; Fausto-Sterling 2000; Nehm and Young 2008).
Personality traits and predispositions are not identical in individuals, but they are also not well captured by the binary system of gender (Witelson 1991)—even in spite of pervasive cognitive schemas that exert pressures toward this pattern. We aren't blank slates, but we also aren't pink and blue notepads.
Brains develop only in interaction; input from the external world, as well as from one's own sensory apparatus, is as critical to development of the brain as food and water are to the entire organism.
Brains change and develop over the lifetime. Few inputs are irreversible.

Even the animal experiments on brain organization showed that the "permanent" effects of early steroid hormone exposures could be eliminated or even reversed by fairly brief interventions in the physical and/or social environment.

Gender relations change, and these are demonstrably related to changes in psychosexual outcomes. For example, structural-level shifts in education (removing barriers to admission for women to colleges and graduate programs, barring gender discrimination in funding, and so on) have quickly reshaped the landscape in terms of the proportion of college graduates who are female, as well as the sex composition of particular programs of study (accounting, law, medicine, biology, and so forth).

APPLYING these observations along with the principle of norms of reaction might lead us to focus on the dynamism of development—understanding *processes* more than *permanent states,* development rather than "essences." A long time ago I was asked to analyze a sex survey conducted by a popular magazine, and the editor was especially interested in comparing the interests and behavior of men and women. There were some fascinating findings that shook up the usual story about women wanting relationships and men wanting sex, but the editor kept pressing me to look for things that would confirm the familiar story. When he finally said to me, "People don't buy this magazine to learn something, they like to confirm what they already know"—I knew it was time to withdraw from the project. When research starts to look too much like an "infomercial" for cherished beliefs, it is no longer science. Brain organization theory is little more than an elaboration of long-standing folk tales about antagonistic male and female essences and how they connect to antagonistic male and female natures. As a folktale, it's a pat answer, a curiosity killer. And the data don't fit into the tidy male–female brain patterns, anyway (as ideas about "mosaicism" already acknowledge). Why keep trying to fit the data into a story about sex?

A wise person once told me, "Life is full of choices that aren't always clear. Err on the side of good narrative." The theory of sexual organization of the brain is getting in the way of the science of human development. And it's also getting in the way of good narrative. What we need now is a way to cultivate and reinvigorate curiosity about how the body really matters in the development of human personality and behavior, because curiosity and skepticism are the real engines of scientific discovery. What good is a science that doesn't tell us anything new?

Notes

1. Sexual Brains and Body Politics

1. Hausman's work has been mentioned in many pieces on recent controversies around women in math, science, and engineering, including a commentary in the *New York Times* criticizing the National Academy of Sciences report on maximizing the potential of women in science and engineering, coverage of the "parity" debate that followed release of an MIT report on discrimination against senior women at MIT, and criticism of a Census Bureau report on wage discrimination (Tierney 2006; Holden 2000; Abel 2001; Smallwood 2001; Ward 2003).

2. Another common feature in the many supportive responses to Summers's comments has been a tendency to misstate the circumstances of both the original controversial speech and the eventual fallout. Quite a few pundits and scientists have claimed that Summers was skewered for simply mentioning the possibility of innate sex differences, and that these comments "cost him his job" (Pinker 2005; Tierney 2006; Baumeister 2007). Summers did not simply "mention the possibility" of innate differences, though. He skipped an entire morning of presentations that were devoted to exploring recent empirical research on structural and social factors that exert differential pressures on male and female faculty members, then delivered a speech in which he proffered the opinion that the most important factors in sex disparities in science, technology, engineering, and mathematics (STEM) careers are probably not social factors like discrimination or structural factors like a needlessly male-oriented tenure clock, but sex differences in interests (what he called the "high powered job hypothesis") and in abilities.

3. John Colapinto's book, in particular, details a myriad of ways that the medical

and psychiatric treatment for this child, under the direction of John Money, was extremely traumatic. Reimer was subjected to repeated physical exams and relentless questioning about his gender identity and development, and was made to simulate (through clothed "humping") sexual intercourse with his twin brother, in order to supposedly break through his extreme discomfort with sexuality and help him learn a "female sexual role." Many aspects of his treatment—particularly the degree to which he was constantly examined and interrogated about his gender and sexual preferences—mirror the experiences of intersex children. A belief that this sort of treatment is traumatic, rather than the idea that one's sense of masculinity comes from early sex hormones, leads me to strongly support the demands of the intersex rights movement for delay of surgical interventions for children with ambiguous genitalia. But I believe it is just as important to take a hard look at the effects of nonsurgical interventions—the constant surveillance of these children's development—which, while not physical, may be just as "irreversible" as sex-reassignment surgeries. I explore these factors at length in Chapter 9. At the same time, it is important to counter some of the more overblown villainizing of Money, such as the widespread suggestion that Reimer's treatment constituted an "experiment" rather than a treatment course that was pursued in good faith, based on the best available information at the time.

4. While some sexual rights advocates and opponents are strongly committed to the idea that evidence of "innate" differences in the brains of lesbian, gay, bisexual, and transgender (LGBT) people would necessitate granting equal rights, historians have pointed out that "innate" difference has often been used as grounds for discrimination against the sexually different. Moreover, scientific findings can't be the neutral arbiters of political questions, because scientific questions are always driven, to some extent, by socially and politically inflected understandings of what portion of any phenomenon is important to understand. Biologist Anne Fausto-Sterling (2000) has given an elegant example of why this is the case in a discussion on research concerning sex differences in the corpus callosum (CC), the fibers that connect the left and right hemispheres of the brain. "If we were to reach social agreement on the politics of gender in education . . . what we believe about CC structure wouldn't matter. We know now, for example, that 'training with spatial tasks will lead to improved achievement on spatial tests.' Let's further suppose that we could agree that schools 'should provide training in spatial ability in order to equalize educational opportunities for boys and girls'" (145). She goes on to describe the various paths that research on the CC might take in order to refine interventions to maximize the abilities of both sexes, noting that "feminists would not object to such studies because the idea of inferiority and immutability would have been severed from the assertion of difference, and they could rest secure in our culture's commitment to a particular form of equal educational opportunity" (145).

5. There is a second reason, too, that I have opted for critique. Although I have learned an immense amount from critical studies of scientific practice, and think this is indeed where the cutting edge of philosophy of science is being

forged, I would argue that current science and technology studies tend to fore-ground scientific practices at the expense of theories. This is possibly because STS scholars have found that scientists' theories are altogether too tidy to be interesting and, as indicators of what "really" goes on in science, aren't as reliable as the evaluation of practices. Or perhaps it is more accurate to say that in current STS accounts, theory fades behind practices in a way that allows science its messiness and doesn't force resolutions or coherence—the "resolution of scientific conflicts" that was so popular to study a few years ago has mostly been abandoned for less-settled pastures (compare Nelkin 1992 to Mol 2002). Again, to someone who values scientific methods as an insider of sorts, theories are important, because they are in a dialectical relationship with scientific practices: it is impossible to maintain that a theory is adequate without examining the practices and evidence that relate to the theory, and it is impossible to choose methods and specify what constitutes data without knowing what theory one hopes to test.

6. Of course, people do not generally experience these domains as separate, either, and there is evidence that the sex/gender/sexuality division works better for analyzing some people's experiences than for analyzing others' (Valentine 2007). For more on the problematic boundaries, especially between gender and sexuality, and between gender and sex, see Kessler 1990 and 1998; Laqueur 1992; Butler 1993; Oudshoorn 1994; Fausto-Sterling 2000 and 2005; and Mak 2004 and 2006.

2. Hormones and Hardwiring

1. This section on early endocrinology and the search for sex differences is based primarily on three texts: Nelly Oudshoorn's *Beyond the Natural Body: An Archaeology of Sex Hormones* (1994); Anne Fausto-Sterling's *Sexing the Body: Gender Politics and the Construction of Sexuality* (2000), esp. chaps. 6 and 7; and Chandak Sengoopta, "Glandular Politics: Experimental Biology, Clinical Medicine, and Homosexual Emancipation in Fin-de-Siècle Central Europe" (1998).

2. Ernest Starling, professor of physiology at the University of London, coined this term in 1905 (Oudshoorn 1994; Fausto-Sterling 2000). Fausto-Sterling explains that the term *hormone* comes "from the Greek 'I excite or arouse'" (150).

3. Bell's comment has a strong echo in current thinking about hormones and "sex-typed interests," which undercuts the current claim that this is a radical new idea. In Chapter 8 I explore the empirical support for the idea that non-erotic interests (also called motivations, preferences, and priorities) are shaped by early hormone exposures. In that chapter, I do not dwell on the strong resonance between nineteenth- and early twentieth-century claims about "internal secretions" and women's "natural" priorities or sphere, but the connections are obvious. I do, however, mention that the kinds of interests that are considered sex-typed, as well as estimates of the size and direction of some differences, have undergone substantial and rapid change recently, which itself chal-

lenges the notion that interests are significantly shaped by hormones. For example, the idea that men would be naturally more drawn to the adversarial profession of law has suffered a real blow, as what was an almost universally male profession has approached gender parity in less than fifty years: in 1960, 2 percent of law degrees in the United States were awarded to women (Smith 1995); by 2002, 50 percent of law school entrants were women; and more than 30 percent of practicing lawyers in the United States are now women (*Charting Our Progress: The Status of Women in the Profession Today* 2006; see also www.abanet.org/legaled/statistics/stats.html).

4. Because this experiment was prior to the biochemical isolation of the hormones, the effect was achieved with "extracts of the testis" and compared to "extracts of the uterus." Oudshoorn explains that this was typical of the sort of assay that was used for experimental investigation of hormones before biochemists became involved in endocrine research. Prior to identification of the chemical structure of hormones, purified extracts were identified as "male sex hormones" or "female sex hormones" by their potential to induce changes in particular sex-specific tissues or characteristics, which included sex-differentiated structures like the comb of a rooster, as well as reproductive structures (Oudshoorn 1994, esp. chap. 2).

5. Biologists and historians of science generally are in agreement that ideas about "sex hormones" were closely related to more general ideas about the nature of the sexes and sex differences that were prominent during the early years of endocrinology. In particular, notions of separate spheres appropriate to men and women were dominant, and perhaps were even more tightly held for being under attack by feminist campaigns such as the drives for suffrage and access to higher education. As Fausto-Sterling (2000) puts it, "We can understand the emergence of scientific accounts of sex hormones only if we see the scientific and the social as part of an inextricable system of ideas and practices—simultaneously social and scientific" (148). Oudshoorn (1994) draws on a similar idea in her contention that the "sex hormones" were literally *invented,* by which she means *not* that there are no material substances like estradiol or testosterone in the body, but that the concept of sex hormones does not neatly map onto these substances. Instead, the concept of "sex hormones" was drawn from what she calls "pre-scientific" ideas about male and female natures and the essences that control them. Empirical evidence about the actual location and function of the hormones that were early designated as "sex hormones" has always had to contend with these preexisting ideas.

6. The gonads and the genital tubercle are "bipotential," meaning that the same structure can develop into either a male-typical or a female-typical form: the gonad will later develop into either a testis or an ovary, and the genital tubercle will develop into either a clitoris or a penis. Each embryo also has both a male duct system (the Wolffian ducts) and a female duct system (the Müllerian ducts). In typical development, in a given individual one of these systems regresses and the other develops. The Wolffian ducts become glands that support semen production (the seminal vesicles and the prostate) and the duct sys-

tem that stores and moves sperm from the testes through to the urethra (for instance, the epidymis and the vas deferens). The Müllerian ducts become the fallopian tubes, uterus, cervix, and top portion of the vagina.

7. It is also the case that subtle differences within the sex chromosomes can alter the usual course of development, such as, for example, the very rare cases in which XX fetuses develop testes rather than ovaries. This results when the sex-determining sequence of the Y chromosome is transposed to an X chromosome during accidental recombination between the X and Y chromosomes (Ferguson-Smith and Affara 1988).

8. Jennifer Terry's superb study *An American Obsession: Science, Medicine, and Homosexuality in Modern Society* documents several medical "treatments" for lesbianism, including abstinence from all sex (the favored approach of Havelock Ellis), clitoridectomy, and removal of the ovaries (Terry 1999, 80, 104). The last of these might have been understood as indirect hormonal therapy.

9. In the classic paper, Young and his colleagues cite a paper by John and Joan Hampson that was still in press at the time as support for the claim that the new brain organization theory would "direct attention to a possible origin of behavioral differences between the sexes which is ipso facto important for psychologic and psychiatric theory" (Phoenix et al. 1959, 370).

10. This dialogue is well documented in the two edited volumes, *Sex and Internal Secretions* (ed. William C. Young, 1961) and *Sex Research: New Developments* (ed. John Money, 1965c).

11. Other passages, from his dissertation and other articles during the same period, indicate that Money was struggling to articulate a position that prioritized the role of socialization without discounting some contribution of biology (Redick 2004).

12. For a taste of Money's larger-than-life personality and his impact on fellow sexologists, see the festschrift commemorating his seventieth birthday (Coleman 1991).

13. In the Reimer case (Money 1975), it does seem that Money suppressed or ignored pertinent evidence that the gender reassignment was not going well (see especially the detailed account of the child's annual visits to Johns Hopkins Medical Center, and the consultations of his local doctors with Money through Reimer's early adolescence, in Colapinto's *As Nature Made Him* (2001). From later publicity by Milton Diamond (1982) and an update by Diamond and Sigmundson (1997), it was clear that this child was not adjusting well by any measure. It will probably never be clear why Money mishandled this case, as he no doubt did. Nonetheless, this case stands in somewhat sharp contrast with his overall research trajectory, which included a serious devotion to the idea that prenatal hormone exposures shape human gender and sexuality in a substantial, though not determining, way. See, for example, his explanation for what he called "gender coding" in Money 1994, p. 190.

14. See the comprehensive bibliography of Money's works through 1991 (Money 1991).

15. The two forms, distinguished by the degree of insensitivity, are complete androgen insensitivity syndrome (CAIS) and partial androgen insensitivity syndrome (PAIS) (Mazur 2005).

16. In *As Nature Made Him: The Boy Who Was Raised as a Girl,* John Colapinto (2001) presents evidence that the gender change actually happened quite a lot later, with his parents perhaps not fully committing to raise the child as a girl until the child was 22 months old.

17. The ERA was ratified by twenty-two states during the same year that it was passed by the U.S. Congress. But Congress had set a time limit on gaining ratification by the necessary thirty-eight states, which expired when the push for ratification was still three states short (Francis n.d.).

18. Oddly, van den Wijngaard declared that the era of organization theory was "over by 1980," but my own research very clearly shows that it was just getting warmed up at that point. I can't be certain what led van den Wijngaard to this conclusion, but I believe it might be an artifact of her disciplinary perspective as a biologist. Indeed, it does seem that fewer and fewer biologists were interested in organization theory as time went on, but the theory not only has retained its power with psychologists but continues to recruit scientists from many other disciplines, as my survey of studies in Chapters 3 and beyond shows. As the theory was more absorbed into the "background assumptions" of studies, it became harder to trace. Van den Wijngaard and others may therefore assume that its day is past, but both my research and my interviews with scientists confirmed that the theory is stronger than ever at present.

19. Because it is never possible to gather the definitive information on exposures that makes causal inference in case-control research relatively strong, it may be somewhat more accurate to call these studies correlational. However, the important distinction in this context is the fact that investigators attempt to begin with starkly different groups and then make inferences about what has made the people in the two groups different. That is the initial structure of case-control studies. Correlational studies do not require an initial sorting of subjects, but they can explore the covariation of two or more continuous variables among subjects who are not seen as qualitatively different. Thus, I believe the term *case-control* better reflects the conceptual approach, if not exactly the available data, in these studies.

3. Making Sense of Brain Organization Studies

1. Of course, for this to really work, the various hormone regimens that were tested would need to involve a restricted range, all within the broad normative values for the genetic sex of any fetus, so that the rearing and social experiences are similar across groups. If some individuals were given dramatically "cross-sex" hormone exposures, they would be born intersex, which would counter the entire point of blinding. Assuming that smaller differences in hormone exposures would result in smaller behavioral effects, there would need to be particularly big experimental groups.

2. This nucleus, INAH3, is generally smaller in women than in men, and some

studies have also found INAH3 to be smaller in gay men compared to hetero-sexual men (LeVay 1991; Byne et al. 2000). Brizendine's theory would predict that gay men have lower, more "female-type" libidos than heterosexual men, but this suggestion would go completely against behavioral evidence. I will have more to say about INAH3 and sexual orientation in Chapter 7.

3. The only difference in lateralization between males and females that Sommer and colleagues found is that men are more likely to be left-handed; there were no sex differences in functional imaging studies that used language tasks, nor in dichotic listening tasks (meant to reflect the relative accuracy with which the two sides of the brain process language inputs), nor in a key structure for language processing, the planum temporale. Although the discrepancy between handedness the other measures of lateralization is interesting, the review by Sommer and colleagues found some evidence that greater left-handedness among men may be a result of greater social pressure on girls and women to conform to the typical right-handed pattern (2008, 82).

4. This calculation is based on the largest meta-analysis of sex differences in spatial abilities (based on 78 studies), which gives an estimate of .56 for the effect size of sex on mental rotation ability (Voyer, Voyer, and Bryden 1995). Here are some examples of how well you could predict the gender of the test taker for scores in a specific range (assuming that the test has a mean score of 100 and scores are normally distributed): scores of 90–95 would be roughly 56 percent women and 44 percent men, scores of 100–105 would be roughly 48 percent women and 51 percent men, and scores of 140–145 would be roughly 24 percent women and 76 percent men. To get a more stable estimate, this example is based on a sample of ten thousand instead of one thousand—with fewer people, it would be even harder to reliably guess gender.

5. Feminists and critical theorists who are concerned with the real-world implications of science vis-à-vis various political projects (sexist and racist oppressions, ecological exploitation in the developing world, environmental racism, and so on) often maintain not only that "objective" is a poor *description* of science, but that objectivity is not necessarily a good *prescription* for science. Rather than aiming for objectivity, these theorists suggest, we should aim for a more thorough understanding of and accountability for the social factors that are always, and inevitably, present in science (Longino 1990; Harding 1998; Hammonds and Subramaniam 2003). That is, they maintain that it is necessary to become aware of how social and political factors operate in science, not in order to "purge" the social from science (which can't be done, in any case), but in order to create more ethical and humane forms of science. The problem, as Donna Haraway has so powerfully stated it, is "how to have *simultaneously* an account of radical historical contingency for all knowledge claims and knowing subjects, a critical practice for recognizing our own 'semiotic technologies' for making meanings, *and* a no-nonsense commitment to faithful accounts of a 'real' world" (Haraway 1991, 187).

6. The fact that science is not, and can never be, a simple mirror of the world also does not imply that science is simply "made up" and is not constrained by material phenomena that actually exist—the material world "pushes back" and

exerts its own effects in science, even if we accept the postmodern premise that we humans have no hope of a direct access to that world that is unmediated by our own practices and culturally determined cognitive and linguistic structures. There is no need to dogmatically insist (against all evidence) that science *really is objective* in order to believe in science as a good and worthwhile endeavor, and even to believe in science as a particularly useful and trustworthy way of learning about the world (see, for example, Haraway 1991; Rouse 2004). I (and others) believe that insights developed in philosophy and social studies of science about scientific practice and the relationship between scientific practice and the "knowable" world are a critical resource for doing excellent science, as well as for appreciating the limits of our knowledge (Barad 2007). For statements of how social and historical studies of science can and should be a resource for science, see Harding 2006, Conkey 2003, and Fausto-Sterling 2000 (esp. chap. 1); for a few studies of scientific practice that are especially good examples of work that can be a resource for both those who seek to understand science and those who practice it, see Latour and Woolgar 1986, Mol 2002, and Barad 1999.

7. Moreover, while a specific measure may differ from a definition, the question of which definition is "correct" is not something that can be resolved by empirical evidence; instead, correct definitions are matters of agreement. Choosing a definition is generally about finding a way to "slice" the phenomena that occur in the natural world in a way that is useful for various human purposes, and for that reason the same definitions won't seem correct for all purposes. As Michael Maraun has explained it, measurement practice is rule-based and therefore is "autonomous with respect to empirical phenomena." Quoting Wittgenstein, Maraun concurs that "whether a phenomenon is a symptom of rain, experience teaches; what counts as a criterion of rain is a matter of agreement, of our determination" (1998, 440). Still, refining measures is an essential pursuit for scientists, because it is what provides us with consistency in our investigations—without which our causal inferences are not valid and none of our observations can be generalized beyond the very specific circumstances of any particular study.

8. In most of the studies prior to 1980, especially those studies conducted under John Money and/or Anke Ehrhardt at Johns Hopkins, the situation is further complicated by the fact that the psychosexual assessments are based on extensive clinical observation and interview with hormone-exposed subjects. The domains or specific items of interest on which their behavior "might" differ from nonexposed controls was first extrapolated from these already-directed observations. Even though, from a clinical perspective of exploring the patterns and range of behavior among people with a particular condition, this is a potentially very rich research method, it is very poorly suited to an investigation of the etiology of group differences between exposed and unexposed subjects.

9. Even though mounting is almost universally treated as the quintessential male-typical sexual behavior, it has been known since before brain organization theory was proposed that mounting is not the exclusive behavioral province of

males. The fact that normal females mount other animals was one of the reasons William Young's team repeated Vera Dantchakoff's experiments to include control animals (Phoenix et al. 1959). Even more interesting information has emerged recently, namely, that in mice (and by implication, possibly other mammals), mounting is not even sexually dimorphic, meaning that it is not a behavior that is "organized" for permanent behavioral patterns by early hormone exposures (Bodo and Rissman 2007). In other words, in at least some species, mounting behavior depends, not on permanent features of the brain, but only on adult circulating hormones. Experiments on mice showed that both normal females and males who were genetically manipulated to be insensitive to androgen were just as likely as untreated males to show mounting behavior in response to estradiol implantation in adulthood. In fact, on five of the six indicators of "masculine mating behavior" (time from mount to pelvic thrusting, number of mounting episodes with pelvic thrusts, number of "successful" mounting episodes, number of pelvic thrusts per mounting episode, and total number of pelvic thrusts displayed during testing), scores were *higher* in the males with "testicular feminization" than in the wild-type males, and the latter three of these indicators were higher in wild-type females than in wild-type males. The only indicator on which wild-type males scored as the "most masculine" was the latency time between approaching and mounting (Bodo and Rissman 2007, 2184–85).

10. Historians of sexuality face a number of terminological and conceptual difficulties because of these shifts. When one wishes to trace "homosexuality," for example, one quickly finds that prior to the last third of the nineteenth century, there was no term to refer to individuals who erotically preferred people of their own sex. More surprising, perhaps, is that historical work convincingly demonstrates that there was not even a common concept that grouped people by their sexual desires for same-sex or other-sex partners; instead, well into the end of the nineteenth century, behaviors, rather than people, were classified (Foucault 1978), and these on the basis of either facilitating reproduction (viewed as "natural" sexuality) or not (viewed as "unnatural" and/or "sinful") (Katz 1995). Thus, it is not just the categories "homosexual" and "heterosexual" that are new; also new is the concept of "sexual orientation" that specifically refers to an enduring, consistent preference for partners of one sex.

11. The late sociologist of science Dorothy Nelkin often emphasized the value of examining scientific controversies for understanding some of the interplay between science and society (for instance, Nelkin 1992). I was working with Professor Nelkin when I first began investigating research on brain organization theory, so I was fascinated by the *lack* of overt controversy among scientists doing that work, especially because I came from public health research where the measurement of sexuality-related variables was hotly debated. As I eventually discovered, the controversies were indeed there, but they were "buried" by the fact that everyone used the same terms and simply assumed that they meant by them what others meant. Examining measures is a way to find this sort of "hidden" disagreement in scientific work. This sort of buried

or ignored disagreement is akin to what Annemarie Mol (2002) calls "distribution" in her study of atherosclerosis. There, she closely examines different practices associated with diagnosing, treating, and preventing "atherosclerosis," thereby revealing how the "object of engagement" for these practices, while going under the common name *atherosclerosis,* is actually a slightly different entity in each case. For example, in the pathology lab, atherosclerosis is a thickening of the lumen in the arteries, but in the clinic it is (among other things) pain in walking. Scientists and laypeople generally assume that these are different "aspects" or levels of the same unified phenomenon, but Mol shows that this is not quite the case—a person may have no pain upon walking and have severe thickening of the lumen, for example. One of the means of reconciling this kind of discontinuity in the ontology of atherosclerosis (what it *is*) is to "distribute" practices to different scientific-medical locations (the clinic, the surgery room, the pathology lab, and so on).

4. Thirteen Ways of Looking at Brain Organization

1. It's worth noting that, despite the fact that comparing hormone-exposed subjects to their unaffected relatives is meant to control for family differences in rearing and other social factors, the exposed and unexposed groups that are compared are often not from the same set of families. This is because the conditions are rare, and the comparisons for these studies are virtually always "within-sex" comparisons, meaning that unexposed females are compared with exposed females, and unexposed males with exposed males. In order to have more statistical power in the analyses, investigators increasingly "pool" the unexposed female relatives of both the exposed males and the exposed females, and compare this pooled group to the exposed females. The same procedure is followed for pooling the unexposed male relatives. The other, in my opinion larger, problem with the approach is the initial assumption that using sibling comparisons is an adequate way to control for social factors, especially when comparing intersex to non-intersex siblings.

2. Two partial exceptions bear mention. In one study, the authors concluded that "the postnatal social system outweighed prenatal hormonal influences" for girls with CAH (Hochberg, Gardos, and Benderly 1987, 498). However, this observation is fairly tangential to the overall report, the main point of which was an emphasis on the difficulties encountered by genetic females with CAH who are assigned and reared as boys (for instance, they are infertile and have very small penises, which interferes with heterosexual relationships). The other report, by Herdt and Davidson (1988), concerned genetic males with the condition 5-alpha reductase deficiency. As I describe below, that report should be read as a rebuttal of the interpretations of the same syndrome published in 1979 by Julianna Imperato-McGinley and colleagues. Herdt and Davidson were more concerned with pointing out the many ways in which individuals with this syndrome were not (as Imperato-McGinley and colleagues had suggested) either reared unambiguously as girls or fully living as men as adults. While the report should be read as a caution against overreaching with brain

organization theory, given the ambiguities in the social context, Herdt and Davidson do not attempt to draw conclusions about the possible organizing effects of hormones.

3. Herdt and Davidson were an especially unusual interdisciplinary team. Their work is all the more interesting when one realizes that one of Davidson's mentors was Frank Beach, one of the most important contributors to psychological research in the twentieth century, especially noted for his animal studies on endocrine–behavior links. In her excellent history of mid-twentieth-century hormone research, Marianne van den Wijngaard (1997) relays that Beach was one of the most important opponents of brain organization theory in its early years. Beach suggested that the role of genital sensation and function was too critical to dismiss, and he felt brain organization theory did dismiss it. With Beach as one of his mentors, it is not surprising that Davidson went on to combine his training in neuroendocrinology and behavioral endocrinology with a focus on social and cross-cultural aspects of sexuality (Morrissette and Myers 2002). One scientist I interviewed suggested that Herdt and Davidson were outsiders to a "scientific" approach to sexuality, and indeed, counted them among people who "don't even believe that hormones exist" (Dr. J interview, September 16, 1998). This is patently false, of course, but this belief may be one of the reasons Herdt and Davidson's research is so infrequently cited in the neuroendocrine literature, in contrast to Imperato-McGinley's work.

4. Technically CAH is not one disorder, but a family of autosomal recessive disorders that impair the ability of the adrenal glands to synthesize cortisol, a steroid hormone that is important for a wide range of metabolic activities, tissue development, and immune function. Most cases of CAH are caused by a deficiency in the 21-hydroxylase enzyme, which causes overproduction of androgen precursors from the adrenals, beginning around the same time as the bipotential genitals are undergoing sex differentiation. Overproduction of androgens continues throughout gestation and at least until diagnosis and treatment with corticosteroids. According to White and Speiser (2000), women with classic CAH continue to produce increased levels of androgens from the ovaries, even when the CAH is well controlled. An excellent, comprehensive review of the etiology and clinical features of CAH can be found in White and Speiser (2000). My summary of clinical features and treatment of CAH is based also on Therrell et al. 1998, Stikkelbroeck, Hermus, et al. 2003, Grosse and van Vliet 2007, and the brain organization studies described in this section.

5. Dessens and colleagues (2005) identified reports on 283 individual genetic females with CAH, including 250 reared as female and 33 reared as male. In addition to the 33 reared as male, 32 of those reared as female were initially assigned as male, and later reassigned ((33 + 32) / 283 = 22.9 percent).

6. The mixed predictions for males are interesting, given that they, too, have androgen levels that are significantly above the norm, as reflected by such somatic problems as premature maturation (pubic hair and body odor in very young children, for instance) (White and Speiser 2000). But an increasing number of studies suggest a nonlinear relationship between androgens and

various cognitive and personality-related variables, such that the most "male-typical" patterns might be associated, not with extremely high androgens, but with levels in the mid-range or low-normal range (Kimura 1999).

7. A recent meta-analysis suggests that there may be, in fact, a modest increase in spatial abilities among women with CAH, but that small sample sizes have obscured this difference (Puts et al. 2008). The meta-analysis has a number of problems. First, there are too many different types of spatial tests included in the analysis; spatial visualization, spatial perception, and mental rotations are three distinct components of spatial ability that should not be confounded, especially because earlier studies have shown very different results for CAH versus unaffected subjects on different kinds of tests (see, for example, table 1 in Hines, Fane, et al. 2003, 1013). Second, the analysis inexplicably omitted one large study (Johannsen et al. 2006) that found significant impairment in spatial abilities. Third, they included only a portion of the samples from one study (Helleday, Bartfai, et al. 1994), in a way that biases the results toward the brain organization hypothesis. Helleday compared 22 women with CAH to 22 unaffected women, and women with CAH scored lower on each of sixteen cognitive tests, including general intelligence and specific subtests (verbal ability, arithmetic, logic, visuospatial, visuomotor, and field dependence); five of these sixteen differences were significant. To determine whether women with CAH have a more "masculine" cognitive profile of superior spatial abilities relative to verbal abilities, Helleday and colleagues created groups that were matched (within one standard deviation) for general intelligence. Because of the significantly lower scores of CAH women, they were able to find a match for only 13 out of the 22 (about 60 percent). Thus, the matched comparisons, which were done for the specific purpose of calculating discrepancy scores between different abilities among subjects of roughly equal overall intelligence, actually involved comparing the *top-scoring* 60 percent of CAH women with the *bottom-scoring* 60 percent of the control women. This is fine for examining discrepancy scores, but it's not appropriate to include these skewed groups in a meta-analysis examining spatial ability in women with CAH compared with unaffected women.

8. Helleday and colleagues (1993) interpret their findings as showing CAH women to be less "female-typical" in this respect, because they were more likely to score low on a scale of indirect aggression, on which women typically score slightly higher than men. However, CAH women were compared to nonaffected women on an entire aggression-hostility inventory, consisting of five separate subscales, each of which has five items. The only differences were found on two items of the "indirect aggression" subscale. Given that this amounts to two significant differences for 25 comparisons, and given that one-tailed tests were used, this cannot properly be considered a significant finding.

9. Scientific debate following the revelation of Reimer's self-reassignment has been intense and sometimes acrimonious, and the public attention given to this case has been enormous. For example, the PBS series *NOVA* based an episode on this case. Titled "Sex: Unknown," the segment aired on October 30,

2001. Oprah Winfrey also devoted a show to Reimer's case in February 2000 (www.oprah.com/tows/pastshows/tows_2000/tows_past_20000209_b.jhtml). An indication of the volume of media attention after the book was first released can be found in the afterword of Colapinto 2001.

10. See especially the interviews with Roger Gorski and Dick Swaab in the *NOVA* documentary "Sex: Unknown" (2001). Also note that the book description and discussion guide on HarperAcademic.com, the site that currently markets Colapinto's biography of Reimer for academic use, implicitly invokes brain organization theory in describing why the case was initially important: "Dr. Money had been asserting that an individual's gender identity is determined not by the hormones that bath[e] a developing fetus' brain and nervous system, but by socialization. David Reimer's tragedy presented him with the perfect opportunity to prove his theory." As with so many accounts, beginning with Colapinto's own work (1997, 2001), this seriously misstates Money's position vis-à-vis brain organization theory: as his many studies cited in this book demonstrate, Money absolutely believed that hormones play a role in gender and sexuality, albeit a more subtle one than many other scientists and laypeople, most notably Milton Diamond (for instance, Diamond 1965, 1974, and 1998; Diamond and Sigmundson 1997).

11. One of the many problems with coverage of this case is the use of the term *experimentation*. In one sense the treatment plan was experimental, in that there is no earlier recorded case of a child born as a normal, healthy male being reassigned to be reared as female for medical reasons. But the treatment course did not constitute an experiment in the usual sense of the word, where the term generally refers to a course of action primarily followed to gain information. The connotation of "experimenting" on this child distorts the motive and intentions of the many actors involved in this case, especially John Money. Although Money seems to have handled this case badly in several respects, the historical record clearly indicates that he was following the course of treatment that he—as well as most other experts in gender development at the time—thought was best for the child's psychological health. See, for example, the comments from psychiatrist Richard Green in the *NOVA* special "Sex: Unknown," or as quoted in Money's *New York Times* obituary (Carey 2006).

12. Recently clinicians and researchers have begun to pay closer attention to the functional outcomes of constructed or reconstructed vaginas in intersex individuals reared female, and the conventional wisdom that good to excellent results are the norm in these cases has been badly shaken. In fact, the construction of "functional" vaginas has often focused only on the capacity of a vagina to "receive a penis" while utterly neglecting sensations, especially pleasure. In a follow-up study of women with CAH, for example, the questions about the "success" of vaginal functioning were limited to these: "Is the vaginal opening adequate for sexual intercourse? Do you have pain or discomfort during intercourse?" (Mulaikal, Migeon, and Rock 1987, 179). No questions were asked about pleasure, lubrication, or plasticity, among other aspects of vaginal function that might be of interest. Researchers have tended to be much more exacting for qualifying a penis as functional (it must have a urethral opening that

permits urination standing up, it must be capable of erections and ejaculation, it must be large enough to permit vaginal intercourse). For more on this, see Kessler 1998 and Karkazis 2008, as well as Chapter 9 of this book.

13. To my knowledge, only one investigator (Slijper 1984) has attempted to account for the effects of the serious illnesses that affect children with CAH. Slijper is also unusual for having referred in her research report on CAH girls to the empirical evidence for differences in parental behavior toward girls versus boys (417), which is critical for understanding how rearing experiences differ for children whose sex has been questioned or reannounced following birth with ambiguous or cross-sex genitalia. As Ruth Bleier (1984) in particular has pointed out, people whose sex has been seriously doubted by medical professionals and families have invariably gone through very unusual rearing experiences. These include parental anxiety over sex reassignment, multiple genital surgeries, and frequent and intensive interactions with medical and psychological professionals. Perhaps most importantly, it's not clear how any subtle, lingering doubts about whether the child is sexually "normal" might affect rearing. Brain organization researchers typically have suggested that the only issue of real interest in this regard is whether parents have been "consistent" in their expectations and reward of gender-typical behavior (see Money and Ehrhardt 1972, but compare Ehrhardt and Baker 1974 for a slightly more complex view). Bernice Hausman's (2000) analysis of the David Reimer (John/Joan) case suggests a more dynamic interplay between rearing experiences and an individual's developing personality, namely, that rebellion as well as conformity might result from heavy-handed gender expectations. This underscores the scientific principle that the only way to be certain that a variable (like unusual rearing) has not affected an outcome (like gender or sexuality patterns) is to either randomly assign the exposure or strictly control for the variable in analysis, neither of which is possible in these human studies of brain organization.

14. Though DES does not cause intersex development, it does affect the development of genitals and the full reproductive tract in a large subset, possibly a majority, of people who were prenatally exposed. Common effects include reproductive tract lesions, as well as smaller vaginal opening and less plasticity of the vagina (Kaufman et al. 1984; Goldberg and Falcone 1999). In large doses, comparable to some of the higher dosing regimens given to pregnant women, it has also been shown to induce brain lesions in laboratory animals. Thus, using DES as a straightforward indicator of sexual differentiation of the brain is questionable, because of the broad spectrum of developmental disruptions that the drug can cause.

15. The experiments suggested that in rats there are two different pathways of brain organization: an androgen pathway and an estrogen pathway. Testosterone *masculinizes* behavior, while estrogen *masculinizes and defeminizes* behavior. But of course, unless aromatization is blocked so that testosterone can't be converted to estrogen, testosterone exposures will work through both pathways. This new model entailed thinking about masculinity and feminin-

ity as separate sets of traits and behaviors. Traditionally psychologists have treated masculinity and femininity as opposite poles on the same scale, so that an increase in masculinity would always result in a decrease in femininity. In the 1970s psychologist Sandra Bem suggested that being more feminine doesn't necessarily mean being less masculine, and vice versa (Bem 1974). Instead of being either masculine or feminine, it is possible to be both, or neither. This same general principle illustrates how McEwen's and Gorski's teams were thinking about sex-related traits and behavior in rats. The distinction they made between "masculinizing" and "defeminizing" hormone exposures followed this principle: *Masculinization* refers to inducing the development of so-called male-typical traits (in rodents, these are mounting, aggression, and so on) without necessarily inhibiting so-called female-typical traits (such as lordosis, cyclic hormone release). *Defeminization* refers to blocking the development of female-typical traits. I take up this "two-dimensional" model of sexual differentiation in later chapters, as it is important for understanding the specific predictions of hormone effects for different studies.

16. This study is a good indication of how "hormone folklore" sometimes competes with the actual prediction of brain organization theory when scientists conceive and interpret their studies. The folk notion of dichotomous, masculine and feminine properties associated directly with androgens and estrogens, respectively, is particularly apparent in this study. While I have never encountered this observation in the literature, there was actually *no reason* in 1973 to expect that estrogens would have a feminizing effect. The animal data and the formal theory of brain organization were, at that time, simply about androgens, and the hypothesis was that feminization occurred by *default*, not by the action of estrogens. Yalom's study hypothesis was based on the idea that estrogens may have an antiandrogenic effect. In other words, the hypothesis was an extension of brain organization theory based on the long-defunct idea that androgens and estrogens are antagonistic compounds, which had not been credible since the 1930s. Yet the study seemed, and continues to seem, a reasonable test of brain organization theory for the scientists who conceived it and the many others who have cited it. This is the earliest example, but a number of other studies that fit neatly with folk thinking about hormones have been published and subsequently cited as evidence of the organizing effects of hormones, even though they do not actually fit the theory (for instance, Ellis and Ames 1987; Rosenstein and Bigler 1987; Lindesay 1987; Holtzen 1994).

17. There are a few studies that concern other hormones used in the mid-twentieth century to treat "problem pregnancies," especially natural and synthetic progestins (for example, Reinisch 1981; Ehrhardt and Money 1967; Ehrhardt, Grisanti, and Meyer-Bahlburg 1977; Kester 1984), but as a recent review noted, these studies are quite small and involve a great variety of specific hormonal compounds, among other methodological problems (Cohen-Bendahan, van de Beek, and Berenbaum 2005, 361). They are less frequently mentioned in current reviews of the evidence for brain organization in hu-

mans, and I have omitted them from this summary for space considerations. I do review them in Chapter 8, on sex-typed interests, as this is the functional area where these studies are most often still considered informative.

18. For example, testosterone levels are known to vary greatly between male and female fetuses. Yet Hines, Golombok, et al. (2002) did not find significant differences in testosterone levels from the maternal blood of women carrying male versus female fetuses, which is one indicator that hormones in maternal blood may not be a good reflection of fetal hormone exposures. Moreover, it is commonly understood that testosterone levels fluctuate significantly over the course of gestation for males, though this might not be the case in females. Yet in the prominent studies by Simon Baron-Cohen's group, amniotic testosterone correlated with gestational age in females, but not in males (Baron-Cohen, Lutchmaya, and Knickmeyer 2004).

19. *Fixing Sex: Intersex, Medical Authority, and Lived Experience,* by medical anthropologist Katrina Karkazis (2008), is illuminating on this point. In interviews with intersex adults, as well as the parents of intersex children (some of whom are now adults), Karkazis found that parents of girls with CAH in particular were routinely counseled to *expect* behavioral masculinization. Many parents report that they were told that the girls would probably be tomboys, but not lesbians. Dismissing this as though it could have no effect on parent–child interactions is a profound shortcoming of the many studies of this syndrome. I address this in Chapter 9, when I turn in detail to issues of context that are neglected in the studies.

20. While it is conventional in such papers to refer to the results as reflecting "fetal hormone levels," note that actual fetal hormone levels have not been measured in these studies. Accurate language is more cumbersome ("children for whom amniotic fluid measures of fetal testosterone were higher" compared to "children with higher fetal testosterone," and so on), but the simpler language conveys more precision than these studies entail. I thank an anonymous reviewer for raising this point.

21. The investigators argue that testosterone may affect different aspects of "male-typical" spatial relations differently, such that some (such as block building) are somehow stimulated by lower testosterone levels in girls, while others (such as mental rotation speed) are stimulated by higher levels. This seems to stretch their data much too far, especially given that the latter finding was apparent only among a subset of the girls. It also doesn't address the very interesting point that among this subset of children who use mental rotation, the usual sex-difference in speed is reversed, with girls performing significantly faster than boys. All of this hints at a conclusion I will explore at length in Chapter 10: attempting to understand biological influences on cognition through the prism of sex differences forces onto the data not just a simplification, but ultimately a crippling distortion. It seems quite likely that we must find better models, perhaps along the lines of the neural "mosaicism" proposed by Sandra Witelson (1991), but again, less shaped by sex than that model.

22. Udry and colleagues proposed that adult as well as prenatal androgens (testos-

terone and androstenedione) would predict the gendered behaviors of women, with prenatal levels "conditioning" the effect of adult levels. In other words, Udry argues that women with higher prenatal androgen levels are more sensitive to androgens in adulthood, so that the most masculine behavior will be seen among women who had both high prenatal exposures *and* higher adult androgen levels. Using SHBG as the indicator of prenatal androgens, Udry did find the predicted relationships.

23. Hines also suggests that the effect in girls but not boys may be explained by the fact that boys have much higher testosterone levels during fetal development, so normal variability in girls may be more meaningful, given the backdrop of relatively lower testosterone levels. In this regard, it is worth noting the findings of Bolton and colleagues (1989) regarding the relative amounts of testosterone and SHBG in male versus female infants' venous cord blood (which, in contrast to the brain organization studies reported here, is linked to fetal, rather than maternal, circulation and should therefore be more relevant for prenatal brain organization). According to Bolton's research, levels of SHBG were significantly higher in the sera of males versus females, rendering the "free" or active testosterone levels of male infants quite low. Bolton and colleagues suggest that "the effect may be that the male foetus in particular is protected from the androgenizing effects of circulating T, if the widely held hypothesis is accepted that only the free T in the circulation is biologically active. It is possible that the androgenizing effects of T are undesirable at this stage of male development" (205).

24. Within specific domains like aggression or spatial relations, for example, there are finer-grained distinctions that must be made in order to correctly specify a pattern of sex difference in that behavior. In spatial relations, the largest and most consistent differences have to do with mental rotations, which tend to favor males. Girls, though, tend to excel at object location memory (Voyer, Voyer, and Bryden 1995; Postma, Izendoorn, and De Haan 1998; Voyer et al. 2007). Likewise, some aspects of aggression (such as indirect aggression) tend to be higher among females, while physical aggression tends to be higher among males (Osterman, Bjorkqvist, and Lagerspetz 1998).

25. If you roll a die 3 times, there are 216 ways it can come out (6 × 6 × 6). The probability of getting at least one 5 is 103/216 = 47.7 percent. My thanks to Jeanne Stellman for pointing this out.

26. Grouping the findings among these studies as pertaining to "sex-typed interests" is the most generous way to perceive congruence across the studies, and this seems warranted. Items in this domain are labeled as "gender role" in the studies of both Hines, Golombok, et al. (2002) and Udry, Morris, and Kovenock (1995), but examining the items in the various studies suggests that an underlying "interest" domain is a reasonable way to group them.

5. Working Backward from "Distinct" Groups

1. Dörner had introduced the theory behind these studies quite a few years earlier and may have reported the studies before 1975 in German-language publi-

cations. However, 1975 was the first year a research report on the LH surge in human subjects was introduced to the English-language literature.

2. In fact, Dörner had discussed homosexuality in rats at least as early as 1968, when he reported the results of neonatal castration of male rats followed by "androgen substitution in adulthood" (Dörner and Hinz 1968). His direct application of rat findings to questions of human homosexuality is part of the legacy that he seems to have assumed indirectly from Eugen Steinach. Dörner traces much of his work to Steinach and is quite proud of the fact that his teacher Walter Hohlweg worked with Steinach for many years. For more discussion of the relevance of this connection, see Dörner et al. (1991), Sengoopta (2006, 319 n. 282), and the discussion of Dörner's use of animal models in Chapters 6 and 7.

3. It's worth noticing that this is the only place in LeVay's book that he entertains the notion of retrospective recall bias. The many other studies that support his reasoning about the etiology of homosexuality (which he attributes largely to prenatal hormones) escape without such close scrutiny.

4. This study is one of the reasons Melissa Hines should be considered one of the most careful and thoughtful researchers doing brain organization studies. She notes that the relationship between prenatal stress and gender-related behavior is smaller than that between gendered behavior and at least nine other variables, and expresses skepticism about the ultimate meaning of the prenatal stress effect. Note that in the quoted passage, she indicates that "prenatal stress appears to have little, *if any,* effect on gender role behavior in girls" (Hines, Johnston, et al. 2002, 126, emphasis added). Too many investigators cling to even very weak findings, looking only for ways their design may have underestimated the effect of hormones. Even the fact that Hines included a wide range of other variables in the study sets her apart from most others, who typically adopt an approach that severely under-assesses various other variables (especially, but not only, including social experiences) that may affect sex-typed outcomes. When these factors do not explain the variance in gender-related behavior, the researchers are tempted to attribute all of the "residual" influence to prenatal hormones. I explain why this is a problem in Chapter 9.

5. At least one very influential scientist who continues to believe that prenatal stress plays a role in sexual orientation seems to miss this discrepancy in timing. Citing the two studies conducted by Lee Ellis and colleagues, the Dutch neuroscientist Dick Swaab has written that "maternal stress is thought to lead to increased occurrence of homosexuality in boys, particularly *when the stress occurs during the first trimester*" (Swaab 2004, 308, emphasis added). As noted above, the initial study by Ellis suggested that the effect of stress was significant for male sexual orientation only when it occurred during the second trimester; first-trimester stress was not significant in his analysis. The lack of significant effect in the first trimester in that study is particularly striking given that Ellis used one-tailed tests and extremely generous levels for the p values to designate findings as "significant" (in some cases mentioning findings as significant at the level of $p < .12$, one-tailed).

6. Geschwind and Behan originally suggested an association between abnormally *high* levels of testosterone and a range of other traits, including dyslexia, migraine, immune dysfunction, and laterality (Geschwind and Behan 1982). When Geschwind and Galaburda extended these associations to include male homosexuality, they used fairly complicated reasoning to suggest a scenario whereby they would expect associations among all of these factors, even though homosexuality in men would be caused by low androgens and the others would be caused by high androgens (Geschwind and Galaburda 1985b). If immune dysfunction, handedness, dyslexia, migraine, and other factors developed during an earlier period (when testosterone levels were high), but sexual orientation developed during a later period (when testosterone levels had dropped), then maternal stress during early pregnancy could account for a correlation between sexual orientation and these other factors. The correlations they predicted, though, haven't been empirically demonstrated (Berenbaum and Denburg 1995).

7. Some studies (such as Putz et al. 2004; Rahman 2005) begin to suggest constraints, but they are neither sufficiently comprehensive nor precise. For example, Putz et al. (2004) design their analysis to reconcile a gap between brain organization theory and empirical evidence to date. According to the theory, various sex-typed traits (including sexual orientation, 2D:4D ratios, spatial abilities, and so on) all differentiate sexually under the influence of early testosterone exposures; yet the traits don't correlate as predicted. Putz and colleagues suggest that the overall pattern of correlations to date indicates that 2D:4D differentiates at the same time as sexual orientation, but not at the same time as other traits that have been examined. Yet even if one accepts their selective review (which is unacceptably slanted, omitting several studies that offer contrary evidence in gay men), they fail to address the fact that 2D:4D is, according to their review, masculinized in gay men and lesbians, while sexual orientation is supposedly feminized in gay men but masculinized in lesbians. How the two traits could differentiate at the same time, under the influence of the same factor, but in opposite directions, is a serious gap in their argument (especially because, as noted elsewhere in this chapter, the prenatal stress hypothesis has not been supported).

8. Lalumiere, Blanchard, and Zucker (2000) take the nice step of acknowledging that there is a bias toward publishing positive results. There is, though, the additional problem that the bias may operate before publication, at the point of data collection or analysis. One obvious problem is that researchers use noncomparable sampling frames for the "heterosexual" versus "homosexual" samples, drawing the former primarily from student and other university-based samples, and the latter primarily from social events and organizations. Thus, differences related to the sampling frames (such as being or not being a university student) may get conflated with being heterosexual vs. homosexual. The size of the "homosexual" versus "heterosexual" samples in relation to the populations from which they're drawn (the homosexual samples constitute a much smaller fraction of the "eligible" populations from which they're drawn) also suggests that the homosexual groups are more selective respond-

ers; this in turn suggests that left-handers, or "non-consistent-right-handers," in the parlance of some researchers, may be more likely to participate.

9. In another small study reported at the 1991 meeting of the International Academy of Sex Research, Tkachuk and Zucker also reported greater left-handedness among lesbians, but that result was never published in a peer-reviewed journal. It has been reported in a number of reviews on the relationship between lateralization and sexual orientation (such as Lalumiere, Blanchard, and Zucker 2000). I do not include it here because I am restricting this analysis to published studies, which have undergone a higher level of methodological scrutiny than conference proceedings.

10. Putz et al. (2004) reported a link between sexual orientation and 2D:4D, but several factors make those findings less than impressive. First, as they note, they found only two significant findings in the predicted direction out of 57 comparisons in that study, and they did not use any statistical corrections for multiple comparisons. Second, the comparisons for sexual orientation are probably among the most unstable in the study, because very few respondents were classified as not heterosexual (3 of 120 women, and 6 of 119 men; D. Putz, personal communication, January 11, 2010).

11. Currently there is no evidence in humans that prenatal hormone exposures do result in the sorts of structural differences that have been observed in other species, particularly in the rat; therefore this design is particularly dependent on animal research for its causal claims. It's also important to remember that hormone exposures, during development or otherwise, are not the only plausible cause for structural–functional correlations between the brain and behavior. For example, correlations between brain structure and particular traits or behaviors could arise by way of a direct genetic influence, without the intervention of hormones. And studies have shown that experience is sometimes directly responsible for differences in the number or density of cells in a particular region. Nonetheless, neuroanatomical investigators themselves typically propose the organizing effect of early hormones as the etiological theory for the differences they explore. This is corroborated in review articles in neuroanatomy, the consistent inclusion of psychoendocrine studies of brain organization in the background literature for neuroanatomical studies, and comments made by scientists during the interviews.

12. In the interviews, the idea that LeVay "scooped" Allen was raised by LeVay, Allen, and Allen's mentor, Roger Gorski. LeVay was almost apologetic about it, Allen was matter-of-fact, and Gorski was a bit rueful, suggesting that it might have caused Allen to miss her "big break" scientifically. I raise it here just because it helps underscore how finding a difference in male versus female brains, and/or gay versus straight brains, is perceived as a "coup" by the scientists themselves. The findings translate into big publicity, which in turn can increase scientific citations, funding, and prestige. This is just one of the many reasons scientists work so diligently to find such differences but may be somewhat less creative in thinking about how the differences they *do* find might be artifactual rather than real.

13. They also noted the interesting finding that the one female-to-male transsex-

ual had a neuron number "in the male range," but because the range of neuron numbers for males and females in the study overlaps, it is not clear what to make of this.

14. Still, the usual criticism—the fact that the subjects who did not die of AIDS were simply presumed to be heterosexual because there was no information on their sexuality—misses its mark. It may be annoying and imprecise to simply assume that people are heterosexual unless proven otherwise, but bad classification confined to the "heterosexual" group would actually decrease, not increase, the chance of finding a difference between the two groups. (This is because the comparison would then be between "homosexual" men and a group of men with mixed sexual orientations, so there would have to be a very large "real" difference between gay and straight men for it to show up in this messy comparison.) The more serious problem is that *both* the homosexual and the heterosexual groups potentially include subjects of uncertain sexual orientation.

6. Masculine and Feminine Sexuality

1. Historians of sexuality (Cook 2004; D'Emilio and Freedman 1988) have observed that this is a general pattern in how sexual ideals and practices have changed over the past two centuries, at least in the United States and United Kingdom. Interestingly, it seems to be true of certain other patterns of change in gendered behavior, as well. In Chapter 8, for example, I describe how some gendered aspects of children's play have shifted in the past few decades, and note that the changes also generally involve girls taking up toys and games that were previously considered appropriate only for boys (such as construction and science toys), without any evidence of boys significantly adopting so-called girls' games and toys. See Blakemore and Centers (2005) for some recent empirical evidence on this.

2. The article has been cited 99 times overall, with 21 of these citations occurring during the years 1999 through 2008, according to a June 20, 2008, search of the ISI Web of Science database.

3. Of course, this was published in 1968, so most scientists writing at the time would refer to an abstract "partner" with male pronouns by default. Note, though, that the rest of the article eliminates any doubt about the intention of that possessive pronoun "his." They state, for example, that "ordinary women report more often dependency on touch and caressing by husband and boyfriend for such a sexual response" (Ehrhardt, Evers, and Money 1968, 121).

4. The habit of excluding larger-scale data on sexual norms and behavior in the population has mostly continued in brain organization research. In her later studies, Ehrhardt has consistently compared a hormone-exposed group to a group with presumably normal hormone exposures: unaffected siblings or other relatives, children from the same school or community, or patients from the same hospitals and clinics whose diagnoses and records suggest that their hormone exposures were normal. Most other investigators use very similar

strategies. Money, on the other hand, preferred a strategy of contrasting different clinical syndromes. Money once explained "the clinical research method, long practised in the Johns Hopkins Psychohormonal Research Unit, [consists] of matching two different syndromes on the criterion that each is admirably suited to be the reciprocal of the other as a control, comparison, or contrast group" (Money and Lewis 1982, 339).

5. This creates a particular problem that I call a "floating norm"—because investigators generally ignore available data about population norms, they take the characteristics of any given "unexposed" comparison group as "normal." Yet a comparison group isn't necessarily representative of "normal," especially because the comparison groups are often even smaller than the groups of women with unusual hormone exposures. For example, Kenneth Zucker and colleagues (1996) studied sexuality in women with CAH and found more same-sex interest and behavior in the CAH women than in the comparison group. But the CAH-affected women still have a *lower* rate of lesbianism than is found in any current population sample—the control group was somewhat unusual in that there was no hint of any same-sex eroticism among them. Likewise, if you map out the findings for various studies to compare percentages of girls in hormone-exposed groups versus control groups who endorse such items as being a tomboy, having high energy expenditure, preferring "functional/androgynous" clothing, wanting to get married, and so on, you'll see that the "norm" in one sample is often at the same level as the "masculinized" in another sample. Compare, for example, Ehrhardt, Grisanti, and Meyer-Bahlburg 1977 with Ehrhardt and Money 1967 and Ehrhardt, Epstein, and Money 1968.

6. The powerful influence of John Money is at least partly responsible for the uniformity of definitions in these studies. Money was the author or co-author of eight reports on early hormone exposures and human sexuality (Ehrhardt and Money 1967; Money, Ehrhardt, and Masica 1968; Ehrhardt, Epstein, and Money 1968; Ehrhardt, Evers, and Money 1968; Money and Alexander 1969; Masica, Money, and Ehrhardt 1971; Money and Ogunro 1974; Money and Daléry 1976); four more were authored by younger investigators who had worked directly with Money: Anke Ehrhardt (Ehrhardt, Grisanti, and Meyer-Bahlburg 1977; Meyer-Bahlburg, Grisanti, and Ehrhardt 1977) and Richard Green (Yalom, Green, and Fisk 1973; Kester et al. 1980). It can be difficult to attribute individual authorship in team-written papers, but certain key phrases found in these articles were definitely penned by Money (see, for instance, Money 1965b, esp. p. 9).

7. In 1977 June Reinisch published an analysis of the effects of DES versus progesterone on personality (with no sexuality-related outcomes) and reported that the progesterone-exposed subjects were "more independent, sensitive, individualistic, self-assured, and self-sufficient" while the estrogen-exposed subjects were "more group-oriented and group-dependent" (562). This fit with the pre-aromatization, folk-derived hypothesis that estrogens would be antagonistic to or opposite in effect from androgens. Progestogens, on the other hand, had already been noted to masculinize genitalia and were therefore hy-

pothesized to masculinize the brain as well. Reinisch didn't label the groups as "masculine" or "feminine," but the implications were quite clear, given her conclusion that these findings accorded with "the investigations with animals and extends them solidly into the human realm" as well as the fact that the personality factors she studied were linked with gendered patterns in school achievement and success. As will become even clearer in Chapter 8, the reversal of hypotheses and findings in terms of the effects of estrogen on brain differentiation has never been properly accounted for, but it is devastating for any claims that DES has masculinizing *or* feminizing effects on human behavior.

8. Curiously, Kester and colleagues (1980) indicated that they began with the hypothesis that all the different "female hormone" regimens they studied would be feminizing, signaling that when they began their study, this team was not yet familiar with the animal data showing that much of the masculinization and/or defeminization from androgen exposure is accomplished through aromatization of androgen to estrogen.

9. Money and Ogunro (1974) do not explicitly use the term *conservative* to denote the feminized or nonmasculinized sexuality of the patients with AIS described in this report, but conservatism is nonetheless implied. See, for example, the discussion on page 194 of the patients' rejection of prosthetic penises as "an adjunctive coital aid," as well as their "negative evaluation" of oral sex.

10. It is possible, of course, to read echoes of Freud in this assessment. Regardless of the origin of the idea, the notion that sexuality is fundamentally about penises—who has them, who doesn't, where they go, and so on—is a spectacularly consistent theme in sexuality research and theory. Van den Wijngaard (1997) astutely observed that both critics and proponents of brain organization theory were unified in a focus on masculinity and male development: "In 1971, the ideas of Beach and most other researchers accorded with Freud's philosophy, which considers the male sexual organ to be the origin of sexuality" (54). I think we may say that a good many researchers, now as well as then, at least implicitly *view sexuality itself as a "masculine" attribute*.

11. Kester and colleagues (1980) imply in their narrative (280) and in table XI (277) that subjects exposed to synthetic progesterone began masturbation later, and masturbate less often, than the other subjects. However, as noted above, the statistical approach in this study was unacceptable, so the finding is not meaningful. In addition to the fact that dozens of independent variables were included, and the cell sizes of the multiple subject and control groups were quite small (ranging from 10 to 22), one-tailed statistical tests were used and the p level was set at .10. This means that even if there had been an appropriately limited number of comparisons, there would already be a 20 percent chance of falsely finding a significant result for any given variable.

12. Though the hypotheses about precocity were usually oblique, one of the most direct statements of the expectation that "masculine" psychosexual development might entail erotic precocity can be read in Money and Ehrhardt's (1972) description of androgens and childhood sexuality, in which they inter-

sperse a discussion about androgenized girls with observations about sexual play in nonhuman primates. Specifically, Money and Ehrhardt noted that among juvenile rhesus monkeys, normal males and androgenized females (but not normal females) "progress" from an ambiguous mounting stance to one that more closely resembles adult male sexual behavior (100). Immediately after this description of juvenile rhesus monkeys, Money and Ehrhardt noted that there was not "a greater degree of verbal curiosity or interest" in sex among androgenized human girls (101). This was consistent with a general pattern wherein investigators noted the lack of precocity in otherwise masculinized girls, much as they reported other "limits" to the masculinizing influence of hormones. Regarding girls with "progestin-induced hermaphroditism," for instance, Ehrhardt and Money (1967) noted that "the oldest girl . . . appeared reticent rather than precocious in romance and boyfriends" (98). Note that they contrasted her reticence with precocity, rather than with normative developmental timing.

13. For example, in a later report comparing the sexuality of women with two different intersex syndromes, Money's team underscored the importance of lesbianism by noting, "No case of exclusive lesbianism, transsexualism, or transvestism was reported from either patient group. Although the two groups differed in sexual and erotic behavior, both were within the range of what in our culture is accepted as feminine" (Masica, Money, and Ehrhardt 1971, 131).

14. Between 1981 and 2008, about two-thirds of the roughly 300 human studies that tested a hypothesis related to brain organization included some information about subjects' sexual behaviors or erotic interests. Every one of these included information about sexual orientation, but fewer than 20 studies (about 12 percent) included information about other aspects of sexuality, such as number of partners, arousal patterns, libido, and so on. Even though only a small *proportion* of studies after 1980 reported on detailed aspects of sexual behavior, however, it is important to note that the *number* of studies including such variables is nearly as high as during the earliest period of research (1967–1980). Thus the conflict between the definitions of feminine sexuality used in earlier and later studies cannot be dismissed as the result of a few anomalous studies.

15. At the risk of repeating myself, I don't take a position on which definition is "correct." Though the latter definition seems to align better with current, Western data on the sexual behavior of adult women, the qualifiers I have already used in this sentence signal that I consider this a contingent alignment. In other words, I do not think there is good reason to treat the construct "feminine sexuality" as if it were valid transculturally and ahistorically. Furthermore, definitions are matters of convention, not "objective" facts of the natural world (this is similar to a point I made in the discussion on measurement in Chapter 3). For both these reasons, I do not attempt to venture a "real" definition for feminine sexuality. Finally, as I also pointed out in Chapter 3, the main concern is that scientists' definitions should relate well to other work in this research network, not to some gold-standard definition.

16. Marianne van den Wijngaard has pointed out that the discovery of the role of aromatization in the masculinization of behavior could have challenged the very designation of "estrogens" versus "androgens," but did not. In other words, as Richard Whalen has noted, "our attribution of a quality of a compound (e.g. defining testosterone as an androgen and not as an estrogen) reflects an often unstated conviction about mechanism of action" (quoted in van den Wijngaard 1997, 43). The obverse, defining estradiol as an estrogen and not an androgen, perhaps provides an even clearer case, because estradiol is so critically important for the development of male-typical brain differentiation in many species.

17. In experiments on rats and mice, which have been the dominant models for studying sexual differentiation (Wallen 2005), the most common indicator of male sexual behavior is mounting, but other indicators may include thrusting, intromission, and ejaculation. Female sexual behavior is most often defined simply as lordosis, the arched-back posture that indicates receptivity to mounting. Some studies also investigate the "appetitive" behaviors, including ear wiggling, darting, hopping, and other things a female rat does to indicate sexual interest.

18. In response to the two-pathway model, experiments were designed to discover which specific behaviors were controlled by the androgen pathway and which through the estrogen pathway. For example, many earlier experiments had focused on inducing or preventing masculinization through administering testosterone to developing females, or castrating developing males during the critical period of development. A newer design involves identifying the particular behaviors under control of the estrogen pathway, which researchers achieve by blocking the metabolic conversion of testosterone through aromatase inhibitors, or through directly implanting or injecting estrogen into the brain. For example, if a rat is exposed to ample androgen but also to an aromatase inhibitor during the critical period, it will not display mounting as an adult. So, in the rat, the development of mounting behavior (an aspect of masculinization) is under the control of the estrogen pathway. Researchers may also look separately at "female typical" behaviors to see whether these are controlled by the same pathway or not. In some species, behaviors designated as "male-typical" versus "female-typical" are largely sex-differentiated under the control of the estrogen pathway, but in primates there is less evidence that aromatization of androgens to estrogens is necessary either to facilitate the development of male-typical traits or to suppress the development of female-typical traits (Wallen 2005).

19. Making a clean break from earlier definitions would have meant a sort of clearing of the "score" in terms of scientific evidence, and starting over in assessing the hypotheses about hormone effects with newer, more appropriate measures of human sexuality. Given my sense of the scientific integrity of Meyer-Bahlburg, Ehrhardt, and their colleagues, I think it is quite possible that this team would have taken that route, had they appreciated how they were departing from earlier definitions.

20. As of March 20, 2008, this study had been cited 111 times, according to the

ISI Web of Science. I have evaluated the context of discussion for more than three dozen of these citations and have not identified any that note the discrepancy between the study and brain organization theory, either at the time when the study was done (when there was no hypothesized role for estrogen) or later (when the hypothesized action of estrogen was, like androgen, to involve masculinization). Yalom and his colleagues were not alone in expecting feminization from estrogen exposures. Even though the idea of sex antagonism had been discarded by pioneering endocrinologists as early as the 1920s, van den Wijngaard (1997) has documented how much the cultural notions of sex-specificity and sex antagonism inflected work on brain organization theory through the 1980s, at least.

21. Historians and philosophers of science recently have become quite interested in this sort of "structured forgetting" or patterned ignorance in science. In a terrifically interesting article on systematic ignorance about female sexuality, and even female sexual physiology, Nancy Tuana (2004) coined the phrase *the epistemology of ignorance* as a way to approach ignorance as a structure rather than as a simple lack of information or knowledge. Like Tuana, Robert Proctor has pointed out that the study of ignorance, what he calls "agnotology," may require different tools than the study of knowledge. Exploring citation rates for articles with negative findings, as opposed to supportive findings in terms of brain organization theory, is one way to highlight how the epistemology of ignorance functions as an invisible support to the theory. See discussions in this chapter, as well as in Chapters 7 and 8, for specific comparisons of citation rates along this line.

7. Sexual Orienteering

1. "Homosexuality: Prenatal Environment May Dictate Sexual Orientation," *Women's Health Weekly*, February 20, 2003, 5, web.lexis-nexis.com.

2. In an earlier analysis that focused exclusively on studies linking prenatal hormone exposures to aspects of adult human sexuality (Young 2000), I identified all such studies that had been published as original, peer-reviewed research in English-language scientific journals between 1967 and 1995. Of these 75 studies, roughly 51 percent (38 studies) focused primarily on sexual orientation. Of that near-total sample of the first three decades of brain organization research on human sexual orientation, just 13 studies (roughly one-third) included an explicit definition of sexual orientation, another 12 studies included an "operational definition" only (meaning that they described the measure or measures they had used), and the final third of studies (13) included no definition of sexual orientation at all. For a full review of the methods and the earlier analysis of sexual orientation measurement, see Young 2000, chap. 7.

3. Some recent studies don't choose people with supposedly "extreme" cross-sex traits, but instead look for correlation (both within-sex and between-sex comparisons) between sex-typed psychological traits such as aggression and physiological markers of early hormone exposure such as the 2D:4D digit-length

ratio (for instance, Benderlioglu and Nelson 2004; Bailey and Hurd 2005). This new addition to the brain organization theory repertoire doesn't materially affect my analysis here.

4. The sexual preference test in Stellflug, Cockett, and Lewis's 2006 study differed very slightly from that used by Resko et al. (1996) but was at least as stringent. Stellflug and colleagues classified rams as male-oriented if they were exclusively sexually active with rams during "2 types of sexual performance tests: 1) a series of nine 30-min serving capacity tests in which individual rams were observed with 3 unrestrained, estrual teaser ewes, and 2) a series of three 30-min sexual partner preference tests in which individual rams were observed with 2 restrained estrual ewes and 2 restrained rams" (464).

5. Bailey has long maintained that sexual orientation is a largely dichotomous phenomenon, with most people being attracted to either men or women, and very few people attracted to both. This characterization is generally favored by scientists who investigate possible biological contributions to sexual orientation, in part because it helps to justify their recruitment techniques for studies. They often seek "pure" homosexual and heterosexual subjects, which makes sense if the "usual" nature of the phenomenon of sexual orientation is a strict dichotomy, but this creates biased samples if the nature of the phenomenon is more continuous. Many of these same researchers agree that sexual orientation is significantly more continuous in women, which is one reason they do fewer studies with female samples.

6. Note that researchers in social science and public health typically use the term *gender* in this formulation, where most brain organization researchers would use the term *sex*. But both groups are interested in whether their subjects are sexually attracted to men, women, or both. Researchers could apply a more nuanced sense of gender, and explore whether people are attracted to more or less feminine, or more or less masculine, partners, but very few researchers take that approach.

7. Note that this group is still not as exclusive as the "perfectly consistent" homosexuals that many brain organization studies include. That is, if the point of intersection were based on people with *only* same-sex partners (instead of at least one such partner), the shaded area reflecting people whose desire, behavior, and identity all perfectly align would be even smaller.

8. In an interview, Dörner indicated that men were recruited for all of his studies in an eclectic fashion that began with referrals from dermatologists, venerologists, sexologists, and other clinicians "who treat sexual problems," but that this main method was also supplemented with self-referral and the "snowball" process of subjects referring additional subjects. He then personally interviewed the men using a "very, very open" interview that "followed the process of Kinsey" in order to identify the sex of actual partners, as well as the sex of fantasy or ideal partners. He indicated that a major objective of the interviews was to ensure that the "homosexual" sample was "pure" and did not include any bisexuals or transsexuals (Dörner interview, September 1998).

9. For a more detailed discussion of this problem, see Young 2000, esp. 401–410.

10. Simon LeVay, one of the few prominent brain organization researchers whose primary training and experience is in neuroscience, has a more subtle view. LeVay (1996) correctly credits the early sexologist and champion of homosexual rights Magnus Hirschfeld with originating the notion of masculine and feminine "sex centers" corresponding to desire for females and males, respectively. But he notes that neural complexity, as well as data from various species on the involvement of multiple brain areas in various aspects of sexual function, suggests that "we should be very cautious about equating the medial preoptic area and the ventromedial nucleus with the neural centers proposed by Hirschfeld. It is probably safer to suggest that what Hirschfeld proposed as 'centers' are actually extended networks, in which the medial preoptic area and the ventromedial nucleus may be elements" (134).

11. Reite and colleagues (1995), for example, cite studies that show gay men to be "atypical" on handedness (Lindesay 1987) but fail to note that the atypicality for that trait is opposite (hypermasculine) to the trait they have investigated (a measure of cerebral laterality that they found to be more "feminine" in gay men). Gladue and Bailey (1995b) do not find any of the hypothesized relationships, which may help to explain their sparse discussion. Still, their framework suggests a simple expectation that "homosexuals" will be more left-handed.

12. A second tension specifically concerns the studies of prenatal stress. As indicated in Chapter 5, the overall evidence does not generally support the hypothesis that either gay men or lesbians are born from pregnancies with higher than usual levels of maternal stress. Nonetheless, many of the studies investigating a link between handedness and homosexuality rely on the idea that male homosexuality may be linked with some "hypermasculine traits" if prenatal stress has caused testosterone to briefly surge and then drop. Of course, this suggests several constraints to the theory, which have never been drawn out in any publication on the topic. First, it suggests that if prenatal stress is associated with female homosexuality, the "stressed period" (when testosterone presumably surges) must be the time when sexual orientation develops, so the timing would need to be precise. Second, the stressed period for gay males should be even earlier in gestation, so that by the critical period for development of sexual orientation, testosterone would have already dropped below typical levels. Third, it suggests that in lesbians for whom prenatal stress influenced orientation, traits that are "feminized" in gay men should also mostly be *hyperfeminized* in these lesbians; the exception would be traits that developed earlier than, or concurrent with, sexual orientation. Thus, it suggests a specific sequence of trait differentiation, which in turn would produce a particular pattern of gender similarities and differences. As I argue below, the pattern is not borne out.

13. Because the issues are so extensive, a few examples will have to suffice. The problem in McCormick, Witelson, and Kingstone (1990) is the framing issue that I described. They missed the point that Geschwind and colleagues use the prenatal stress scenario to predict that gay men have a lower testosterone level early in development but a higher level later in development. Thus, they characterize Geschwind and Galaburda as suggesting that high levels of testoster-

one cause male homosexuality. Because his study involved only male subjects, it is not clear if Lindesay (1987) appreciated the importance of framing, but he, too, failed to grasp that Geschwind and colleagues had suggested a testosterone surge-then-trough. Lindesay specifically wrote that Geschwind's hypothesis is at odds with brain organization theory (which it is not; but its connection is elaborate, to say the least). Becker and colleagues (1992), too, cite Geschwind and Galaburda as if they offer a straightforward hypothesis of higher testosterone leading to homosexuality.

14. For more detailed discussion of these problems, see Young 2000, chap. 8.

15. Holtzen (1994) implicitly acknowledged the problem by running his analyses in both the homosexual/heterosexual and the androphile/gynephile frame, but he did not comment on how one frame versus the other would be required to connect his theory with other psychoendocrine research. McCormick, Witelson, and Kingstone (1991) noted the contradictions between Lindesay's (1987) interpretation that excess left-handedness among gay men signals hypermasculinity, but they incorrectly ascribed this same view to Geschwind and colleagues. These are examples of two early reports in which investigators indicated some awareness of internal theoretical tension in studies linking left-handedness and male homosexuality (specifically, the notion that left-handedness is caused by excess testosterone, and the traditional brain organization notion that orientation to men is caused by low testosterone). But this was as close as any of the studies came to acknowledging the complex framing problems I've identified here.

16. Imperato-McGinley and colleagues (1979, 1991) are exceptional in this regard. Yet, because they also argue a much stronger role for hormones than is typical of investigators who study intersex people, it fits with their overall approach.

17. Note that I have indicated that the investigators "adopt" either a continuous or a categorical approach in their brain organization studies, not that they necessarily would define sexual orientation in this way if they were asked for an abstract definition. There are some investigators whose writings have left no question that they believe sexual orientation to be a categorical matter (for example, LeVay 1996; Wilson and Rahman 2005; Bailey 2003), and others who seem to believe it is a matter of degree (for example, Ehrhardt et al. 1985; Meyer-Bahlburg et al. 2008), but the issue at hand is the nature of the measures they adopt and the analytic strategies they use to interpret these measures.

18. In some studies, the way investigators divide and refer to the groups may appear to support a notion of sexual orientation as continuous, but on closer inspection this is often simply a matter of choosing a lower cutpoint for the "nonheterosexual" group. For example, van Anders and Hampson (2005) divided their subjects into groups that were either "heterosexual (HS)" (scoring 0 or 1 on Kinsey scales for behavior and fantasy) or "not strictly heterosexual (NHS)" (scoring 2 or more on either behavior or fantasy). Laudably, these investigators avoid the common problem of calling their "not strictly heterosexual" group "homosexual." But they do not give the sort of information that

would allow us to conclude that their measure reflects a background assumption of sexual orientation as continuous, such as providing a frequency distribution of subjects' sexual orientation scores, nor do they conduct their analyses with sexual orientation as a continuous variable. However, because they have set the cutpoint for the NHS group so low (just a Kinsey 2 score on either fantasy or behavior), one might argue that they have not sought to create the most starkly contrasting groups possible. It would be useful to know how they settled on those cutpoints: Do they believe that Kinsey scores of 0–1 versus 2 or more actually correspond to meaningfully different groups? Did this division seem to make sense from the frequency distribution of their respondents' scores? Or—and this would be the most problematic reason—was this division the one that yielded a significant difference between groups on the dependent variable of spatial relations? Without more information in the methods section, we can't eliminate this last possibility.

19. "In an earlier study of the responses provided, we found that a substantial minority of offspring provided apparently erroneous responses regarding their sexual orientation, based on contrary answers given elsewhere in the questionnaire regarding their being attracted to members of the same sex or fantasizing about sexually interacting with the same sex. Consequently, for the present study, these subjects were excluded from the present analysis" (Ellis and Cole-Harding 2001, 217).

20. The study by Kraemer and colleagues (2006) is interesting in this regard. This is one of the very few studies of a nonclinical sample of adults that uses a continuous measure of sexual orientation in the search for evidence of a correlation between sexual orientation and other indicators of neurohormonal sexual differentiation of the brain. In women, they found a very small but significant correlation between the finger-length ratio (2D:4D) and the continuous measure of sexual orientation, and the direction of the correlation was consistent with the usual brain organization prediction (a more female-directed sexual orientation was associated with a more "masculine" 2D:4D ratio). Interestingly, though, they found no such effects in men. Although they do believe the effects indicate the influence of early testosterone on sexual orientation in women, they interpret their evidence as inconsistent with the "predominant suggestion of *fixed organizational effects* of androgens in the brain and a *categorical* sexual orientation" in women (210; emphasis added).

21. Lisa Diamond's book *Sexual Fluidity* (2008) provides a lovely qualitative exploration of the complexity and changeability of sexual desire in women. On a related note, recent limited evidence suggests that both heterosexual and lesbian women show similar physiological responses to erotic material. That is, genital arousal can be detected in women who are exposed to erotic images regardless of whether the image "matches" their preferred sexual partners—in fact, even images of nonhuman primates having sex stimulated genital arousal in a small group of women, but not men, who were tested (Chivers and Bailey 2005). On the one hand, this study raises interpretive problems similar to those I mentioned regarding this team's other study examining the relation between physiological response and sexual orientation (Rieger, Chivers, and

Bailey 2005). On the other hand, it echoes some "real world" data on the independence of various aspects of sexual orientation in women, suggesting that constructing "pure" homosexual or heterosexual groups for studies just doesn't make sense.

22. I've opted to omit analyses of "fluctuating asymmetry" from this review because these studies have so consistently yielded negative results (for example, Rahman and Wilson 2003).

23. Of course, it is likely that additional studies that fail to find differences have been conducted but not published, given the publication bias toward studies that find difference. Gladue and Bailey (1995), for example, cited a study by Tkachuck and Zucker that was presented as a poster at the 1991 meeting of the International Academy of Sex Research. That study, which apparently found no differences in spatial abilities between sexual orientation groups, was never published.

24. The two largest studies do report that gay men score in a feminine direction relative to straight men on some cognitive traits. These include Collaer, Riemers, and Manning's (2007) analysis of the massive BBC Internet survey, with more than 11,000 gay men, and Qazi Rahman's multipublication study of a broad range of psychological outcomes (e.g., Rahman and Wilson 2003; Rahman, Abrahams, and Wilson 2003; Rahman, Wilson, and Abrahams 2003, 2004; Rahman, Andersson, and Govier 2005), which included data on 60 gay men. The next three largest studies, though, do *not* find any association between sexual orientation and cognitive abilities (Cohen 2002; Gladue and Bailey 1995; Tuttle and Pillard 1991). Although small samples do limit the ability to find group differences, it is important to remember that sampling biases might well *inflate* the probability of finding differences, possibly accounting for many of the gay versus straight differences in cognition that have been reported. Most studies use different sampling strategies for gay versus heterosexual subjects, with the gay samples typically being recruited from community-based political or social organizations and events, while the heterosexual samples are overwhelmingly drawn from university students and staff. Sampling bias cannot even be ruled out in the BBC Internet survey. Logically, Internet-based activity may follow social network patterns, such that people with certain common interests or characteristics might be overrepresented if they are more likely to participate, and to pass the Internet link to their personal contacts more readily, than others who are potential research subjects. For example, sexual minorities (gay men, lesbians, bisexuals, and transgender people) may have a greater interest in studies that investigate sexual orientation, and sexual minorities who endorse biological causation theories may be especially likely to participate in studies that investigate factors that are popularly associated with sexual orientation via such theories (such as handedness, the 2D:4D ratio, and childhood gender nonconformity). The possibility that differential social network participation influenced the composition of the BBC Internet survey is hinted at by the extremely different proportions of heterosexual versus bisexual or lesbian women who completed the survey in various regions around the globe (Lippa 2008, 181).

25. Iris Sommer, the Dutch neuroscientist who recently completed multiple meta-analyses that debunked the usual presumption of greater lateralization among men for language tasks, has offered an intriguing observation about the sex difference in handedness that might be relevant. Sommer notes that several lines of evidence suggest that "social influences may be a better candidate" than early testosterone exposures to explain the sex difference in handedness: more women report having been forced to switch handedness from left to right; among right-handers, more females than males use the left hand for some tasks; and the size of the sex difference in handedness is generally larger in non-Western countries, which Sommer and colleagues take as suggesting "more sex-specific social pressure in non-Western cultures" (Sommer et al. 2008, 82).

26. On the face of it, Martin and Nguyen's (2004) analysis of anthropometric estimates of bone size seems to support the idea that lesbians and gay men have "cross-sex" hormone exposures in early development. There are two major reasons to be cautious about accepting their analysis, however. The most important reason is related to sampling. Due to recruitment of participants at sites with highly unbalanced gay/straight populations (gay pride parades, gay bars vs. Brigham Young student union, churches, and so on), and the fact that measurements were taken at the recruitment site, it is not accurate to conclude that the body measurements were done blindly. Though the researchers indicated that they made attempts to avoid recruitment bias by approaching everyone who was in the recruitment area, my own field research experience suggests that this is a very difficult criterion to discharge: given the density of population at sites such as street corners, bars, and gay pride parades, it is virtually impossible to approach everyone in a delineated vicinity. It is sometimes a better procedure to use systematic selection (for instance, every fourth person who enters an area from the point where a recruiter is available). People were recruited at the various sites by researchers who knew the hypothesis, and either knew or could guess with high accuracy the sexual orientation of a given subject. Thus it is possible that recruitment was not unbiased. The second reason is that the only true probability sample that offers information on anthropometry and sexual orientation (Bogaert and Friesen 2002) does not find any such differences.

27. See the note in Chapter 5 regarding my decision to omit another study, Putz et al. 2004, from this analysis.

28. Digit ratios might not be a good candidate for signaling early hormone exposures, in any case. Although the 2D:4D ratio is sometimes reported as masculinized in genetic females with CAH, the balance of evidence seems to tilt away from this interpretation (Brown et al. 2002 and Okten, Kalyoncu, and Yaris 2002 did report such masculinization, but the much larger study reported in Buck et al. 2003 did not). The ratio is highly variable across different ethnic groups, as is the degree of sex difference (Manning et al. 2004). Moreover, as Martin and Nguyen (2004) observe, "the anatomical development of the hand is complex, dynamic and continues throughout life. Unlike other long bones, the distal phalanges may continue to grow in adulthood, whereas

the metacarpals appear to uniformly decrease in length as one ages, and these changes appear to occur more in males than females. . . . Hand use, that is, load stress, may play a larger role in phalangeal length than is commonly assumed. Clearly the type of work performed by male and female homosexuals and heterosexuals may not be the same, and future studies should be designed to control for this variable" (36). This line of reasoning even offers a potential (if highly speculative) explanation for the conflicting 2D:4D evidence. Recall that studies in the United States found gay men to have more masculine digit ratios, while studies of European men found the opposite. There has been a major emphasis on body building and masculine physique among American gay men over the past couple of decades, and body building might be an important activity that would influence growth of the phalanges (finger bones). If body building were less central to gay culture and aesthetics among European gay men during this time (or if it took different precise forms, or entered the subculture later), then the observed 2D:4D ratios could represent "real" homosexual versus heterosexual differences (that is, not due to sampling biases) that arise from local "biocultures" in adulthood, rather than from prenatal influences.

29. Following are some of the numerous statistical problems with this study:
Dittmann and colleagues (1990, "Congenital Adrenal Hyperplasia I") include in a table of significant results any single finding that reaches even $p < .20$—and this with one-tailed tests, and *twelve* or more comparisons on each item: sisters versus patients, simple virilizers versus salt wasters, simple virilizers versus sisters, salt wasters versus sisters. Each of these four groups is compared for three different age groups of decreasing size ($> 11; > 16; > 21$); plus for some items, the responses are divided in different ways, and both analyses are reported. The chance for type II error (failing to reject the null hypothesis of no difference between groups) is therefore staggering. With a 2-tailed test and a p level set at .05, there is a 5 percent chance of type II error for each comparison. If the p level is set at .20 to begin with, this increases that chance to 20 percent, and using a one-tailed test doubles that to 40 percent. Doing multiple comparisons on the same item correspondingly multiplies that chance by the number of comparisons. Cutting the data this many ways, they literally cannot fail to find differences. Given that there are only 14 sisters in the entire comparison group, 12 women with SW, 20 with SV, and 34 patients total, the entire exercise becomes absurd.

In addition to the fact that some item responses were dichotomized in multiple ways, and reported separately, there is an additional problem with the way they have grouped responses (for instance, sometimes they compare responses of "yes" to "no + not applicable"; sometimes they also add "unsure" to "no" and "not applicable"). But responses of "not applicable" and "unsure/don't know" should generally never be added to a substantive response, at least not without a clearly explained rationale for doing so.

There are 43 separate topic areas listed (see appendix 1, p. 166), which include an unspecified number of individual items. It appears that every single item was subjected to four or more comparisons of the sort mentioned above.

Of the 47 supposedly separate single items that showed some statistical significance, only 17 are independent. Ten items are subject to the kind of "multiple cuts" at dichotomization described in item 2, above, so 20 of the 47 results actually represent just 10 items. Another problem is that two clusters of four items each are nonindependent, because the items are specific versus more global versions of the same question (such as "romantic or erotic fantasies involving a partner" and "romantic or erotic fantasies, all kinds"); another 3 sets of two items each are nonindependent because they involve comparisons of the entire subject group vs. sisters in one case, and a specific age range (> 13 years) vs. sisters in the second case. Again, nonindependent items should be treated as multiple comparisons, so this further exacerbates the problem listed in item 1, above.

In spite of these problems, the study has been influential. As of July 2, 2008, this study was cited at least 95 times in other journal articles, according to the ISI Web of Science.

8. Sex-Typed Interests

1. A difficulty in characterizing Baron-Cohen's work is the fuzziness of the conceptual category "systems." In measurement terms, the idea that systems comprise "phenomena that are . . . lawful, finite, and deterministic" does not act as a sensitive and specific guide to the sorts of interests that he classifies, in practice, as masculine. Does the law comprise a "lawful" system? If so, what do we make of the rapid change in this field from one where men dominated for centuries, to one where there is now slight female dominance in new practitioners? Accounting provides another example. Surely accountants deal with "systems" that are "lawful, finite, and deterministic." But Wootton and Kemmerer (2000) document that "the gender composition of the accounting profession changed from 10% female in 1930 to 53% female by 1990." Differences in the extent to which "systemizing" characterizes specific sciences have been used to explain the fact that biology is an increasingly female-dominated science but physics is not. This relies on an old-fashioned, Newtonian view of physics, though. Physicist Karen Barad's recent book *Meeting the Universe Halfway: Quantum Physics and the Entanglement of Matter and Meaning* (2007) offers an elegant rebuke to this characterization. Reflecting deeply on the phenomenon of diffraction and Niels Bohr's principle of complementarity, Barad introduces readers to the irreducibly context-dependent nature of phenomena.

2. Scientists have justified this characterization with reference to observations and experiments with nonhuman animals, studies that have themselves been shaped by midcentury ideas about female domesticity and masculine industriousness (see van den Wijngaard 1997 for an excellent extended critique). Van den Wijngaard and other scientists have also pointed out that the species generally selected for examples fit the expectations for finding maternal/domestic females and adventurous/aggressive males (Young and Balaban 2003; Parish and de Waal 2000).

3. The theoretical model has always suggested that interest in one's own appearance is a "feminine" trait. But this is a terrific example of a sort of "sex-typed interest" that is extremely time- and culture-bound, and it has not even been well explored among North American populations. Occasionally investigators will offer suggestive data to counter the characterization. Ehrhardt and Baker (1974) observed, for example, that roughly half of both CAH-affected and unaffected boys reported moderate or strong interest in clothing and appearance, leading the researchers to conclude that "interest in appearance *per se* is thus not specifically feminine but also a noticeable part of boys' behavior in childhood and adolescence" (45).

4. In fact, hormone exposures are not "female-typical" in genetic males with androgen insensitivity syndrome, because the testes secrete normal amounts of androgens from fetal life onward. But because target tissues are insensitive to androgen, due to the receptor dysfunction, the functional exposure to hormones is presumed to be at the extreme opposite end of the spectrum from male-typical, with little response to androgens (in the case of partial androgen insensitivity, or PAIS) or no response to androgens (in the case of complete androgen insensitivity, or CAIS).

5. The primary difference between the more specific "activities and interests" scales (like the PSAI or the CGPQ) and gender scales (like the RCGQ) is that gender scales typically also include items more closely related to gender identity than to gender related interests. An example of such items would be asking whether an adult recalls having "felt masculine versus feminine" or their "frequency of feeling good about being a girl (or a boy)" as a child (Meyer-Bahlburg, Sandberg, Dolezal, and Yager 1994; Meyer-Bahlburg, Sandberg, Yager, et al. 1994; Meyer-Bahlburg et al. 2006).

6. Note, though, that I could not identify any brain organization studies that specifically employed the BSRI to comment on aggression. This might be because, compared to other measures, the BSRI is used more consistently as a composite scale only, without individual item analyses.

7. In an interview, Dr. Berenbaum (June 2, 2008) suggested that the aggression scale of the MPQ is not as reliable a measure as the Reinisch aggression inventory. Still, she agreed that the findings regarding aggression in CAH were weak, at best.

8. This again brings to mind recent studies in agnotology, or "epistemology of ignorance," concerning how knowledge is lost, ignored, or forgotten (Tuana 2004; Proctor 2008). I do not have the space to devote detailed attention to this matter here, but my analyses throughout this book have revealed an unmistakable pattern: data that don't support the theory are far more likely to be lost than data that do support it (or *seem* to). See, in particular, how smaller and methodologically weaker studies that seem to support brain organization theory are routinely cited much more often than studies that fail to support the theory. (For just two examples, compare Yalom, Green, and Fisk 1973 and Ehrhardt et al. 1985, on the one hand, with Titus-Ernstoff et al. 2003 on the other; and Meyer-Bahlburg and Ehrhardt 1982 and Reinisch 1981, on the one hand, with Jaffe et al. 1989, on the other).

9. I have tried to include all studies that meet three criteria: the findings were reported in a peer-reviewed journal (prior to January 2008); group comparisons involved statistical analyses (so case reports and studies doing simple numerical comparisons are not included); and at least one of the dependent variables involves sex-typed but non-erotic interests. Note that each study may have resulted in either a single publication (to date) or several publications, and that I list only those specific publications that report findings related to sex-typed interests, excluding eroticism.

10. It's worth noting that "interests" as they are measured might capture both perceptual differences and motivation in early childhood. In older children and adults, motivation may, in turn, be affected by both habituation and skills, meaning that individuals may find greater reward in activities in which they excel, and so direct even more of their interests toward such activities.

11. There are a few other thing to take into account when evaluating this model. It is extremely cheap and easy to investigate 2D:4D on large numbers of people in a variety of settings, which at least partly explains the rush of recent interest. That also should cause us to wonder about how many other investigations have ended up in the file drawer because they found no link between digit ratio and sex-typed traits. A recent comprehensive analysis of publication patterns also found that a large proportion of 2D:4D research is conducted by a single team (John Manning and colleagues), and that positive studies appear more likely to be published and cited (Voracek and Loibl 2009). They also noted that most of the research is done by psychologists, and that so far the research has circulated primarily among scientists with an interest in this work. Thus, much of the research hasn't yet been subjected to broad evaluation by scientists in the fields of anatomy, physiology, and biology.

12. I made an attempt to clarify some of the inconsistencies, but as Dr. Udry no longer had access to the data files and his data analyst had retired, he was unable to clear up the confusion (Udry, personal communication, January 14, 2008). Note that, while Cohen-Bendahan, van de Beek, and Berenbaum (2005) detail inconsistencies between Udry's two reports, their review misstates his actual findings; for instance, he did not, as they report, find any significant correlation between testosterone in maternal blood and the "gendered behavior" of adult offspring (see, for example, Udry, Morris, and Kovenock 1995, 364).

13. It's worth noting two other things. Although the correlation could not be explained by any of five "background" factors that were analyzed (mother's education, having older brothers at home, having older sisters at home, the mother having a male partner living in the home, or parental adherence to traditional sex roles), there is ample room for the influence of other background factors that were not measured (such as whether or not the mother was employed, or the status of her occupation). Also, Hines's study was large (342 male and 337 female children) and involved an oversampling of children at the extreme ends of masculine and feminine behavior, to increase the study's power to detect small effects (and to avoid the extreme expense of conducting bioassays on blood samples from nearly 14,000 women). This is a smart sam-

pling plan for maximizing the chance of unearthing any correlation between variables. It does mean, however, that very small effects, without much real-world meaning, can show up as statistically significant.

14. Although both of the latter studies were based on relatively small samples (60 and 53 children, respectively, including males and females), it seems unlikely that larger samples would have revealed a relationship between prenatal testosterone and play patterns. Knickmeyer, Wheelwright, and colleagues (2005), for example, noted that "the extremely low ß and r values found in the current study suggest that even in a larger sample, no stronger relationship between [fetal] T and play would have been observed" (522). Both studies used modified versions of the Bates and Bentler (1973) inventory of child play behaviors, and both found significant differences in play between boys and girls in the groups tested, as well as a large effect size for sex. Grimshaw and colleagues suggested that the lack of correlation may reflect a weakness in their measure of sex-typed play, but the potential weakness they perceived related to their interest in play experiences, rather than interests per se. Given that they selected nineteen activities that required object manipulation, and asked parents to rate their child's level of preference for such activities, their measure seems especially well suited to the question of androgen effects on sex-typed interests.

15. Psychologists Cordelia Fine and Giordana Grossi point out that it is frequently impossible to specify on the basis of any clear, *a priori* criterion whether specific skills are examples of systemizing or empathizing. This gives Baron-Cohen and colleagues too much leeway to drop evidence that doesn't fit their theory (Fine, personal communication, June 3, 2010).

16. Baron-Cohen's study is a longitudinal analysis of sex-typed traits among children whose mothers underwent amniocentesis Therefore they can never recruit more subjects, and specific reports are based on the participation of subsets of children whose parents respond to requests for particular information or task participation. Thus, the analyses are necessarily based on fairly small samples, which in turn results in low statistical power. The team deals with this in a variety of ways, which may simply reflect the fact that various analyses are "spearheaded" by different members of this team, often as doctoral research projects. Still, a shift in theoretical or statistical approach is important because it can have a dramatic effect on the shape of results, especially with a small sample; it's important to be consistent with these methodological decisions, in order to avoid accidentally choosing the analysis that yields "more congenial" results. But the team is frequently inconsistent in ways that could affect their results. Consider how they treated statistical outliers in two different studies published in the same year, with the same lead author. In the study on gender-typed play, Knickmeyer, Wheelwright, et al. (2005) drop an outlier, a girl whose testosterone (T) is two standard deviations above the mean, based on the fact that there is not supposed to be a change in fetal T among girls during this same period. Nonetheless, once she is dropped, there is still a near-significant relationship between gestational age and T among girls ($p = .07$), but not among boys ($p = .33$), even though T supposedly *does*

change for boys during this same period. (As an aside, this is a signal that there is at play either measurement error or a problem with current knowledge about the biological process that they are trying to study.) In the study on social relationships and restricted interests (Knickmeyer, Baron-Cohen et al. 2005), they followed a different strategy. Here, they did *not* drop the two girls with the highest fetal T from further analyses, even though the linear relationship between gestational age and T among girls (a relationship that they questioned before) was no longer significant when they were excluded. They explained, "We did not exclude these girls from further analysis because they were within 3 *standard deviations of the mean score* for fT and gestational age and we wished to keep the sample size as large as possible" (202, emphasis added). There is no clear reason for them to apply different standards in the two studies.

Another contrast, again between two different analyses published in the same year with the same author, concerns the way they model the relationship between estradiol and testosterone. In an analysis of fetal testosterone and vocabulary size, Lutchmaya and colleagues (2001) wrote that "oestradiol . . . is known to have masculinising as well as feminizing effects on development"(420), which was in keeping with their finding that estradiol was inversely related to the (female-favoring) trait of vocabulary size. (Estradiol was, in fact, the *strongest* predictor of vocabulary size at the first time point, though the authors never pointed this out, perhaps because their overall research model focuses on testosterone.) Contrast this with the way they describe estradiol in a study of testosterone and eye contact: "Oestradiol in the amniotic fluid . . . has feminising effects on development, so could be considered as an opposing influence to testosterone" (Lutchmaya, Baron-Cohen, and Raggatt 2002, 329). In this analysis, estradiol was not related to eye contact. Given that estradiol supposedly worked in tandem with T on one trait they studied, it is unwarranted that they should simply hypothesize that it would be "an opposing influence" to T on another trait. And it is troubling that they do not point out the discrepancy between the two studies in terms of the findings regarding estradiol.

A third example, from the same pair of studies, is that this team used a quadratic equation for the model of eye contact, but not vocabulary size, though they suggested that there is probably a nonlinear relationship between T and both of these sex-typed traits. It seems that the eye contact data were collected first (at 12 months), but reported later than the vocabulary analysis (the data for which were collected at 18 and 24 months). In the eye contact study, they entered the quadratic T term along with the predictor variables based on Grimshaw, Bryden, and Finegan 1995 (Lutchmaya, Baron-Cohen, and Raggatt 2002, 331), but in the vocabulary study they never did include the quadratic T term, even though they had suggested that there would probably be a nonlinear relationship between T and vocabulary size (see esp. p. 421). This leaves the unfortunate impression that they may have chosen the model for the different analyses based not on their theories, but on the results that the specific method yielded.

17. The group includes progesterone (also called "natural progestogen") and synthetic progestins, which in turn include the "true" progestins such as medroxyprogesterone acetate (MPA), and more "androgenic progestins" such as those derived from testosterone (it may come as a surprise that this latter group is the sort of progestin used in most oral contraceptives) (Darney 1995). Progesterone and the true progestins can act as "antiandrogens" (for example, MPA is sometimes used with estrogen to feminize physical characteristics like facial hair in male-to-female transsexuals). But most of the synthetic progestogens are significantly *androgenic,* meaning that they promote the development or maintenance of male-typical characteristics (like facial hair growth, or deepening of the voice).

18. A note on the progestin-exposure studies that are and are not included in this section is in order. A small study of genetic females who were born with masculinized genitalia after prenatal exposure to various progestins (Ehrhardt and Money 1967) is not included because they used no control group. Katharina Dalton and colleagues presented several other reports of sex-typed interests among progestin-exposed people at conferences in the 1970s, but these are not included in this review because they were never published.

19. Citation counts in this section are based on a search of the ISI Web of Science on June 10, 2008. Note that the main focus of each article citing Jaffe et al. (1989) was the safety and efficacy of MPA (also known as "Provera") as a hormonal contraceptive.

20. Several scientists whom I interviewed mentioned the difficulties of interpreting studies that fail to find group differences. Some of them (Drs. D, R, and T, for example) generally thought that steps could and should be taken to push for more publication of "null" or "negative" findings. Others were more equivocal on this point, because they often saw many reasons that studies might fail to find a group difference, even if there is a "true" underlying difference present. Dr. U, for example, noted that using measures without good sensitivity for capturing the constructs of interest, as well as the necessarily small sample sizes in these studies, both would tend to underestimate differences between hormone-exposed and unexposed groups. But the most obvious strategy for increasing the power of these studies, namely, using larger unexposed control groups, is almost never used. The neglected study by Jaffe and colleagues (1989), ironically, is one of the very few that did use this strategy: they compared 74 boys and 98 girls who had been exposed to 459 boys and 546 girls who had not been exposed (351). Thus, this study had excellent power for detecting even small group differences.

21. Dalton did publish some other results from the progesterone-exposure study, but these were limited to analyses of IQ, general school performance, and developmental milestones (for example, Dalton 1976). The findings on specifically sex-typed interests and personality traits (including "tomboyism") were never published, which makes subsequent citation and mention difficult to trace (for instance, the conference paper does not show up at all in the ISI Web of Science), but my impression is that the unpublished sex-typed interests findings are cited even more frequently than the IQ findings of the published

report—probably because the idea that early hormones affect overall IQ went out of favor a very long time ago.

22. This has sometimes been acknowledged in reviews. For example, regarding both of the studies by Richard Green and colleagues, Hines, Golombok, et al. (2002) remarked that "the studies assessed a large number of variables, raising the possibility that the few significant findings resulted from chance" (1684).

23. Again, the authors of the report are more ready to see a consistent pattern of findings, suggesting that the progesterone-only regimen was especially linked with more feminine traits. Still, they noted that "in a study with many dependent variables, and a limited number of subjects, caution must be exercised in the interpretation of 'significant' findings" (Kester et al. 1980, 281).

24. It is fair to characterize the statistical approach in this study as overly generous, for a number of reasons. First, the authors themselves acknowledged the complexity of effects on sexual differentiation from the hormones involved, and noted that animal and human studies had suggested both masculinizing and demasculinizing effects. They even wrote that "since little is known about the behavioral effects of the individual estrogens and progestogens, it is impossible to predict with any certainty their interactive effects on sex-dimorphic behavior" (Ehrhardt et al. 1984, 458). Yet they employed one-tailed statistical tests, even though the groups they studied were exposed to a variety of compounds. Second, they conducted a very large, though unreported, number of comparisons, which dramatically increases the odds of finding significant results simply by chance. This is a special problem when so many of the dependent variables are interrelated. For example, in the domain of "parenting rehearsal," they separately analyzed six different items, each assessed both by an interview with the child and one with the mother. Out of these twelve comparisons, finding one significant result is hardly unexpected. A similar problem affects their conclusions regarding girls' interest in feminine appearance (two marginal differences out of ten comparisons, again using one-tailed tests).

25. A third report, appearing in *Nature* (Reinisch 1977), framed the analysis of the psychological effects of prenatal progestin in terms of "IQ and personality tests," but did not mention masculinization or feminization. For that reason, and because it does not apparently present new data, I omit detailed discussion of that article in this section. The two different frameworks for presenting the same findings are interesting, though. Marianne van den Wijngaard (1997) has speculated that Reinisch was increasingly disposed to interpret the findings in terms of the hypothesis of sex differentiation, as brain organization theory was becoming an important touchstone for research on hormones during this period. It seems more likely to me that the difference between the *Nature* article and the *Archives of Sexual Behavior* article (Reinisch and Karow 1977) was simply a matter of space: the latter allowed the authors much more room to discuss etiologic theories and related studies.

26. It might be premature to suggest that the DES-exposure hypothesis is dead,

given the disproportionate attention given to studies finding effects of DES on psychosexual development, relative to those studies that find none. For example, based on a check of citations in the Web of Science as of June 10, 2008, average citations for the studies that seem to support "organizing effects" of DES are 2.25 per year, nearly twice the rate (1.16) of those studies counter to the hypothesis.

27. Additional data on (non-erotic) sex-typed interests in CAH have been reported, but they are not included here because of insufficient detail regarding methods and measures, lack of control groups, and/or no statistical analysis of group differences (for example, Ehrhardt, Epstein, and Money 1968; Lev-Ran 1974a, 1974b; Money and Daléry 1976; Money and Schwartz 1977; Money, Schwartz, and Lewis 1984; Slijper et al. 1998; Gupta et al. 2006).

28. Of the three, note that only Hines and Kaufman (1994) made direct observations of children's behavior. In interviews, two researchers indicated that the usual method, parental reports, is especially prone to bias regarding children's level of activity (Drs. R and U).

29. According to later analysis of a larger sample of Sheri Berenbaum's CAH study, in which she used a standardized, validated instrument (the Self-Image Questionnaire for Young Adolescents, SIQYA, in Petersen et al. 1984), CAH-affected subjects and controls did not differ in vocational/educational goals (Berenbaum et al. 2004). During an interview, Dr. Berenbaum suggested that the SIQYA is a less sensitive measure of sex-typed occupational preferences, so this finding probably does not temper the earlier finding that girls with CAH have more masculine career interests.

30. In one study, women with CAH scored lower on a measure of indirect aggression (a scale on which women typically score higher), but were no different on the other four aggression scales, even though each of these scales typically shows sex differences, and liberal statistical tests were used (Helleday et al. 1993).

31. In a recent review chapter on prenatal hormones and sex-typed interests, Berenbaum and Resnick (2007) write that females with CAH are "more likely to report that they would use physical aggression in hypothetical conflict situations" (151). Because only one study (Pasterski et al. 2007) has found females with CAH to be more aggressive than controls, whereas many other studies have failed to find such differences (Ehrhardt, Epstein, and Money 1968; Ehrhardt and Baker 1974; Slijper 1984; Gordon et al. 1986; Dittmann et al. 1990, "Congenital Adrenal Hyperplasia I"), including the combined sample of Berenbaum's own study (Berenbaum et al. 2004), this interpretation is misleading—though again, I think the difficulty is that Berenbaum and Resnick overlook the data from the 2004 study, as their focus in that analysis wasn't on aggression. I do not think it involves a deliberate distortion of the evidence. Still, the totality of evidence strongly suggests that CAH is not associated with increased aggression in either males or females.

32. It's probably worth noting again that some of the scientists who do this work disagree that it is a good idea to look at the evidence from all the study de-

signs, because they consider certain subsets of studies to be illogical or otherwise unconvincing. In particular, several scientists who do cohort studies on people with known unusual hormone exposures were critical of the case-control studies of sexual minorities, especially studies that rely on the Geschwind-Behan-Galaburda hypothesis (interviews with Drs. D, R, T, and U). Nonetheless, the different designs do offer distinct strengths and benefits (the case-control studies don't have to contend with the issue of intersexuality, for example). Most importantly, these studies are all used as a package to support the theory in a great many scientific reviews and in the popular press.

33. The "present attitudes and future expectations" items (1990, part I, 407) in Dittmann and colleagues' studies (1990, parts I and II) illustrate how extremely dated norms are presented as the "feminine" expectation in these studies: "giving up work (if living with a partner), subordinate role (if living with a partner), perceived advantages in life for men vs. women, and future preference for career vs. family." Among other problems, these measures give no possibility for girls and women to be affected by political awareness rather than early hormones when evaluating, say, the relative advantages in life for men versus women (treating that as a potential indicator of "brain masculinization" seems extraordinarily similar to early twentieth-century conservative physicians who suggested that "sex bitterness" among feminists was a biological condition). It is also somewhat amazing that Dittmann, like most other scientists, does not consider that difficult, medically managed childhoods and (often multiple) genital surgeries may affect women's attitudes toward childbearing and childrearing.

34. We must be thankful for the studies, like this one, that give us the nitty-gritty details. One reason so few studies do is that very few of them rely on observation of behavior, but instead ask children and parents to complete questionnaires. The questionnaires, in turn, are more susceptible to stereotyped responses, especially if the child's sex has already been made salient by anything in the questionnaires.

35. Additional data are available from the recent *Digest of Education Statistics* published by the National Center for Education Statistics (2007). See table 9, for example, which lists the number of persons age 18 and over, by highest level of education attained, age, sex, and race/ethnicity for 2007. Notice that the male-to-female ratio of people over 18 who hold a doctoral degree (more than 2 to 1) is reversed for the youngest age group for whom this statistic is available. Among people aged 25 to 29 years old, more than 2.15 times as many women as men hold doctorates.

36. For example, the vervets were tested in groups, which means that the choices each vervet made were not independent, but were dependent on group factors such as what other vervets (especially more dominant vervets) were already playing with. Out of 44 males and 44 females, only 33 males and 30 females made contact with any of the six toys, which was the necessary criterion for getting a "percent contact score" and being included in the analysis. Males showed a trend toward dominance ranking being correlated with a greater

amount of toy contact. In addition to dominance, choices would also have been affected by more mundane facts such as things that were already going on in the enclosure before the toy was introduced (so that some vervets were engaged and distracted by other tasks when the toy was put in), the spatial distribution of the vervets before the toys were put in the cages, and so on. There was also a near-significant trend for male vervets to get higher scores for "approach" and "contact" with toys in general. It would also be interesting to see the shape of the data. Were a few vervets, of either sex, playing with the toys much more often than the others? How did the passing of toys from one vervet to another work—did toys get "abandoned," "taken," "given"? Were they ever shared?

37. The hypothesis of sex differences in color preference got quite a lot of airtime after another study supposedly showed that, as a headline in the *Independent* put it, "Boys Like Blue, Girls Like Pink—It's in Our Genes" (Connor 2007). Right off the bat, the article gives a terrific sense of the successive sleights of hand that neatly link stereotypes to science: "Scientists have found that women tend to prefer pink—or at least a redder shade—while men prefer blue, and that the gender difference may be down to genes rather than upbringing." Move down a bit, and you learn the structure of the study: "As fast as they could, each young man and woman had to choose their preferred colour from a series of paired, coloured rectangles shown on a computer screen. The universally preferred colour for both sexes was blue, but females also showed a distinct preference for reddish colours, Dr Hurlbert said." To complete the slippage, notice that the study was done on young adults—who have already been through years of color-coding by gender—though a quote in the article from the study's lead scientist (echoing the headline) alludes to "little girls" and "boys." In the end, a study showing that young men and women prefer blue, and women slightly prefer a more reddish-blue, getting boiled down to a claim that "little" (read: not yet socialized) girls have a preference for pink (notice this is an altogether new element in the story). That's not science, it's folklore. I thank an anonymous reviewer for bringing this example to my attention.

38. Alexander and Hines (2002) reported that "in some instances, . . . vervet monkeys contacted toys in ways that appeared to resemble children's contact with them, such as moving the car along the ground. They also interacted with the doll in ways that resembled female vervet contact with infants, such as inspecting it physically" (472). They also include two photographs in the article as "examples of a female and a male animal contacting toys" (473). The authors do not provide data on whether there were any sex differences in the way that vervets contacted toys. Given the ubiquitous nature of stories about children engaging in sex-typed play even with "inappropriate" objects (a girl cuddling a fire truck, a boy using a Barbie as a gun), it would be important to know if vervets were more likely to "contact" specific toys in ways that resemble how their same-sex human counterparts would contact them. In any case, the picture and the emphasis on those instances (presumably not the majority

of the time, or they would have said so) when vervets appeared to use toys in ways similar to how human children use them lend credibility to the evolutionary story, though the data don't neatly support it.

9. Taking Context Seriously

1. I restrict consideration for now to the classical form of CAH, because the vast majority of brain organization studies focus on this form rather than the milder "nonclassical" form, which does not masculinize the genitals. I will return later to consider how aspects of context might be important for evaluating girls and women with nonclassical CAH.

2. We know so little about how variation in physical sexual sensations might contribute to specific activity and partner preferences—maybe women for whom penetration is somewhat less comfortable or pleasurable are more likely to fantasize about other activities, especially cunnilingus? This harkens back to the important midcentury experimental psychologist Frank Beach, who resisted brain organization theory for years, famously shouting, "All is in the penis!" prompting brain organization proponent Roger Gorski to retort, "No, all is in the brain!" (van den Wijngaard 1997, 39).

3. Remarkably, the effect of a serious, genetically transmitted condition on one's plans for parenthood does not seem to be much studied. On December 27, 2008, a search of the ISI Web of Science database for articles using the term combination ("huntington chorea" or "hemophilia" or "tay-sachs" or "epilepsy" or "sickle-cell anemia" or "muscular dystrophy") and ("effect on fertility" or "wish for children" or "maternalism") turned up 0 results.

4. Judith Butler's commentary on this case (2001) is similar in the sense that she, too, points to the "narrative of gender essentialism" (632) and to the extraordinary scrutiny under which Reimer developed a sense of self. I read Butler's account of Reimer's own subjective narrative as an exploration of how it is possible to "do justice" to him, to honor his account of himself, even as we pursue a critical stance toward gender as an institution, structure, and social process. My emphasis is different and involves a simpler point that Butler did not directly make: that relentless, overt scrutiny is itself a technology for *destabilizing* gender, for affirming that it is not "true" or "natural."

5. In a sense, proving either theory with reference to evidence could be a circular exercise. In the case of naturalized gender, the circularity is this: stable gender typically develops because everyone believes that it reflects the natural state of affairs—the "truth" of the developing person; if it didn't work, they must not have believed. In the case of brain organization theory, the circularity goes like this: gender identity develops under the influence (or absence) of androgens; if masculine identity develops, there *must have been* "sufficient" androgens.

6. This description of embodied gender has been influenced and informed by a number of developmental biologists who are thoughtful about gender, most especially Evan Balaban and Anne Fausto-Sterling. Fausto-Sterling has recently begun to conduct empirical research on how systematic differences in early interactions with infants could be evaluated in terms of a cascade-and-

feedback model of gender development (Fausto-Sterling 2006). I explore the embodiment of gender in much more detail in Chapter 10.

7. There is a sort of "chicken-and-egg" quality to this relationship. Did same-sex-oriented people form visible communities because they were distinct "types," including being relatively gender-nonconforming? Or did manner-isms cultivated through membership in or even knowledge of gay and lesbian subcultures sediment into patterns that connected same-sex desires with gen-der nonconformity? It's not possible to definitely answer this question, but the latter explanation fits better with several observations from history and con-temporary sexual cultures. The history of "the closet," for example, serves as a compelling argument against the notion that erotically atypical people are "by nature" gender-atypical in other ways. How would it be possible for so many people to conceal their sexual orientation if significant gender atypi-cality were universal among people with same-sex desires? And there is am-ple evidence that the link between sexual orientation and gender is self-reinforcing. Scores of anthropological and historical works document that gender-atypical lesbians and gay men have long been more readily recognized (as well as more punished and demonized) than their more gender-conforming counterparts. (Two excellent examples are Bérubé 1990, especially chapter 2 on "fitting in" in military life, and Chauncey 1994, especially regarding differ-ent men's relationship to the identity of "fairy.")

8. Here, though, to foreshadow some of the argument in Chapter 10, it is impor-tant to recognize that steroid levels alone don't really tell us much, because hormones operate within a complex and dynamic system that involves multi-ple chemical and neurological actors, feedback loops, receptor activity, and so on. One of my early suspicions about the effect of early hormone exposures was that these initial exposures "calibrate" an individual's system as initial preparation for these later dynamic functions. In this way, for example, fe-males become far more sensitive and responsive to testosterone at much lower levels than males are. This could show up empirically as a result that females do not show the same responses to testosterone fluctuations, or show the same testosterone responses to emotion, that males show. But those results may re-flect the fact that researchers typically examine the higher levels of production and response that are characteristic of males. These assays may not be sensi-tive enough to reveal small variations in the much lower levels found in females (which are more technically demanding and therefore expensive to de-tect). Biologically important fluctuations in females may therefore remain un-discovered and therefore unanalyzed.

10. Trading Essence for Potential

1. Pinto-Correia has fervently and convincingly argued that neither Leeuwen-hoek nor the other "spermists" of the seventeenth century used the specific term *homunculus*, which had far too much negative baggage at that time. I can't begin to do justice to the tale in the short space of a note, but the whole story is laid out in Pinto-Correia's *The Ovary of Eve* (1997).

2. Both Lewontin (2000) and Balaban (2006) explain one important source of stochasticity, which is the uneven distribution in space of molecules within developing cells, and the fact that there are limited temporal windows for particular kinds of molecular interactions to take place. "The consequence of there being a very small number of chemical units processed by spatially constrained intracellular machines is that there is considerable variation from cell to cell in the rate and number of molecules that are synthesized" (Lewontin 2000, 36).

3. Though they suggest the familiar story of the "female-typical" pattern (reliance on local markers) as the default strategy, this conclusion fits awkwardly with their data. First, the notion that use of local markers is a "female specialization" implies that females have this skill but are not skilled at spatial location, which is false. Thus, while blocking androgens does improve males' use of local markers, it seems unwarranted to conclude that androgens "eliminate the female specialization" (ability to use local markers).

4. The tone of popular coverage on this trend has sometimes been dire: "Newly released statistics show there are about 2 million more women than men on the nation's college campuses, a disparity that could alter the fabric of marriage in America, experts say. More middle-aged men lacking a college education are single. Marriage rates among professional women have flattened" (Emerson and Hall 2006, 1). The new trend has also prompted worries that efforts to keep gender balance in college have raised admissions standards even higher for women (Levy-Prewitt 2007).

5. Again, it's worth noting how the "ecotype NOR" is different from a strict NOR that maps development of a single genotype. The assumption in this comparison between the sex difference seen in the U.S. population in 1930 versus 1990 is that there is no plausible reason to believe that there has been any large-scale shift in the sex distribution of relevant genes in that sixty-year period. In terms of the brain organization story, it's important to point out that there is also no reason to believe that sex differences in prenatal hormone exposures have changed for some nongenetic reason. Conversely, there *has* been a discernible change in the relevant environment, namely, a series of measures to eliminate systematic sex discrimination in education and employment.

6. This study is especially useful because they used a control group who practiced a different kind of video game, and did not improve the spatial skills. As the authors explain: "Some previous studies found that video-game practice had little or no effect . . . , implying that improved performance on spatial tasks after video-game practice may depend on the kinds of spatial abilities that are needed in the game and in the spatial task" (853).

References

AAMC. See American Association of Medical Colleges.

Abel, D. 2001. MIT Bias Claims Challenged; Merit Cited as Reason for Pay, Lab Disparity. *Boston Globe*, February 8.

Achenbach, T. M., and C. S. Edelbrock. 1979. Child-Behavior Profile 2. Boys Aged 12–16 and Girls Aged 6–11 and 12–16. *Journal of Consulting and Clinical Psychology* 47 (2): 223–233.

Adkins-Regan, E. 2002. Development of Sexual Partner Preference in the Zebra Finch: A Socially Monogamous, Pair-Bonding Animal. *Archives of Sexual Behavior* 31 (1): 27–33.

Alexander, G. M. 2003. An Evolutionary Perspective of Sex-Typed Toy Preferences: Pink, Blue, and the Brain. *Archives of Sexual Behavior* 32 (1): 7–14.

Alexander, G. M., and M. Hines. 2002. Sex Differences in Response to Children's Toys in Nonhuman Primates *(Cercopithecus aethiops sabaeus)*. *Evolution and Human Behavior* 23 (6): 467–479.

Alexander, G. M., and B. S. Peterson. 2004. Testing the Prenatal Hormone Hypothesis of Tic-Related Disorders: Gender Identity and Gender Role Behavior. *Development and Psychopathology* 16 (2): 407–420.

Allen, L. S., and R. A. Gorski. 1992. Sexual Orientation and the Size of the Anterior Commissure in the Human Brain. *Proceedings of the National Academy of Sciences of the United States of America* 89 (August): 7199–7202.

Allen, L. S., M. Hines, J. E. Shryne, and R. A. Gorski. 1989. 2 Sexually Dimorphic Cell Groups in the Human Brain. *Journal of Neuroscience* 9 (2): 497–506.

American Association of Medical Colleges. 2010. *Women Enrollment and Graduates in U.S. Medical Schools, 1961–2008.* American Association of Medical

Colleges 2009. www.Aamc.Org/Data/Facts/Enrollmentgraduate/Table31-Women-Count.Htm.

Ammini, A. C., R. Gupta, A. Kapoor, A. Karak, A. Kriplani, D. K. Gupta, and K. Kucheria. 2002. Etiology, Clinical Profile, Gender Identity and Long-Term Follow Up of Patients with Ambiguous Genitalia in India. *Journal of Pediatric Endocrinology and Metabolism* 15 (4): 423–430.

Anders, S. M. Van, and E. Hampson. 2005. Testing the Prenatal Androgen Hypothesis: Measuring Digit Ratios, Sexual Orientation, and Spatial Abilities in Adults. *Hormones and Behavior* 47 (1): 92–98.

Arnold, A. P., and R. A. Gorski. 1984. Gonadal-Steroid Induction of Structural Sex-Differences in the Central Nervous System. *Annual Review of Neuroscience* 7:413–442.

Bailey, J. M. 2003. *The Man Who Would Be Queen: The Science of Gender-Bending and Transsexualism*. Washington, DC: Joseph Henry Press.

Bailey, A. A., and P. L. Hurd. 2005. Finger Length Ratio (2D:4D) Correlates with Physical Aggression in Men but not in Women. *Biological Psychology* 68 (3): 215–222.

Bailey, J. M., L. Willerman, and C. Parks. 1991. A Test of the Maternal Stress Theory of Human Male Homosexuality. *Archives of Sexual Behavior* 20 (3): 277–293.

Balaban, E. 1998. Eugenics and Individual Phenotypic Variation: To What Extent Is Biology a Predictive Science? *Science in Context* 11 (3–4): 331–356.

———. 2001. Behavior Genetics: Galen's Prophecy or Malpighi's Legacy? In *Thinking about Evolution: Philosophical and Political Perspectives*, ed. R. S. Singh, C. B. Krimbas, D. B. Paul, and J. Beatty. Cambridge: Cambridge University Press.

———. 2006. Cognitive Developmental Biology: History, Process and Fortune's Wheel. *Cognition* 101 (2): 298–332.

Barad, K. 1999. Agential Realism: Feminist Interventions in Understanding Scientific Practices. In *The Science Studies Reader*, ed. M. Biagioli. New York: Routledge.

———. 2007. *Meeting the Universe Halfway: Quantum Physics and the Entanglement of Matter and Meaning*. Durham, NC: Duke University Press.

Barker, M., A. Iantaffi, and C. Gupta. 2006. Bisexuality Research Questioned. *Psychologist* 19 (2): 81.

Barnett, R., and C. Rivers. 2004. *Same Difference: How Gender Myths Are Hurting Our Relationships, Our Children, and Our Jobs*. New York: Basic Books.

Baron-Cohen, S. 2003a. *The Essential Difference: The Truth about the Male and Female Brain*. New York: Basic Books.

———. 2003b. They Just Can't Help It. *Guardian*, April 17, 4.

———. 2007. Sex Differences in Mind: Keeping Science Distinct from Social Policy. In *Why Aren't More Women in Science?* ed. S. J. Ceci and W. M. Williams. Washington, DC: American Psychological Association.

Baron-Cohen, S., S. Lutchmaya, and R. Knickmeyer. 2004. *Prenatal Testosterone in Mind: Amniotic Fluid Studies*. Cambridge, MA: MIT Press.

Barres, B. A. 2006. Does Gender Matter? *Nature* 442:133–136.

Bates, J. E., and P. M. Bentler. 1973. Play Activities of Normal and Effeminate Boys. *Developmental Psychology* 9 (1): 20–27.

Baumeister, R. F. 2007. Is There Anything Good about Men? (invited address). American Psychological Association Annual Meeting, San Francisco. www .psy.fsu.edu/~baumeistertice/goodaboutmen.htm.

Bazelon, E. 2008. Hormones, Genes and the Corner Office. Review of "The Gender Paradox: Men, Women, and the Real Gender Gap" by Susan Pinker. *New York Times,* March 9.

Beach, F. A. 1971. Hormonal Factors Controlling the Differentiation, Development, and Display of Copulatory Behaviors in the Ramstergig (!) and Related Species. In *The Biopsychology of Development,* ed. E. Tobach, L. R. Aronson, and E. Shaw. New York: Academic Press.

Becker, J. T., S. M. Bass, M. A. Dew, L. E.Kingsley, O. A. Selnes, and K. Sheridan. 1992. Hand Preference, Immune System Disorder and Cognitive Function among Gay/Bisexual Men: The Multicenter AIDS Cohort Study (MACS). *Neuropsychologia* 30 (3): 229–235.

Bekker, M. H. J., G. L. Vanheck, and A. Vingerhoets. 1996. Gender-Identity, Body-Experience, Sexuality and the Wish for Having Children in DES-Daughters. *Women and Health* 24 (2): 65–82.

Bem, S. L. 1974. The Measurement of Psychological Androgyny. *Journal of Consulting and Clinical Psychology* 42 (2): 155–162.

Benderlioglu, Z., and R. J. Nelson. 2004. Digit Length Ratios Predict Reactive Aggression in Women, but Not in Men. *Hormones and Behavior* 46 (5): 558–564.

Beral, V., and L. Colwell. 1981. Randomised Trial of High Doses of Stilboestrol and Ethisterone Therapy in Pregnancy: Long-Term Follow-Up of the Children. *Journal of Epidemiology and Community Health* 35:155–160.

Berenbaum, S. A. 1998. How Hormones Affect Behavioral and Neural Development. Introduction to Special Issue, "Gonadal Hormones and Sex Differences in Behavior." *Developmental Neuropsychology* 14 (2–3): 175–196.

———. 1999. Effects of Early Androgens on Sex-Typed Activities and Interests in Adolescents with Congenital Adrenal Hyperplasia. *Hormones and Behavior* 35:102–110.

———. 2008. *Information for Families with CAH.* www.personal.psu.edu/faculty/ s/a/sab31/cah_info.html.

Berenbaum, S. A., and J. M. Bailey. 2003. Effects on Gender Identity of Prenatal Androgens and Genital Appearance: Evidence from Girls with Congenital Adrenal Hyperplasia. *Journal of Clinical Endocrinology and Metabolism* 88 (3): 1102–1106.

Berenbaum, S. A., K. K. Bryk, S. C. Duck, and I. M. Resnick. 2004. Psychological Adjustment in Children and Adults with Congenital Adrenal Hyperplasia. *Journal of Pediatrics* 144 (6): 741–746.

Berenbaum, S. A., K. K. Bryk, N. Nowak, C. A. Quigley, and S. Moffat. 2009. Fingers as a Marker of Prenatal Androgen Exposure. *Endocrinology* 150 (11): 5119–24.

Berenbaum, S. A., and S. D. Denburg. 1995. Evaluating the Empirical Support for the Role of Testosterone in the Geschwind-Behan-Galaburda Model of Cerebral Lateralization: Commentary. *Brain and Cognition* 27 (1): 79–83.

Berenbaum, S. A., S. C. Duck, and K. Bryk. 2000. Behavioral Effects of Prenatal versus Postnatal Androgen Excess in Children with 12-Hydroxylase-Deficient Congenital Adrenal Hyperplasia. *Journal of Clinical Endocrinology and Metabolism* 85 (2): 727–733.

Berenbaum, S. A., and M. Hines. 1992. Early Androgens Are Related to Childhood Sex-Typed Toy Preferences. *Psychological Science* 3 (3): 203–206.

Berenbaum, S. A., and S. M. Resnick. 1997. Early Androgen Effects on Aggression in Children and Adults with Congenital Adrenal Hyperplasia. *Psychoneuroendocrinology* 22 (7): 505–515.

———. 2007. The Seeds of Career Choices: Prenatal Sex Hormone Effects on Psychological Sex Differences. In *Why Aren't More Women in Science?* ed. S. J. Ceci and W. M. Williams. Washington, DC: American Psychological Association.

Berenbaum, S. A., and E. Snyder. 1995. Early Hormonal Influences on Childhood Sex-Typed Activity and Playmate Preferences: Implications for the Development of Sexual Orientation. *Developmental Psychology* 31 (1): 31–42.

Berger, J., and T. L. Conner. 1969. Performance Expectations and Behavior in Small Groups. *Acta Sociologica* 12 (4): 186–198.

Bérubé, A. 1990. *Coming Out under Fire: The History of Gay Men and Women in World War Two.* New York: Free Press.

Blakemore, J. E. O., and Renee E. Centers. 2005. Characteristics of Boys' and Girls' Toys. *Sex Roles* 53 (9/10): 619–633.

Blanchard, R., and R. A. Lippa. 2007. Birth Order, Sibling Sex Ratio, Handedness, and Sexual Orientation of Male and Female Participants in a BBC Internet Research Project. *Archives of Sexual Behavior* 36:163–176.

Bleier, R. 1984. *Science and Gender: A Critique of Biology and Its Theories on Women.* 2nd ed. New York: Pergamon Press.

Blum, D. 1997. *Sex on the Brain: The Biological Differences between Men and Women.* New York: Penguin Books.

Bodo, C., and E. F. Rissman. 2007. Androgen Receptor Is Essential for Sexual Differentiation of Responses to Olfactory Cues in Mice. *European Journal of Neuroscience* 25 (7): 2182–2190.

Bogaert, A. F. 1998. Physical Development and Sexual Orientation in Women: Height, Weight, and Age of Puberty Comparisons. *Personality and Individual Differences* 24 (1): 115–121.

———. 2003. The Interaction of Fraternal Birth Order and Body Size in Male Sexual Orientation. *Behavioral Neuroscience* 117 (2): 381–384.

Bogaert, A. F., and R. Blanchard. 1996. Handedness in Homosexual and Heterosexual Men in the Kinsey Interview Data. *Archives of Sexual Behavior* 25 (4): 373–378.

Bogaert, A. F., and C. Friesen. 2002. Sexual Orientation and Height, Weight, and Age of Puberty: New Tests from a British National Probability Sample. *Biological Psychology* 59 (2): 135–145.

Bogaert, A. F., C. Friesen, and P. Klentrou. 2002. Age of Puberty and Sexual Orientation in a National Probability Sample. *Archives of Sexual Behavior* 31 (1): 73–81.

Bogaert, A. F., and S. Hershberger. 1999. The Relation between Sexual Orientation and Penile Size. *Archives of Sexual Behavior* 28 (3): 213–221.

Bogart, L. M., H. Cecil, D. A. Wagstaff, S. D. Pinkerton, and P. R. Abramson. 2000. Is It "Sex"?: College Students' Interpretations of Sexual Behavior Terminology. *Journal of Sex Research* 37 (2): 108–116.

Bolton, N. J., J. Tapanainen, M. Koivisto, and R. Vihko. 1989. Circulating Sex Hormone-Binding Globulin and Testosterone in Newborns and Infants. *Clinical Endocrinology* 31:201–207.

Bostwick, J. M., and K. A. Martin. 2007. A Man's Brain in an Ambiguous Body: A Case of Mistaken Gender Identity. *American Journal of Psychiatry* 164 (10): 1499–1505.

Bowman, R. E., N. J. Maclusky, Y. Sarmiento, M. Frankfurt, M. Gordon, and V. N. Luine. 2004. Sexually Dimorphic Effects of Prenatal Stress on Cognition, Hormonal Responses, and Central Neurotransmitters. *Endocrinology* 145 (8): 3778–3787.

Boyar, R. M., and J. Aiman. 1982. The 24-Hour Secretory Pattern of LH and the Response to LHRH in Transsexual Men. *Archives of Sexual Behavior* 11 (2): 157–169.

Bradley, S. J., G. D. Oliver, A. B. Chernick, and K. J. Zucker. 1998. Experiment of Nurture: Ablatio Penis at 2 Months, Sex Reassignment at 7 Months, and a Psychosexual Follow-Up in Young Adulthood. *Pediatrics* 102 (1): e9.

Brizendine, L. 2006. *The Female Brain*. New York: Morgan Road Books.

Brooks, David. 2006. Is Chemistry Destiny? *New York Times,* September 17, 14.

Brown, W., M. Hines, B. A. Fane, and S. M. Breedlove. 2002. Masculinized Finger Length Patterns in Human Males and Females with Congenital Adrenal Hyperplasia. *Hormones and Behavior* 42 (4): 380–386.

Buchmann, C., and T. A. Diprete. 2006. The Growing Female Advantage in College Completion: The Role of Family Background and Academic Achievement. *American Sociological Review* 71 (4): 515–541.

Buck, J. J., R. M. Williams, I. A. Hughes, and C. L. Acerini. 2003. In-Utero Androgen Exposure and 2nd to 4th Digit Length Ratio: Comparisons between Healthy Controls and Females with Classical Congenital Adrenal Hyperplasia. *Human Reproduction* 18 (5): 976–979.

Bullough, V. L. 1994. *Science in the Bedroom*. New York: Basic Books.

Butler, J. 1990. *Gender Trouble: Feminism and the Subversion of Identity*. New York: Routledge.

———. 1993. *Bodies That Matter*. New York: Routledge.

———. 2001. Doing Justice to Someone: Sex Reassignment and Allegories of Transsexuality. *GLQ: A Journal of Lesbian and Gay Studies* 7 (4): 621–636.

Byne, W. 2006. Developmental Endocrine Influences on Gender Identity: Implications for Management of Disorders of Sex Development. *Mount Sinai Journal of Medicine* 73 (7): 950–959.

Byne, W., M. S. Lasco, E. Kemether, A. Shinwari, M. A. Edgar, S. Morgello, L. B.

Jones, and S. Tobet. 2000. The Interstitial Nuclei of the Human Anterior Hypothalamus: An Investigation of Sexual Variation in Volume and Cell Size, Number and Density. *Brain Research* 856 (1–2): 254–258.

Byne, W., and B. Parsons. 1993. Human Sexual Orientation: The Biologic Theories Reappraised. *Archives of General Psychiatry* 50 (3): 228–239.

Byne, W., S. Tobet, L. A. Mattiace, M. S. Lasco, E. Kemether, M. A. Edgar, S. Morgello, M. S. Buchsbaum, and L. B. Jones. 2001. The Interstitial Nuclei of the Human Anterior Hypothalamus: An Investigation of Variation with Sex, Sexual Orientation, and HIV Status. *Hormones and Behavior* 40 (2): 86–92.

Cahill, L. 2003. Sex- and Hemisphere-Related Influences on the Neurobiology of Emotionally Influenced Memory. *Progress in Neuro-Psychopharmacology and Biological Psychiatry* 27 (8): 1235–1241.

———. 2005. His Brain, Her Brain. *Scientific American* 292 (5): 40–47.

Cameron, D. 2007. *The Myth of Mars and Venus: Do Men and Women Really Speak Different Languages?* Oxford: Oxford University Press.

CARES Foundation. 2008. www.caresfoundation.org/productcart/pc/index.html.

Carey, B. 2005. Straight, Gay or Lying? Bisexuality Revisited. *New York Times,* July 5.

———. 2006. John William Money, 84, Sexual Identity Researcher. *New York Times,* July 11.

Carmines, E. G., and R. A. Zeller. 1979. *Reliability and Validity Assessment.* Newbury Park, CA: Sage.

Carvel, J. 2002. Britons Stand Tall, if Slightly Heavy, in Europe. *Guardian,* August 28.

Cavelaars, A., A. E. Kunst, J. J. M. Geurts, R. Crialesi, L. Grotvedt, U. Helmert, E. Lahelma, O. Lundberg, A. Mielck, N. K. Rasmussen, E. Regidor, T. Spuhler, and J. P. Mackenbach. 2000. Persistent Variations in Average Height between Countries and between Socio-Economic Groups: An Overview of 10 European Countries. *Annals of Human Biology* 27 (4): 407–421.

Ceci, S. J., and W. M. Williams, eds. 2007. *Why Aren't More Women in Science?* Washington, DC: American Psychological Association.

Charting Our Progress: The Status of Women in the Profession Today. 2006. Chicago: Commission on Women in the Profession.

Chase, C., and ISNA. 1998. Amicus Brief on Intersex Genital Surgery, submitted to the Constitutional Court of Colombia. Intersex Society of North America. www.isna.org/node/97.

Chauncey, G. 1994. *Gay New York: Gender, Urban Culture, and the Making of the Gay Male World 1890–1940.* New York: Basic Books.

Chivers, M. L., and J. M. Bailey. 2005. A Sex Difference in Features That Elicit Genital Response. *Biological Psychology* 70 (2): 115–120.

Clausen, J., and W. M. Hiesey. 1960. The Balance between Coherence and Variation in Evolution. *Proceedings of the National Academy of Sciences of the United States of America* 46 (4): 494–506.

Clausen, J., D. D. Keck, and W. M. Hiesey. 1947. Heredity of Geographically and Ecologically Isolated Races. *American Naturalist* 81 (797): 114–133.

Clemens, L. G., and B. A. Gladue. 1978. Feminine Sexual-Behavior in Rats Enhanced by Prenatal Inhibition of Androgen Aromatization. *Hormones and Behavior* 11 (2): 190–201.

Clemens, L. G., M. Hiroi, and R. A. Gorski. 1969. Induction and Facilitation of Female Mating Behavior in Rats Treated Neonatally with Low Doses of Testosterone Propionate. *Endocrinology* 84 (6): 1430.

Cohen, K. M. 2002. Relationships among Childhood Sex-Atypical Behavior, Spatial Ability, Handedness, and Sexual Orientation in Men. *Archives of Sexual Behavior* 31 (1): 129–143.

Cohen-Bendahan, C. C. C., J. K. Buitelaar, S. H. M. Van Goozen, and P. T. Cohen-Kettenis. 2004. Prenatal Exposure to Testosterone and Functional Cerebral Lateralization: A Study in Same-Sex and Opposite-Sex Twin Girls. *Psychoneuroendocrinology* 29 (7): 911–916.

Cohen-Bendahan, C. C. C., J. K. Buitelaar, S. H. M. Van Goozen, J. F. Orlebeke, and P. T. Cohen-Kettenis. 2005. Is There an Effect of Prenatal Testosterone on Aggression and Other Behavioral Traits? A Study Comparing Same-Sex and Opposite-Sex Twin Girls. *Hormones and Behavior* 47 (2): 230–237.

Cohen-Bendahan, C. C. C., C. van de Beek, and S. A. Berenbaum. 2005. Prenatal Sex Hormone Effects on Child and Adult Sex-Typed Behavior: Methods and Findings. *Neuroscience and Biobehavioral Reviews* 29 (2): 353–384.

Cohen-Bendahan, C. C. C., S. H. M. Van Goozen, J. K. Buitelaar, and P. T. Cohen-Kettenis. 2005. Maternal Serum Steroid Levels Are Unrelated to Fetal Sex: A Study in Twin Pregnancies. *Twin Research and Human Genetics* 8 (2): 173–177.

Cohen-Kettenis, P. T. 2005. Gender Change in 46,XY Persons with 5 Alpha-Reductase-2 Deficiency and 17 Beta-Hydroxysteroid Dehydrogenase-3 Deficiency. *Archives of Sexual Behavior* 34 (4): 399–410.

Cohen-Kettenis, P. T., S. H. M. Van Goozen, C. D. Doorn, and L. J. G. Gooren. 1998. Cognitive Ability and Cerebral Lateralisation in Transsexuals. *Psychoneuroendocrinology* 23 (6): 631–641.

Colapinto, J. 1997. The True Story of John/Joan. *Rolling Stone,* December 11, 54–97.

———. 2001. *As Nature Made Him: The Boy Who Was Raised as a Girl.* New York: HarperCollins.

Coleman, E., ed. 1991. *John Money: A Tribute.* Binghamton: Haworth Press.

Collaer, M. L., S. Reimers, and J. T. Manning. 2007. Visuospatial Performance on an Internet Line Judgment Task and Potential Hormonal Markers: Sex, Sexual Orientation, and 2D:4D. *Archives of Sexual Behavior* 36 (2): 177–192.

Condry, J., and S. Condry. 1976. Sex Differences: A Study of the Eye of the Beholder. *Child Development* 47:812–819.

Conkey, M. 2003. Has Feminism Changed Archaeology? *Signs* 28 (3): 867–883.

Connellan, J., S. Baron-Cohen, S. Wheelwright, A. Batki, and J. Ahluwalia. 2000. Sex Differences in Human Neonatal Social Perception. *Infant Behavior and Development* 23 (1): 113–118.

Connor, S. 2007. Boys Like Blue, Girls Like Pink: It's in Our Genes. August 21. The Independent (online), Science Section. www.independent.co.uk/news/science/boys-like-blue-girls-like-pink—its-in-our-genes-462390.html.

Cook, H. 2004. *The Long Sexual Revolution: English Women, Sex, and Contraception, 1800–1975.* Oxford: Oxford University Press.

Cook, T. D., and D. T. Campbell. 1979. *Quasi-Experimentation: Design and Analysis Issues for Field Settings.* Boston: Houghton Mifflin.

Cornean, R. E., P. C. Hindmarsh, and C. G. D. Brook. 1998. Obesity in 21-Hydroxylase Deficient Patients. *Archives of Disease in Childhood* 78 (3): 261–263.

Costa, P. T., and R. R. McCrae. 1992. 4 Ways 5 Factors Are Basic. *Personality and Individual Differences* 13 (6): 653–665.

Coyne, S. M., J. T. Manning, L. Ringer, and L. Bailey. 2007. Directional Asymmetry (Right-Left Differences) in Digit Ratio (2D:4D) Predict Indirect Aggression in Women. *Personality and Individual Differences* 43 (4): 865–872.

Crouch, N. S., and S. M. Creighton. 2007. Long-Term Functional Outcomes of Female Genital Reconstruction in Childhood. *British Journal of Urology International* 100 (2): 403–406.

Crouch, N. S., C. L. Minto, L. M. Laio, C. R. J. Woodhouse, and S. M. Creighton. 2004. Genital Sensation after Feminizing Genitoplasty for Congenital Adrenal Hyperplasia: A Pilot Study. *BJU International* 93 (1): 135–138.

Csatho, A., A. Osvath, E. Bicsak, K. Karadi, J. Manning, and J. Kallai. 2003. Sex Role Identity Related to the Ratio of Second to Fourth Digit Length in Women. *Biological Psychology* 62 (2): 147–156.

Culp, R. E., A. S. Cook, and P. C. Housley. 1983. A Comparison of Observed and Reported Adult-Infant Interactions: Effects of Perceived Sex. *Sex Roles* 9 (4): 475–479.

D'Emilio, J. 1983. *Sexual Politics, Sexual Communities: The Making of a Homosexual Minority in the United States, 1940–1970.* Chicago: University of Chicago Press.

D'Emilio, J., and E. B. Freedman. 1988. *Intimate Matters: A History of Sexuality in America.* New York: Harper and Row.

Dalton, K. 1968. Ante-Natal Progesterone and Intelligence. *British Journal of Psychiatry* 114 (516): 1377–1382.

———. 1976. Prenatal Progesterone and Educational Attainments. *British Journal of Psychiatry* 129:438–442.

Darney, P. D. 1995. The Androgenicity of Progestins. *American Journal of Medicine* 98:S104–S110.

Davis, K. B. 1924–1925. *A Study of Certain Auto-Erotic Practices.* New York: National Committee for Mental Hygiene.

Davis, P. G., C. V. Chaptal, and B. S. McEwen. 1979. Independence of the Differentiation of Masculine and Feminine Sexual-Behavior in Rats. *Hormones and Behavior* 12 (1): 12–19.

Decoster, J., and H. M. Claypool. 2004. A Meta-Analysis of Priming Effects on Impression Formation Supporting a General Model of Informational Biases. *Personality and Social Psychology Review* 8 (1): 2–27.

Dessens, A. B., P. T. Cohen-Kettenis, G. J. Mellenbergh, N. Van Der Poll, J. G. Koppe, and K. Boer. 1999. Prenatal Exposure to Anticonvulsants and Psychosexual Development. *Archives of Sexual Behavior* 28 (1): 31–44.

Dessens, A. B., F. M. E. Slijper, and S. L. S. Drop. 2005. Gender Dysphoria and Gender Change in Chromosomal Females with Congenital Adrenal Hyperplasia. *Archives of Sexual Behavior* 34 (4): 389–397.

Diamond, L. M. 2008. *Sexual Fluidity.* Cambridge, MA: Harvard University Press.

Diamond, L. M., and R. C. Savin-Williams. 2000. Explaining Diversity in the Development of Same-Sex Sexuality among Young Women. *Journal of Social Issues* 56 (2): 297–313.

Diamond, M. 1965. A Critical Evaluation of the Ontogeny of Human Sexual Behavior. *Quarterly Review of Biology* 40:147–175.

———. 1974. Comments on the Review Article "Fetal Hormones, the Brain, and Human Sex Differences." *Archives of Sexual Behavior* 3 (5): 485–486.

———. 1982. Sexual Identity, Monozygotic Twins Reared in Discordant Sex Roles and a BBC Follow-Up. *Archives of Sexual Behavior* 11 (2): 181–186.

———. 1998. Foreword: Sexual Development: Nature's Substrate for Nurture's Influence. In *Males, Females, and Behavior: Toward Biological Understanding,* ed. L. Ellis and L. Ebertz. Westport, CT: Praeger.

Diamond, M., and H. K. Sigmundson. 1997. Sex Reassignment at Birth: Long-Term Review and Clinical Implications. *Archives of Pediatrics and Adolescent Medicine* 151 (3): 298–304.

Dittmann, R. W., M. F. Kappes, and M. H. Kappes. 1992. Sexual Behavior in Adolescent and Adult Females with Congenital Adrenal-Hyperplasia. *Psychoneuroendocrinology* 17 (2–3): 153–170.

Dittmann, R. W., M. H. Kappes, M. E. Kappes, D. Borger, H. Stegner, R. H. Willig, and H. Wallis. 1990. Congenital Adrenal Hyperplasia I: Gender-Related Behavior and Attitudes in Female Patients and Sisters. *Psychoneuroendocrinology* 15 (5–6): 401–420.

Dittmann, R. W., M. E. Kappes, M. H. Kappes, D. Borger, H. F. L. Meyer-Bahlburg, H. Stegner, R. H. Willig, and H. Wallis. 1990. Congenital Adrenal Hyperplasia II: Gender-Related Behavior and Attitudes in Female Salt-Wasting and Simple-Virilizing Patients. *Psychoneuroendocrinology* 15 (5–6): 421–434.

Doell, R., and H. Longino. 1988. Sex Hormones and Human Behavior: A Critique of the Linear Model. *Journal of Homosexuality* 15 (3–4): 55–78.

Donovan, W., N. Taylor, and L. Leavitt. 2007. Maternal Sensory Sensitivity and Response Bias in Detecting Change in Infant Facial Expressions: Maternal Self-Efficacy and Infant Gender Labeling. *Infant Behavior and Development* 30 (3): 436–452.

Dörner, G., T. Geier, L. Ahrens, L. Krell, G. Münx, H. Sieler, E. Kittner, and H. Müller. 1980. Prenatal Stress as Possible Aetiogenetic Factor of Homosexuality in Human Males. *Endokrinologie* 75 (3): 365–368.

Dörner, G., and G. Hinz. 1968. Induction and Prevention of Male Homosexuality by Androgen. *Journal of Endocrinology* 40:387–388.

Dörner, G., I. Poppe, F. Stahl, J. Kolzsch, and R. Uebelhack. 1991. Gene-Dependent and Environment-Dependent Neuroendocrine Etiogenesis of Homosexuality and Transsexualism. *Experimental and Clinical Endocrinology* 98 (2): 141–150.

Dörner, G., W. Rohde, F. Stahl, L. Krell, and W. Masius. 1975. A Neuroendocrine Predisposition for Homosexuality in Men. *Archives of Sexual Behavior* 4 (1): 1–8.

Dörner, G., W. Rohde, K. Seidel, W. Haas, and G. Scholl. 1976. On the Evocability of a Positive Oestrogen Feedback Action on LH Secretion in Transsexual Men and Women. *Endokrinologie* 67 (1): 20–25.

Dörner, G., W. Rohde, G. Schott, and C. Schnabl. 1983. On the LH Response to Oestrogen and LH-RH in Transsexual Men. *Experimental and Clinical Endocrinology* 82 (3): 257–267.

Dörner, G., B. Schenk, B. Schmiedel, and L. Ahrens. 1983. Stressful Events in Prenatal Life of Bi- and Homosexual Men. *Experimental and Clinical Endocrinology* 81 (1): 83–87.

Dreger, A. D. 1998. *Hermaphrodites and the Medical Invention of Sex.* Cambridge, MA: Harvard University Press.

Driskell, J. E., 1982. Personal Characteristics and Performance Expectations. *Social Psychology Quarterly* 45 (4): 229–237.

Driskell, J. E., and B. Mullen. 1990. Status, Expectations, and Behavior: A Meta-Analytic Review and Test of the Theory. *Personality and Social Psychology Bulletin* 16 (3): 541–553.

Dubey, R. K., E. K. Jackson, D. G. Gillespie, L. C. Zacharia, D. Wunder, B. Imthurn, and M. Rosselli. 2008. Medroxyprogesterone Abrogates the Inhibitory Effects of Estradiol on Vascular Smooth Muscle Cells by Preventing Estradiol Metabolism. *Hypertension* 51 (4): 1197–1202.

Ehrenreich, B., and D. English. 1978. *For Her Own Good: 150 Years of the Experts' Advice to Women.* Garden City, NY: Anchor Press.

Ehrhardt, A. A. 1985. Gender Differences: A Biosocial Perspective. *Psychology and Gender* 32:37–57.

———. 1988. Laudatio for John Money on the Occasion of the 1988 Masters and Johnson Award for the Society of Sex Therapy and Research. *Journal of Sex and Marital Therapy* 14:173–176.

Ehrhardt, A. A., and S. Baker. 1974. Fetal Androgen, Human Central Nervous System Differentiation, and Behavior Sex Differences. In *Sex Differences in Behavior,* ed. R. C. Friedman, R. M. Richart, and R. L. Vande Weile. New York: John Wiley.

Ehrhardt, A. A., R. Epstein, and J. Money. 1968. Fetal Androgens and Female Gender Identity in the Early-Treated Adrenogenital Syndrome. *Johns Hopkins Medical Journal* 122:160–167.

Ehrhardt, A. A., K. Evers, and J. Money. 1968. Influence of Androgen and Some Aspects of Sexually Dimorphic Behavior in Women with the Late-Treated Adrenogenital Syndrome. *Johns Hopkins Medical Journal* 123:115–122.

Ehrhardt, A. A., G. C. Grisanti, and H. F. L. Meyer-Bahlburg. 1977. Prenatal Exposure to Medroxyprogesterone Acetate (MPA) in Girls. *Psychoneuroendocrinology* 2:391–398.

Ehrhardt, A. A., and H. F. L. Meyer-Bahlburg. 1979. Prenatal Sex Hormones and the Developing Brain: Effects on Psychosexual Differentiation and Cognitive Function. *Annual Review of Medicine* 30:417–430.

———. 1981. Effects of Prenatal Sex Hormones on Gender-Related Behavior. *Science* 211 (March 20): 1312–1318.

Ehrhardt, A. A., H. F. L. Meyer-Bahlburg, J. F. Feldman, and S. E. Ince. 1984. Sex-Dimorphic Behavior in Childhood Subsequent to Prenatal Exposure to Exogenous Progestogens and Estrogens. *Archives of Sexual Behavior* 13 (5): 457–477.

Ehrhardt, A. A., H. F. L. Meyer-Bahlburg, L. R. Rosen, J. F. Feldman, N. P. Veridiano, I. Zimmerman, and B. S. McEwen. 1985. Sexual Orientation after Prenatal Exposure to Exogenous Estrogen. *Archives of Sexual Behavior* 14 (1): 57–77.

Ehrhardt, A. A., H. F. L. Meyer-Bahlburg, L. R. Rosen, J. F. Feldman, N. P. Veridiano, E. J. Elkin, and B. S. McEwen. 1989. The Development of Gender-Related Behavior in Females following Prenatal Exposure to Diethylstilbestrol (DES). *Hormones and Behavior* 23:526–541.

Ehrhardt, A. A., and J. Money. 1967. Progestin-Induced Hermaphroditism: IQ and Psychosexual Identity in a Study of Ten Girls. *Journal of Sex Research* 3 (1): 83–100.

Ellis, H. 1925. *The Theory of Sexual Inversion,* 3rd ed. Philadelphia: F. A. Davis.

Ellis, L., and M. A. Ames. 1987. Neurohormonal Functioning and Sexual Orientation: A Theory of Homosexuality-Heterosexuality. *Psychological Bulletin* 101 (2): 233–258.

Ellis, L., M. A. Ames, W. Peckham, and D. Burke. 1988. Sexual Orientation of Human Offspring May Be Altered by Severe Maternal Stress during Pregnancy. *Journal of Sex Research* 25 (1): 152–157.

Ellis, L., D. Burke, and M. A. Ames. 1987. Sexual Orientation as a Continuous Variable: A Comparison between the Sexes. *Archives of Sexual Behavior* 16 (6): 523–529.

Ellis, L., and S. Cole-Harding. 2001. The Effects of Prenatal Stress, and of Prenatal Alcohol and Nicotine Exposure, on Human Sexual Orientation. *Physiology and Behavior* 74 (1–2): 213–226.

Emerson, A., and H. Hall. 2006. It's Not Easy to Find an (Good) Educated Man: Fewer Men on Campus Leaves Fewer Options. *Tampa Tribune,* September 10, 1.

Esgate, A., and M. Flynn. 2005. The Brain-Sex Theory of Occupational Choice: A Counterexample. *Perception and Motor Skills* 100 (1): 25–37.

Eugster, E. A., L. A. Dimeglio, J. C. Wright, G. R. Freidenberg, R. Seshadri, and O. H. Pescovitz. 2001. Height Outcome in Congenital Adrenal Hyperplasia Caused by 21-Hydroxylase Deficiency: A Meta-Analysis. *Journal of Pediatrics* 138 (1): 26–32.

Fagot, B. I. 1978. Influence of Sex of Child on Parental Reactions to Toddler Children. *Child Development* 49 (2): 459–465.

Fausto-Sterling, A. 1985. *Myths of Gender: Biological Theories about Women and Men.* New York: Basic Books.

———. 2000. *Sexing the Body.* New York: Basic Books.

———. 2005. The Bare Bones of Sex: Part 1: Sex and Gender. *Signs* 30 (2): 1491–1527.

———. 2006. Rethinking Nature vs. Nurture. Helen Pond McIntyre Lecture, September 14. Barnard College, New York NY.

Fazio, R. H., and M. A. Olson. 2003. Implicit Measures in Social Cognition Research: Their Meaning and Use. *Annual Review of Psychology* 54:297–327.

Federman, D. D. 1987. Psychosexual Adjustment in Congenital Adrenal Hyperplasia. *New England Journal of Medicine* 316 (4): 209–210.

Feng, J., I. Spence, and J. Pratt. 2007. Playing an Action Video Game Reduces Gender Differences in Spatial Cognition. *Psychological Science* 18 (10): 850–855.

Ferguson-Smith, M. A., and N. A. Affara. 1988. Accidental X-Y Recombination and the Etiology of XX Males and True Hermaphrodites. *Philosophical Transactions of the Royal Society of London, Series B: Biological Sciences* 322 (1208): 133–143.

Fewer Men on Campus. 2006. *USA Today,* July 12.

Finegan, J. A. K., G. A. Niccols, and G. Sitarenios. 1992. Relations between Prenatal Testosterone Levels and Cognitive Abilities at 4 Years. *Developmental Psychology* 28 (6): 1075–1089.

Fink, B., J. T. Manning, and N. Neave. 2004. Second to Fourth Digit Ratio and the 'Big Five' Personality Factors. *Personality and Individual Differences* 37 (3): 495–503.

Fink, B., N. Neave, K. Laughton, and J. T. Manning. 2006. Second to Fourth Digit Ratio and Sensation-Seeking. *Personality and Individual Differences* 41 (7): 1253–1262.

Fischer, A. H. 1993. Sex-Differences in Emotionality: Fact or Stereotype? *Feminism and Psychology* 3 (3): 303–318.

Fischer, A. H., and P. M. Rodriguez Mosquera. 2001. What Concerns Men? Women or Other Men? A Critical Appraisal of the Evolutionary Theory of Sex Differences in Aggression. *Psychology, Evolution and Gender* 3 (1): 5–25.

Forest, M. G. 2004. Recent Advances in the Diagnosis and Management of Congenital Adrenal Hyperplasia due to 21-Hydroxylase Deficiency. *Human Reproduction Update* 10 (6): 469–485.

Foucault, M. 1978. *The History of Sexuality: An Introduction,* vol. 1, trans. R. Hurley. New York: Pantheon.

———. 1990. *The Use of Pleasure: The History of Sexuality,* vol. 2, trans. R. Hurley. New York: Random House.

Francis, R. W. n.d. *The History behind the Equal Rights Amendment.* Alice Paul Institute. www.equalrightsamendment.org/era.htm.

Fredriks, A. M., S. Van Buuren, R. J. F. Burgmeijer, J. F. Meulmeester, R. J. Beuker, E. Brugman, M. J. Roede, S. P. Verloove-Vanhorick, and J. M. Wit. 2000. Continuing Positive Secular Growth Change in The Netherlands, 1955–1997. *Pediatric Research* 47 (3): 316–323.

Freud, S. 1965. *New Introductory Lectures on Psychoanalysis.* New York: Norton.

Gagnon, J. H., W. Simon, and A. J. Berger. 1970. Some Aspects of Sexual Adjustment in Early and Later Adolescence. *Proceedings of the Annual Meeting of the American Psychopathology Association* 59:275–298.

Gamson, W. A., and A. Modigliani. 1987. The Changing Culture of Affirmative Action. In *Research in Political Sociology,* ed. R. A. Braumgart. Greenwich, CT: JAI Press.

Garfinkel, H. 1967. *Studies in Ethnomethodology.* Englewood Cliffs, NJ: Prentice-Hall.

Gastaud, F., C. Bouvattier, L. Duranteau, R. Brauner, E. Thibaud, F. Kutten, and P. Bougneres. 2007. Impaired Sexual and Reproductive Outcomes in Women with Classical Forms of Congenital Adrenal Hyperplasia. *Journal of Clinical Endocrinology and Metabolism* 92 (4): 1391–96.

Geschwind, N., and P. Behan. 1982. Left-Handedness: Association with Immune Disease, Migraine, and Developmental Learning Disorder. *Proceedings of the National Academy of Sciences of the United States of America: Biological Sciences* 79 (16): 5097–5100.

———. 1984. Hormones, Handedness and Immunity. *Immunology Today* 5 (7): 190–191.

Geschwind, N., and A. M. Galaburda. 1985a. Cerebral Lateralization: Biological Mechanisms, Associations, and Pathology: 1. A Hypothesis and a Program for Research. *Archives of Neurology* 42 (5): 428–459.

———. 1985b. Cerebral Lateralization: Biological Mechanisms, Associations, and Pathology: 2. A Hypothesis and a Program for Research. *Archives of Neurology* 42 (6): 521–552.

———. 1985c. Cerebral Lateralization: Biological Mechanisms, Associations, and Pathology: 3. A Hypothesis and a Program for Research. *Archives of Neurology* 42 (7): 634–654.

Gherardi, S., and B. Poggio. 2001. Creating and Recreating Gender Order in Organizations. *Journal of World Business* 36 (3): 245–259.

Gieryn, T. F. 1999. *Cultural Boundaries of Science.* Chicago: University of Chicago Press.

Gladue, B. A., and J. M. Bailey. 1995a. Aggressiveness, Competitiveness, and Human Sexual Orientation. *Psychoneuroendocrinology* 20 (5): 475–485.

———. 1995b. Spatial Ability, Handedness, and Human Sexual Orientation. *Psychoneuroendocrinology* 20 (5): 487–497.

Gladue, B. A., W. W. Beatty, J. Larson, and R. D. Staton. 1990. Sexual Orientation and Spatial Ability in Men and Women. *Psychobiology* 18 (1): 101–108.

Gladue, B. A., R. Green, and R. E. Hellman. 1984. Neuroendocrine Response to Estrogen and Sexual Orientation. *Science* 225:1496–1498.

Glantz, S. A. 2002. *Primer of Biostatistics.* 5th ed. New York: McGraw-Hill Medical.

Glass, S. J., and R. H. Johnson. 1944. Limitations and Complications of Organotherapy in Male Homosexuality. *Journal of Clinical Endocrinology* 4 (11): 540–544.

Goffman, E. 1963. *Stigma: Notes on the Management of Spoiled Identity.* Englewood Cliffs, NJ: Prentice-Hall.

Goh, H. H., S. S. Ratnam, and D. R. London. 1984. The Feminisation of Gonadotrophin Responses in Intact Male Transsexuals. *Clinical Endocrinology* 20:591–596.

Goldberg, C. 1999. MIT Issues Report Acknowledging Sex Discrimination. *New York Times,* March 23.

Goldberg, J. M., and T. Falcone. 1999. Effect of Diethylstilbestrol on Reproductive Function. *Fertility and Sterility* 72 (1): 1–7.

Golombok, S., and J. Rust. 1993. The Measurement of Gender Role Behaviour in Pre-School Children: A Research Note. *Journal of Child Psychology and Psychiatry* 34 (5): 805–811.

Goncalves, E. M., S. H. V. De Lemos-Marini, M. P. De Mello, M. T. M. Baptista, L. F. R. D'Souza-Li, A. D. Baldin, W. R. G. Carvalho, E. S. Farias, and G. Guerra. 2009. Impairment in Anthropometric Parameters and Body Composition in Females with Classical 21-Hydroxylase Deficiency. *Journal of Pediatric Endocrinology and Metabolism* 22 (6): 519–529.

Gooren, L. 1990. The Endocrinology of Transsexualism: A Review and Commentary. *Psychoneuroendocrinology* 15 (1): 3–14.

Gooren, L., and P. T. Cohen-Kettenis. 1991. Development of Male Gender Identity/Role and a Sexual Orientation towards Women in a 46,XY Subject with an Incomplete Form of the Androgen Insensitivity Syndrome. *Archives of Sexual Behavior* 20 (5): 459–470.

Gooren, L., B. R. Rao, H. Van Kessel, and W. Harmsen-Louman. 1984. Estrogen Positive Feedback on LH Secretion in Transsexuality. *Psychoneuroendocrinology* 9 (3): 249–259.

Gordon, A. H., P. A. Lee, M. K. Dulcan, and D. N. Finegold. 1986. Behavioral Problems, Social Competency, and Self Perception among Girls with Congenital Adrenal Hyperplasia. *Child Psychiatry and Human Development* 17 (2): 129–138.

Gorski, R. A. 1979. Neuroendocrinology of Reproduction: Overview. *Biology of Reproduction* 20 (1): 111–127.

Götestam, K. O., T. J. Coates, and M. Ekstrand. 1992. Handedness, Dyslexia and Twinning in Homosexual Men. *International Journal of Neuroscience* 63:179–186.

Gould, S. J. 1996. *The Mismeasure of Man.* 2nd ed. New York: W. W. Norton.

Goy, R. W., F. B. Bercovitch, and M. C. McBrair. 1988. Behavioral Masculinization Is Independent of Genital Masculinization in Prenatally Androgenized Female Rhesus Macaques. *Hormones and Behavior* 22:552–571.

Goy, R. W., and D. A. Goldfoot. 1975. Neuroendocrinology: Animal Models and Problems of Human Sexuality. *Archives of Sexual Behavior* 4 (4): 405–420.

Green, R., and R. Young. 2001. Hand Preference, Sexual Preference, and Transsexualism. *Archives of Sexual Behavior* 30 (6): 565–574.

Grimshaw, G. M., G. Sitarenios, and J. A. K. Finegan. 1995. Mental Rotation at 7 Years: Relations with Prenatal Testosterone Levels and Spatial Play Experiences. *Brain and Cognition* 29 (1): 85–100.

Grosse, S. D., and G. Van Vliet. 2007. How Many Deaths Can Be Prevented by Newborn Screening for Congenital Adrenal Hyperplasia? *Hormone Research* 67 (6): 284–291.

Gupta, D. K., S. Shilpa, A. C. Ammini, M. Gupta, G. Aggarwal, G. Deepika, and

K. Kamlesh. 2006. Congenital Adrenal Hyperplasia: Long-Term Evaluation of Feminizing Genitoplasty and Psychosocial Aspects. *Pediatric Surgery International* 22 (11): 905–909.

Hall, J. A. Y. and D. Kimura. 1995. Sexual Orientation and Performance on Sexually Dimorphic Motor Tasks. *Archives of Sexual Behavior* 24 (4): 395–407.

Hall, L. S. 2000. Dermatoglyphic Analysis of Total Finger Ridge Count in Female Monozygotic Twins Discordant for Sexual Orientation. *Journal of Sex Research* 37 (4): 315–320.

Hamer, D., and P. Copeland. 1994. *The Science of Desire: The Search for the Gay Gene and the Biology of Behavior.* New York: Simon and Schuster.

Hammonds, E., and B. Subramaniam. 2003. A Conversation on Feminist Science Studies. *Signs* 28 (3): 923–944.

Hampson, E., C. L. Ellis, and C. M. Tenk. 2008. On the Relation between 2D:4D and Sex-Dimorphic Personality Traits. *Archives of Sexual Behavior* 37:133–144.

Hampson, E., J. F. Rovet, and D. Altmann. 1998. Spatial Reasoning in Children with Congenital Adrenal Hyperplasia due to 21-Hydroxylase Deficiency. *Developmental Neuropsychology* 14 (2–3): 299–320.

Haraway, D. 1978. Animal Sociology and a Natural Economy of the Body Politic, Part II: The Past Is the Contested Zone: Human Nature and Theories of Production and Reproduction in Primate Behavior Studies. *Signs* 4 (1). 21–36.

——. 1989. *Primate Visions: Gender, Race and Nature in the World of Modern Science.* New York: Routledge.

——. 1991. *Simians, Cyborgs, and Women: The Reinvention of Nature.* New York: Routledge.

Harding, S. 1998. *Is Science Multicultural? Postcolonialisms, Feminisms, and Epistemologies.* Bloomington: Indiana University Press.

——. 2006. *Science and Social Inequality: Feminist and Postcolonial Issues.* Urbana: University of Illinois Press.

Harris, A. 2004. *Gender as Soft Assembly.* New York: Routledge.

Harter, S. 1982. The Perceived Competence Scale for Children. *Child Development* 53:87–97.

Hausman, B. L. 2000. Do Boys Have to Be Boys? Gender, Narrativity, and the John/Joan Case. *NWSA Journal* 12 (3): 114–138.

Hausman, P. 2000. A Tale of Two Hormones. Paper presented at the National Academy of Engineering, Southeast Regional Meeting, Atlanta, April 26. www.patriciahausman.com/speeches.html

Helleday, J., A. Bartfai, E. M. Ritzen, and M. Forsman. 1994. General Intelligence and Cognitive Profile in Women with Congenital Adrenal-Hyperplasia (CAH). *Psychoneuroendocrinology* 19 (4): 343–356.

Helleday, J., G. Edman, E. M. Ritzen, and B. Siwers. 1993. Personality Characteristics and Platelet MAO Activity in Women with Congenital Adrenal-Hyperplasia (CAH). *Psychoneuroendocrinology* 18 (5–6): 343–354.

Helleday, J., B. Siwers, E. M. Ritzen, and K. Hugdahl. 1994. Normal Lateralization for Handedness and Ear Advantage in a Verbal Dichotic-Listening Task in

Women with Congenital Adrenal-Hyperplasia (CAH). *Neuropsychologia* 32 (7): 875–880.

Heller, C. G., and W. O. Maddock. 1947. The Clinical Uses of Testosterone in the Male. *Vitamins and Hormones: Advances in Research and Applications* 5:393–432.

Henderson, B. A., and S. A. Berenbaum. 1997. Sex-Typed Play in Opposite-Sex Twins. *Developmental Psychobiology* 31 (2): 115–123.

Hendricks, S. E., B. Graber, and J. F. Rodriguez-Sierra. 1989. Neuroendocrine Responses to Exogenous Estrogen: No Differences between Heterosexual and Homosexual Men. *Psychoneuroendocrinology* 14 (3): 177–185.

Hendricks, S. E., J. R. Lehman, and G. Oswalt. 1982. Responses to Copulatory Stimulation in Neonatally Androgenized Female Rats. *Journal of Comparative and Physiological Psychology* 96 (5): 834–845.

Herdt, G. H., and J. Davidson. 1988. The Sambia "Turnim-Man": Sociocultural and Clinical Aspects of Gender Formation in Male Pseudohermaphrodites with 5-Alpha-Reductase Deficiency in Papua New Guinea. *Archives of Sexual Behavior* 17 (1): 33–56.

Herman, R. A., and K. Wallen. 2007. Cognitive Performance in Rhesus Monkeys Varies by Sex and Prenatal Androgen Exposure. *Hormones and Behavior* 51 (4): 496–507.

Hewlett, S. A. 2002. *Creating a Life: Professional Women and the Quest for Children.* New York: Hyperion.

Hiesey, W. M., J. Clausen, and D. D. Keck. 1942. Ecological Aspects of Evolution: Relations between Climate and Intraspecific Variation in Plants. *American Naturalist* 76:5–22.

Hines, M. 1982. Prenatal Gonadal Hormones and Sex Differences in Human Behavior. *Psychological Bulletin* 92 (1): 56–80.

———. 2003. Sex Steroids and Human Behavior: Prenatal Androgen Exposure and Sex-Typical Play Behavior in Children. In *Steroids and the Nervous System,* ed. G. Panzica and R. C. Melcangi. *Annals of the New York Academy of Sciences* 1007:272–282.

———. 2004. *Brain Gender.* New York: Oxford University Press.

———. 2007. Do Sex Differences in Cognition Cause the Shortage of Women in Science? In *Why Aren't More Women in Science? Top Researchers Debate the Evidence,* ed. S. J. Ceci and W. M. Williams. Washington, DC: American Psychological Association.

Hines, M., S. F. Ahmed, and I. A. Hughes. 2003. Psychological Outcomes and Gender-Related Development in Complete Androgen Insensitivity Syndrome. *Archives of Sexual Behavior* 32 (2): 93–101.

Hines, M., C. Brook, and G. S. Conway. 2004. Androgen and Psychosexual Development: Core Gender Identity, Sexual Orientation, and Recalled Childhood Gender Role Behavior in Women and Men with Congenital Adrenal Hyperplasia (CAH). *Journal of Sex Research* 41 (1): 75–81.

Hines, M., B. A. Fane, V. L. Pasterski, G. A. Mathews, G. S. Conway, and C. Brook. 2003. Spatial Abilities following Prenatal Androgen Abnormality: Targeting and Mental Rotations Performance in Individuals with Congenital Adrenal Hyperplasia. *Psychoneuroendocrinology* 28 (8): 1010–1026.

Hines, M., S. Golombok, J. Rust, K. J. Johnston, and J. Golding. 2002. Testosterone during Pregnancy and Gender Role Behavior of Preschool Children: A Longitudinal, Population Study. *Child Development* 73 (6): 1678–1687.

Hines, M., K. J. Johnston, S. Golombok, J. Rust, M. Stevens, J. Golding, and the ALSPAC Study Team. 2002. Prenatal Stress and Gender Role Behavior in Girls and Boys: A Longitudinal, Population Study. *Hormones and Behavior* 42 (2): 126–134.

Hines, M., and F. R. Kaufman. 1994. Androgen and the Development of Human Sex-Typical Behavior: Rough-and-Tumble Play and Sex of Preferred Playmates in Children with Congenital Adrenal-Hyperplasia (CAH). *Child Development* 65 (4): 1042–53.

Hines, M., and E. C. Sandberg. 1996. Sexual Differentiation of Cognitive Abilities in Women Exposed to Diethylstilbestrol (DES) Prenatally. *Hormones and Behavior* 30 (4): 354–363.

Hines, M., and C. Shipley. 1984. Prenatal Exposure to Diethylstilbestrol (DES) and the Development of Sexually Dimorphic Cognitive Abilities and Cerebral Lateralization. *Developmental Psychology* 20 (1): 81–94.

Hochberg, Z., R. Chayen, N. Reiss, Z. Falik, A. Makler, M. Munichor, A. Farkas, H. Goldfarb, N. Ohana, and O. Hiort. 1996. Clinical, Biochemical, and Genetic Findings in a Large Pedigree of Male and Female Patients with 5 Alpha-Reductase 2 Deficiency. *Journal of Clinical Endocrinology and Metabolism* 81 (8): 2821–2827.

Hochberg, Z., M. Gardos, and A. Benderly. 1987. Psychosexual Outcome of Assigned Females and Males with 46,XX Virilizing Congenital Adrenal Hyperplasia. *European Journal of Pediatrics* 146:497–499.

Holden, C. 2000. Parity as a Goal Sparks Bitter Battle. *Science* 289 (5478): 380.

Hollibaugh, A., and C. Moraga. 1983. What We're Rollin' around in Bed With: Sexual Silences in Feminism. In *Powers of Desire*, ed. A. Snitow, C. Stansell, and S. Thompson. New York: New Feminist Library.

Holtzen, D. W. 1994. Handedness and Sexual Orientation. *Journal of Clinical and Experimental Neuropsychology* 16 (5): 702–712.

Huang, G., N. Taddese, and E. Walter. 2000. Entry and Persistence of Women and Minorities in College Science and Engineering Education. Washington, DC: U.S. Department of Education, National Center for Education Statistics.

Hughes, I. A. 2004. Female Development: All by Default? *New England Journal of Medicine* 351 (8): 748–750.

Hunt, K., H. Lewars, C. Emslie, and G. D. Batty. 2007. Decreased Risk of Death from Coronary Heart Disease amongst Men with Higher "Femininity" Scores: A General Population Cohort Study. *International Journal of Epidemiology* 36 (3): 612–620.

Hurtig, A. L., J. Radhakrishnan, H. M. Reyes, and I. M. Rosenthal. 1983. Psychological Evaluation of Treated Females with Virilizing Congenital Adrenal-Hyperplasia. *Journal of Pediatric Surgery* 18 (6): 887–893.

Hurtig, A. L., and I. M. Rosenthal. 1987. Psychological Findings in Early Treated Cases of Female Pseudohermaphroditism Caused by Virilizing Congenital Adrenal Hyperplasia. *Archives of Sexual Behavior* 16 (3): 209–223.

Hyde, J. S. 2005. The Gender Similarities Hypothesis. *American Psychologist* 60 (6): 581–592.

Imperato-McGinley, J., R. E. Peterson, T. Gautier, and E. Sturla. 1979. Androgens and the Evolution of Male-Gender Identity among Male Pseudohermaphrodites with 5a-Reductase Deficiency. *New England Journal of Medicine* 300 (22): 1233–37.

Imperato-McGinley, J., R. E. Peterson, M. Leshin, J. E. Griffin, G. Cooper, S. Draghi, M. Berenyi, and J. D. Wilson. 1980. Steroid 5α-Reductase Deficiency in a 65-Year Old Male Pseudohermaphrodite: The Natural History, Ultrastructure of the Testes and Evidence for Inherited Enzyme Heterogeneity. *Journal of Clinical Endocrinology and Metabolism* 50:15–22.

Imperato-McGinley, J., M. Pichardo, T. Gautier, D. Voyer, and M. P. Bryden. 1991. Cognitive Abilities in Androgen-Insensitive Subjects: Comparison with Control Males and Females from the Same Kindred. *Clinical Endocrinology* 34 (5): 341–347.

Irvine, J. M. 1990. *Disorders of Desire: Sex and Gender in Modern American Sexology.* Philadelphia: Temple University Press.

Jacklin, C. N., E. E. Maccoby, and C. H. Doering. 1983. Neonatal Sex-Steriod Hormones and Timidity in 6–18-Month-Old Boys and Girls. *Developmental Psychobiology* 16 (3): 163–168.

Jacklin, C. N., K. T. Wilcox, and E. E. Maccoby. 1988. Neonatal Sex-Steroid Hormones and Cognitive-Abilities at 6 Years. *Developmental Psychobiology* 21 (6): 567–574.

Jackson, S. 1999. *Heterosexuality in Question.* Thousand Oaks, CA: Sage.

Jaffe, B., D. Shye, S. Harlap, M. Baras, and A. Lieblich. 1989. Aggression, Physical-Activity Levels and Sex-Role Identity in Teenagers Exposed in Utero to MPA. *Contraception* 40 (3): 351–363.

Johannsen, T. H., C. P. L. Ripa, J. M. Reinisch, M. Schwartz, E. L. Mortensen, and K. M. Main. 2006. Impaired Cognitive Function in Women with Congenital Adrenal Hyperplasia. *Journal of Clinical Endocrinology and Metabolism* 91 (4): 1376–81.

Jost, A. 1953. Problems of Fetal Endocrinology: The Gonadal and Hypophyseal Hormones. *Recent Progress in Hormone Research* 8:379–418.

———. 1970. Hormonal Factors in the Sex Differentiation of the Mammalian Foetus. *Philosophical Transactions of the Royal Society of London* 259:119–130.

Jurgensen, M., O. Hiort, P. M. Holterhus, and U. Thyen. 2007. Gender Role Behavior in Children with XY Karyotype and Disorders of Sex Development. *Hormones and Behavior* 51 (3): 443–453.

Kaiser, A., S. Haller, S. Schmitz, and C. Nitsch. 2009. On Sex/Gender Related Similarities and Differences in fMRI Language Research. *Brain Research Reviews* 61 (2): 49–59.

Kaplan, N. M. 1959. Male Pseudohermaphrodism: Report of a Case, with Observations on Pathogenesis. *New England Journal of Medicine* 261:641–644.

Karkazis, K. 2008. *Fixing Sex: Intersex, Medical Authority, and Lived Experience.* Durham, NC: Duke University Press.

Katz, J. N. 1995. *The Invention of Heterosexuality.* New York: Plume/Penguin Books.

Kaufman, R. H., K. Noller, E. Adam, J. Irwin, M. Gray, J. A. Jefferies, and J. Hilton. 1984. Upper Genital-Tract Abnormalities and Pregnancy Outcome in Diethylstilbestrol-Exposed Progeny. *American Journal of Obstetrics and Gynecology* 148 (7): 973–984.

Kay, J. F., and A. Jackson. 2008. *Sex, Lies, and Stereotypes: How Abstinence-Only Programs Harm Women and Girls.* New York: Legal Momentum.

Keller, E. F. 2000. *The Century of the Gene.* Cambridge, MA: Harvard University Press.

Kelso, W. M., M. E. R. Nicholls, G. L. Warne, and M. Zacharin. 2000. Cerebral Lateralization and Cognitive Functioning in Patients with Congenital Adrenal Hyperplasia. *Neuropsychology* 14 (3): 370–378.

Kenen, S. H. 1997. Who Counts When You're Counting Homosexuals? Hormones and Homosexuality in Mid-Twentieth Century America. In *Science and Homosexualities,* ed. V. A. Rosario. New York: Routledge.

Kessler, S. J. 1990. The Medical Construction of Gender: Case Management of Intersex Infants. *Signs* 16 (1): 3–26.

———. 1998. *Lessons from the Intersexed.* New Brunswick, NJ: Rutgers University Press.

Kessler, S., and W. McKenna. 1978. *Gender: An Ethnomethodological Approach.* New York: John Wiley.

Kester, P. A. 1984. Effects of Prenatally Administered 17a-Hydroxyprogesterone Caproate on Adolescent Males. *Archives of Sexual Behavior* 13 (5): 441–455.

Kester, P., R. Green, S. J. Finch, and K. Williams. 1980. Prenatal "Female Hormone" Administration and Psychosexual Development in Human Males. *Psychoneuroendocrinology* 5:269–285.

Kimura, D. 1999. *Sex and Cognition.* Boston: MIT Press.

———. 2002. Sex Differences in the Brain. *Scientific American,* 32–37.

Kinsey, A. C. 1941. Homosexuality: Criteria for a Hormonal Explanation of the Homosexual. *Journal of Clinical Endocrinology* 1:424–428.

Kinsey, A. C., W. B. Pomeroy, and C. E. Martin. 1948. *Sexual Behavior in the Human Male.* Philadelphia: Saunders.

Kinsey, A. C., W. B. Pomeroy, C. E. Martin, and P. Gebhard. 1953. *Sexual Behavior in the Human Female.* Philadelphia: Saunders.

Kite, M. E., and K. Deaux. 1987. Gender Belief Systems—Homosexuality and the Implicit Inversion Theory. *Psychology of Women Quarterly* 11 (1):83–96.

Klein, F., B. Sepekoff, and T. J. Wolf. 1985. Sexual Orientation: A Multi-Variable Dynamic Process. *Journal of Homosexuality* 11:35–49.

Klienfeld, J. 2005. Truth to Power: Summers of Academic Discontent. *National Review Online,* January 25.

Knickmeyer, R., S. Baron-Cohen, P. Raggatt, and K. Taylor. 2005. Foetal Testosterone, Social Relationships, and Restricted Interests in Children. *Journal of Child Psychology and Psychiatry* 46 (2): 198–210.

Knickmeyer, R., S. Baron-Cohen, B. A. Fane, S. Wheelwright, G. A. Mathews, G. S. Conway, C. G. D. Brook, and M. Hines. 2006. Androgens and Autistic Traits: A Study of Individuals with Congenital Adrenal Hyperplasia. *Hormones and Behavior* 50 (1): 148–153.

Knickmeyer, R., S. Baron-Cohen, P. Raggatt, K. Taylor, and G. Hackett. 2006. Fetal Testosterone and Empathy. *Hormones and Behavior* 49 (3): 282–292.

Knickmeyer, R. C., S. Wheelwright, K. Taylor, P. Raggatt, G. Hackett, and S. Baron-Cohen. 2005. Gender-Typed Play and Amniotic Testosterone. *Developmental Psychology* 41 (3): 517–528.

Koedt, A. 1976. The Myth of the Vaginal Orgasm. In *Female Psychology: The Emerging Self*, ed. S. Cox. Chicago: Science Research Associates. Original edition, 1970.

Konrad, A. M., J. E. Ritchie, P. Lieb, and E. Corrigall. 2000. Sex Differences and Similarities in Job Attribute Preferences: A Meta-Analysis. *Psychological Bulletin* 126 (4): 593–641.

Koutcherov, Y., J. K. Mai, K. W. S. Ashwell, and G. Paxinos. 2002. Organization of Human Hypothalamus in Fetal Development. *Journal of Comparative Neurology* 446 (4): 301–324.

Kraemer, B., T. Noll, A. Delsignore, G. Milos, U. Schnyder, and U. Hepp. 2006. Finger Length Ratio (2D:4D) and Dimensions of Sexual Orientation. *Neuropsychobiology* 53 (4): 210–214.

Kraft-Ebing, Dr. R. V. 1930. *Psychopathia Sexualis: With Especial Reference to the Antipathic Sexual Instinct—A Medico-Forensic Study.* Brooklyn: Physicians and Surgeons Book Co.

Kruijver, F. P. M., J. N. Zhou, C. W. Pool, M. A. Hofman, L. J. G. Gooren, and D. F. Swaab. 2000. Male-to-Female Transsexuals Have Female Neuron Numbers in a Limbic Nucleus. *Journal of Clinical Endocrinology and Metabolism* 85 (5): 2034–41.

Kuhnle, U., and M. Bullinger. 1997. Outcome of Congenital Adrenal Hyperplasia. *Pediatric Surgery International* 12 (7): 511–515.

Kuhnle, U., M. Bullinger, and H. P. Schwarz. 1995. The Quality-of-Life in Adult Female Patients with Congenital Adrenal Hyperplasia: A Comprehensive Study of the Impact of Genital Malformations and Chronic Disease on Female Patients' Life. *European Journal of Pediatrics* 154 (9): 708–716.

Kuhnle, U., M. Bullinger, H. P. Schwarz, and D. Knorr. 1993. Partnership and Sexuality in Adult Female Patients with Congenital Adrenal-Hyperplasia: 1st Results of a Cross-Sectional Quality-of-Life Evaluation. *Journal of Steroid Biochemistry and Molecular Biology* 45:123–126.

Kuhnle, U., and W. Krahl. 2002. The Impact of Culture on Sex Assignment and Gender Development in Intersex Patients. *Perspectives in Biology and Medicine* 45:85–103.

Lalumiere, M. L., R. Blanchard, and K. J. Zucker. 2000. Sexual Orientation and

Handedness in Men and Women: A Meta-Analysis. *Psychological Bulletin* 126 (4): 575–592.

Lancaster, R. 2003. *The Trouble with Nature: Sex in Science and Popular Culture.* Berkeley: University of California Press.

Laqueur, T. 1986. Orgasm, Generation, and the Politics of Reproductive Biology. *Representations* 14 (1): 1–41.

———. 1992. *Making Sex: Body and Gender from the Greeks to Freud.* Cambridge, MA: Harvard University Press.

Lasco, M. S., T. J. Jordan, M. A. Edgar, C. K. Petito, and W. Byne. 2002. A Lack of Dimorphism of Sex or Sexual Orientation in the Human Anterior Commissure. *Brain Research* 936 (1–2): 95–98.

Latour, B., and S. Woolgar. 1986. *Laboratory Life: The Construction of Scientific Facts.* Princeton: Princeton University Press.

Laumann, E. O., J. H. Gagnon, R. T. Michael, and S. Michaels. 1994. *The Social Organization of Sexuality: Sexual Practices in the United States.* Chicago: University of Chicago Press.

Leboucher, G. 1989. Maternal Behavior in Normal and Androgenized Female Rats: Effect of Age and Experience. *Physiology and Behavior* 45 (2): 313–319.

Lefkowitz, E. S., and P. B. Zeldow. 2006. Masculinity and Femininity Predict Optimal Mental Health: A Belated Test of the Androgyny Hypothesis. *Journal of Personality Assessment* 87 (1): 95–101.

Lev-Ran, A. 1974a. Gender Role Differentiation in Hermaphrodites. *Archives of Sexual Behavior* 3 (5): 391–424.

———. 1974b. Sexuality and Educational Levels of Women with the Late-Treated Adrenogenital Syndrome. *Archives of Sexual Behavior* 3 (1): 27–32.

LeVay, S. 1991. A Difference in Hypothalamic Structure between Heterosexual and Homosexual Men. *Science* 253:1034–37.

———. 1996. *Queer Science: The Use and Abuse of Research into Homosexuality.* Cambridge, MA: MIT Press.

Leveroni, C. L., and S. A. Berenbaum. 1998. Early Androgen Effects on Interest in Infants: Evidence from Children with Congenital Adrenal Hyperplasia. *Environmental Neuropsychology* 14 (2–3): 321–340.

Levy-Prewitt, J. 2007. College Bound. *SF Gate,* September 16.

Lewontin, R. 2000. *The Triple Helix: Gene, Organism, and Environment.* Cambridge, MA: Harvard University Press.

Lindesay, J. 1987. Laterality Shift in Homosexual Men. *Neuropsychologia* 25 (6): 965–969.

Link, B., J. Monahan, A. Stueve, and F. Cullen. 1999. Real in Their Consequences: A Sociological Approach to Understanding the Association between Psychotic Symptoms and Violence. *American Sociological Review* 64:316–331.

Link, B., and J. C. Phelan. 2001. Conceptualizing Stigma. *Annual Review of Sociology* 27:363–385.

Lippa, R. A. 2002. Gender-Related Traits of Heterosexual and Homosexual Men and Women. *Archives of Sexual Behavior* 31 (1): 83–98.

―――. 2003. Are 2D:4D Finger-Length Ratios Related to Sexual Orientation? Yes for Men, No for Women. *Journal of Personality and Social Psychology* 85 (1): 179–188.

―――. 2005. *Gender, Nature, and Nurture*. 2nd ed. New York: Routledge.

―――. 2008. Sex Differences and Sexual Orientation Differences in Personality: Findings from the BBC Internet Survey. *Archives of Sexual Behavior* 37 (1): 173–187.

―――. 2009. Sex Differences in Sex Drive, Sociosexuality, and Height across 53 Nations: Testing Evolutionary and Social Structural Theories. *Archives of Sexual Behavior* 38 (5): 631–651.

Lish, J. D., A. A. Ehrhardt, H. F. L. Meyer-Bahlburg, L. R. Rosen, R. S. Gruen, and N. P. Veridiano. 1991. Gender-Related Behavior Development in Females Exposed to Diethylstilbestrol (DES) in Utero: An Attempted Replication. *Journal of the American Academy of Child and Adolescent Psychiatry* 30 (1): 29–37.

Lish, J. D., H. F. L. Meyer-Bahlburg, A. A. Ehrhardt, B. G. Travis, and N. P. Veridiano. 1992. Prenatal Exposure to Diethylstilbestrol (DES): Childhood Play Behavior and Adult Gender-Role Behavior in Women. *Archives of Sexual Behavior* 21 (5): 423–441.

Lloyd, E. A. 2005. *The Case of the Female Orgasm: Bias in the Science of Evolution*. Cambridge, MA: Harvard University Press.

Loder, N. 2000. US Science Shocked by Revelations of Sexual Discrimination. *Nature* 405:713–714.

Loehlin, J. C., D. McFadden, S. E. Medland, and N. G. Martin. 2006. Population differences in finger-length ratios: Ethnicity or latitude? *Archives of Sexual Behavior* 35:739–742.

Longino, H. E. 1990. *Science as Social Knowledge: Values and Objectivity in Scientific Inquiry*. Princeton: Princeton University Press.

Longino, H. E., and R. Doell. 1983. Body, Bias, and Behavior: A Comparative Analysis of Reasoning in Two Areas of Biological Science. *Signs: Journal of Women in Culture and Society* 9 (2): 206–227.

Lucas, T. W., C. A. Wendorf, E. O. Imamoglu, J. Shen, M. R. Parkhill, C. C. Weisfeld, and G. E. Weisfeld. 2004. Marital Satisfaction in Four Cultures as a Function of Homogamy, Male Dominance and Female Attractiveness. *Sexualities, Evolution and Gender* 6 (2–3): 97–130.

Lutchmaya, S., S. Baron-Cohen, and P. Raggatt. 2001. Foetal Testosterone and Vocabulary Size in 18- and 24-Month-Old Infants. *Infant Behavior and Development* 24 (4): 418–424.

―――. 2002. Foetal Testosterone and Eye Contact in 12-Month-Old Human Infants. *Infant Behavior and Development* 25 (3): 327–335.

Lutchmaya, S., S. Baron-Cohen, P. Raggatt, R. Knickmeyer, and J. T. Manning. 2004. 2nd to 4th Digit Ratios, Fetal Testosterone and Estradiol. *Early Human Development* 77 (1–2): 23–28.

Lyons, J. A., and L. A. Serbin. 1986. Observer Bias in Scoring Boys' and Girls' Aggression. *Sex Roles* 14 (5–6): 301–313.

Lytton, H., and D. M. Romney. 1991. Parents' Differential Socialization of Boys and Girls: A Meta-analysis. *Psychological Bulletin* 109 (2): 267–296.

Maccoby, E. E., C. H. Doering, C. N. Jacklin, and H. Kraemer. 1979. Concentrations of Sex Hormones in Umbilical-Cord Blood: Their Relation to Sex and Birth Order of Infants. *Child Development* 50 (3): 632–642.

Maccoby, E. E., and C. N. Jacklin. 1974. *The Psychology of Sex Differences*. Stanford: Stanford University Press.

MacLusky, N. J., I. Lieberburg, and B. S. McEwen. 1979. Development of Estrogen-Receptor Systems in the Rat Brain: Perinatal Development. *Brain Research* 178 (1): 129–142.

Mak, G. A. 2004. Sandor/Sarolta Vay: From Passing Woman to Sexual Invert. *Journal of Women's History* 16 (1): 54–77.

———. 2006. Doubting Sex from Within: A Praxiographic Approach to a Late Nineteenth-Century Case of Hermaphroditism. *Gender and History* 18 (2): 332–356.

Malouf, M. A., C. J. Migeon, K. A. Carson, L. Petrucci, and A. B. Wisniewski. 2006. Cognitive Outcome in Adult Women Affected by Congenital Adrenal Hyperplasia due to 21-Hydroxylase Deficiency. *Hormone Research* 65 (3): 142–150.

Manning, J. T., A. Stewart, P. E. Bundred, and R. L. Trivers. 2004. Sex and Ethnic Differences in 2nd to 4th Digit Ratio of Children. *Early Human Development* 80:161–168.

Maraun, M. D. 1998. Measurement as a Normative Practice: Implications of Wittgenstein's Philosophy for Measurement in Psychology. *Theory and Psychology* 8 (4): 435–461.

Marchant-Haycox, S. E., I. C. McManus, and G. D. Wilson. 1991. Left-Handedness, Homosexuality, HIV Infection and AIDS. *Cortex* 27:49–56.

Marcus, J., E. E. Maccoby, C. N. Jacklin, and C. H. Doering. 1985. Individual Differences in Mood in Early Childhood: Their Relation to Gender and Neonatal Sex Steroids. *Developmental Psychobiology* 18 (4): 327–340.

Martin, J. T., and D. H. Nguyen. 2004. Anthropometric Analysis of Homosexuals and Heterosexuals: Implications for Early Hormone Exposure. *Hormones and Behavior* 45 (1): 31–39.

Masica, D. N., J. Money, and A. A. Ehrhardt. 1971. Fetal Feminization and Female Gender Identity in the Testicular Feminizing Syndrome of Androgen Insensitivity. *Archives of Sexual Behavior* 1 (2): 131–142.

Masters, W. H., and V. E. Johnson. 1965. The Sexual Response Cycle of the Human Female. In *Sex Research: New Developments*, ed. J. Money. New York: Holt, Rinehart and Winston.

———. 1966. *Human Sexual Response*. Boston: Little, Brown.

Mathews, G. A., B. A. Fane, et al. 2004. Androgenic Influences on Neural Asymmetry: Handedness and Language Lateralization in Individuals with Congenital Adrenal Hyperplasia. *Psychoneuroendocrinology* 29 (6): 810–822.

May, B., M. Boyle, and D. Grant. 1996. A Comparative Study of Sexual Experiences: Women with Diabetes and Women with Congenital Adrenal Hyperplasia. *Journal of Health Psychology* 1 (4): 479–492.

Mayhew, T. M., L. Gillam, R. McDonald, and F. J. P. Ebling. 2007. Human 2D (Index) and 4D (Ring) Digit Lengths: Their Variation and Relationships during the Menstrual Cycle. *Journal of Anatomy* 211 (5): 630–638.

Mayrent, S. L., C. H. Hennekens, and J. E. Buring, eds. 1987. *Epidemiology in Medicine*. Philadelphia: Lippincott Williams and Wilkins.

Mazur, T. 2005. Gender Dysphoria and Gender Change in Androgen Insensitivity or Micropenis. *Archives of Sexual Behavior* 34 (4): 411–421.

McCarty, B. M., C. J. Migeon, H. F. L. Meyer-Bahlburg, H. Zacur, and A. B. Wisniewski. 2006. Medical and Psychosexual Outcome in Women Affected by Complete Gonadal Dysgenesis. *Journal of Pediatric Endocrinology and Metabolism* 19 (7): 873–877.

McClintock, A. 1995. *Imperial Leather: Race, Gender, and Sexuality in the Colonial Contest*. New York: Routledge.

McCormick, C. M., and S. F. Witelson. 1991. A Cognitive Profile of Homosexual Men Compared to Heterosexual Men and Women. *Psychoneuroendocrinology* 16 (6): 459–473.

McCormick, C. M., S. F. Witelson, and Edward Kingstone. 1990. Left-Handedness in Homosexual Men and Women: Neuroendocrine Implications. *Psychoneuroendocrinology* 15 (1): 69–76.

McEwen, B. S., I. Lieberburg, C. Chaptal, and L. C. Krey. 1977. Aromatization: Important for Sexual Differentiation of Neonatal Rat-Brain. *Hormones and Behavior* 9 (3): 249–263.

McEwen, B. S., I. Lieberburg, N. MacLusky, and L. Plapinger. 1977. Do Estrogen Receptors Play a Role in Sexual Differentiation of the Rat Brain? *Journal of Steroid Biochemistry and Molecular Biology* 8 (5): 593–598.

McFadden, D. 2002. Masculinization Effects in the Auditory System. *Archives of Sexual Behavior* 31 (1): 99–111.

McFadden, D., and C. A. Champlin. 2000. Comparison of Auditory Evoked Potentials in Heterosexual, Homosexual, and Bisexual Males and Females. *Jaro* 1 (1): 89–99.

McFadden, D., J. C. Loehlin, S. M. Breedlove, R. A. Lippa, J. T. Manning, and Q. Rahman. 2005. A Reanalysis of Five Studies on Sexual Orientation and the Relative Length of the 2nd and 4th Fingers (the 2D:4D Ratio). *Archives of Sexual Behavior* 34 (3): 341–356.

McFadden, D., and E. Shubel. 2003. The Relationships between Otoacoustic Emissions and Relative Lengths of Fingers and Toes in Humans. *Hormones and Behavior* 43 (3): 421–429.

McFarlane, J., C. L. Martin, and T. M. Williams. 1988. Mood Fluctuations: Women versus Men and Menstrual versus Other Cycles. *Psychology of Women Quarterly* 12:201–223.

McGuire, L. S., K. O. Ryan, and G. S. Omenn. 1975. Congenital Adrenal Hyperplasia. II: Cognitive and Behavioral Studies. *Behavior Genetics* 5 (2): 175–188.

McIntyre, M. H. 2003. Digit Ratios, Childhood Gender Role Behavior, and Erotic Role Preferences of Gay Men. *Archives of Sexual Behavior* 32:495–497.

McIntyre, M. H., E. S. Barrett, R. McDermott, D. D. P. Johnson, J. Cowden, and S. P. Rosen. 2007. Finger Length Ratio (2D:4D) and Sex Differences in Aggression during a Simulated War Game. *Personality and Individual Differences* 42(4): 755–764.

McManus, I. C., and M. P. Bryden. 1991. Geschwind's Theory of Cerebral Lateralization: Developing a Formal, Causal Model. *Psychological Bulletin* 110 (2): 237–253.

Meek, L. R., K. M. Schulz, and C. A. Keith. 2006. Effects of Prenatal Stress on Sexual Partner Preference in Mice. *Physiology and Behavior* 89 (2): 133–138.

Mendez, J. P., A. Ulloa-Aguirre, J. Imperato-McGinley, A. Brugmann, M. Delfin, B. Chavez, C. Shackleton, S. Kofman-Alfaro, and G. Perez-Palacios. 1995. Male Pseudohermaphroditism due to Primary 5-Alpha-Reductase Deficiency: Variation in Gender Identity Reversal in 7 Mexican Patients from 5 Different Pedigrees. *Journal of Endocrinological Investigation* 18 (3): 205–213.

Mendonca, B. B., M. Inacio, E. M. F. Costa, I. J. P. Arnhold, F. A. Q. Silva, W. Nicolau, W. Bloise, D. W Russell, and J. D. Wilson. 1996. Male Pseudohermaphroditism due to Steroid 5 Alpha-Reductase 2 Deficiency. *Medicine* 75 (2): 64–76.

Merke, D. P., and S. R. Bornstein. 2005. Congenital Adrenal Hyperplasia. *Lancet* 365 (9477): 2125–2136.

Meyer-Bahlburg, H. F. L. 1977. Sex Hormones and Male Homosexuality in Comparative Perspective. *Archives of Sexual Behavior* 6 (4): 297–323.

———. 1999. What Causes Low Rates of Child-Bearing in Congenital Adrenal Hyperplasia? *Journal of Clinical Endocrinology and Metabolism* 84 (6): 1844–47.

———. 2002. Gender Assignment and Reassignment in Intersexuality: Controversies, Data, and Guidelines for Research. In *Pediatric Gender Assignment: A Critical Reappraisal*, ed. S. Zderic, D. A. Canning, M. C. Carr, and H. McC. Snyder. New York: Kluwer Academic/Plenum.

———. 2005. Gender Identity Outcome in Female-Raised 46,XY Persons with Penile Agenesis, Cloacal Exstrophy of the Bladder, or Penile Ablation. *Archives of Sexual Behavior* 34 (4): 423–438.

Meyer-Bahlburg, H. F. L., C. Dolezal, S. W. Baker, A. D. Carlson, J. S. Obeid, and M. I. New. 2004. Prenatal Androgenization Affects Gender-Related Behavior but Not Gender Identity in 5–12-Year-Old Girls with Congenital Adrenal Hyperplasia. *Archives of Sexual Behavior* 33 (2): 97–104.

Meyer-Bahlburg, H. F. L., C. Dolezal, S. W. Baker, A. A. Ehrhardt, and M. I. New. 2006. Gender Development in Women with Congenital Adrenal Hyperplasia as a Function of Disorder Severity. *Archives of Sexual Behavior* 35 (6): 667–684.

Meyer-Bahlburg, H. F. L., C. Dolezal, S. W. Baker, and M. I. New. 2008. Sexual Orientation in Women with Classical or Non-Classical Adrenal Hyperplasia as a Function Degree of Prenatal Androgen Excess. *Archives of Sexual Behavior* 37 (1): 85–99.

Meyer-Bahlburg, H. F. L., and A. A. Ehrhardt. 1982. Prenatal Sex Hormones and Human Aggression: A Review, and New Data on Progestogen Effects. *Aggressive Behavior* 8:39–62.

————. 1983. Sexual Behavior Assessment Schedule: Adult (SEBAS-A). New York: Columbia University.

Meyer-Bahlburg, H. F. L., A. A. Ehrhardt, J. F. Feldman, L. R. Rosen, N. P. Veridiano, and I. Zimmerman. 1985. Sexual Activity Level and Sexual Functioning in Women Prenatally Exposed to Diethylstilbestrol. *Psychosomatic Medicine* 47 (6): 497–511.

Meyer-Bahlburg, H. F. L., A. A. Ehrhardt, L. R. Rosen, J. F. Feldman, N. P. Veridiano, I. Zimmerman, and B. S. McEwen. 1984. Psychosexual Milestones in Women Prenatally Exposed to Diethylstilbestrol. *Hormones and Behavior* 18:359–366.

Meyer-Bahlburg, H. F. L., A. A. Ehrhardt, L. R. Rosen, R. S. Gruen, N. P. Veridiano, F. H. Vann, and H. F. Neuwalder. 1995. Prenatal Estrogens and the Development of Homosexual Orientation. *Developmental Psychology* 31 (1): 12–21.

Meyer-Bahlburg, H. F. L., G. C. Grisanti, and A. A. Ehrhardt. 1977. Prenatal Effects of Sex Hormones on Human Male Behavior: Medroxyprogesterone Acetate (MPA). *Psychoneuroendocrinology* 2:383–390.

Meyer-Bahlburg, H. F. L., R. S. Gruen, M. I. New, et al. 1996. Gender Change from Female to Male in Classical Congenital Adrenal Hyperplasia. *Hormones and Behavior* 30 (4): 319–332.

Meyer-Bahlburg, H. F. L., D. E. Sandberg, C. L. Dolezal, and T. J. Yager. 1994. Gender-Related Assessment of Childhood Play. *Journal of Abnormal Child Psychology* 22 (6): 643–660.

Meyer-Bahlburg, H. F. L., D. E. Sandberg, T. J. Yager, C. L. Dolezal, and A. A. Ehrhardt. 1994. Questionnaire Scales for the Assessment of Atypical Gender Development in Girls and Boys. *Journal of Psychology and Human Sexuality* 6 (4): 19–39.

Michael, R. T., J. Wadsworth, J. A. Feinleib, A. M. Johnson, E. O. Laumann, and K. Wellings. 2001. Private Sexual Behavior, Public Opinion, and Public Health Policy Related to Sexually-Transmitted Diseases: A U.S.–British Comparison. In *Sex, Love, and Health in America*, ed. E. O. Laumann and R. T. Michael. Chicago: University of Chicago Press.

Migeon, C. J., A. B. Wisniewski, J. P. Gearhart, H. F. L. Meyer-Bahlburg, J. A. Rock, T. R. Brown, S. J. Casella, A. Maret, K. M. Ngai, J. Money, and G. D. Berkovitz. 2002. Ambiguous Genitalia with Perineoscrotal Hypospadias in 46,XY Individuals: Long-Term Medical, Surgical, and Psychosexual Outcome. *Pediatrics* 110 (3): e31.

MIT. 1999. *A Study on the Status of Women Faculty in Science at MIT*. Cambridge, MA: Massachusetts Institute of Technology.

Moir, A. 2003. *Why Men Don't Iron: The Fascinating and Unalterable Differences between Men and Women*. New York: Kensington Press Corp.

Mol, A. 2002. *The Body Multiple: Ontology in Medical Practice*. Durham, N.C.: Duke University Press.

Money, J. 1952. Hermaphroditism: An Inquiry into the Nature of a Human Paradox. Ph.D. diss., Harvard University, Boston.

————. 1965a. Influence of Hormones on Sexual Behavior. *Annual Review of Medicine* 16:67–82.

———. 1965b. Psychosexual Differentiation. In *Sex Research: New Developments,* ed. J. Money. New York: Holt, Rinehart and Winston.

———, ed. 1965c. *Sex Research: New Developments.* New York: Holt, Rinehart and Winston.

———. 1975. Ablatio Penis: Normal Male Infant Sex Re-Assigned as a Girl. *Archives of Sexual Behavior* 4:65–71.

———. 1980. *Love and Love Sickness: The Science of Sex, Gender Difference, and Pair-Bonding.* Baltimore: Johns Hopkins University Press.

———. 1987. Sin, Sickness, or Status? Homosexual Gender Identity and Psychoneuroendocrinology. *American Psychologist* 42 (4): 384–399.

———. 1991. Bibliography of Works by John Money. In *John Money: A Tribute,* ed. E. Coleman. Binghamton, UK: Haworth Press.

———. 1994. *Reinterpreting the Unspeakable: Human Sexuality 2000.* New York: Continuum.

Money, J., and D. Alexander. 1969. Psychosexual Development and Absence of Homosexuality in Males with Precocious Puberty. *Journal of Nervous and Mental Disease* 148 (2): 111–123.

Money, J., and J. Daléry. 1976. Iatrogenic Homosexuality: Gender Identity in Seven 46,XX Chromosomal Females with Hyperadrenocortical Hermaphroditism Born with a Penis, Three Reared as Boys, Four Reared as Girls. *Journal of Homosexuality* 1 (4): 357–371.

Money, J., and A. A. Ehrhardt. 1972. *Man and Woman, Boy and Girl.* Baltimore: Johns Hopkins University Press.

Money, J., A. A. Ehrhardt, and D. N. Masica. 1968. Fetal Feminization Induced by Androgen Insensitivity in the Testicular Feminizing Syndrome: Effect on Marriage and Maternalism. *Johns Hopkins Medical Journal* 123:105–114.

Money, J., J. G. Hampson, and J. L. Hampson. 1955a. An Examination of Some Basic Sexual Concepts: The Evidence of Human Hermaphroditism. *Bulletin of the Johns Hopkins Hospital* 97:301–319.

———. 1955b. Hermaphroditism: Recommendations Concerning Assignment of Sex, Change of Sex, and Psychologic Management. *Bulletin of the Johns Hopkins Hospital* 97:284–300.

Money, J., and V. Lewis. 1966. IQ, Genetics, and Accelerated Growth: Adrenogenital Syndrome. *Bulletin of the Johns Hopkins Hospital* 118:365–373.

———. 1982. Homosexual/Heterosexual Status in Boys at Puberty: Idiopathic Adolescent Gynecomastia and Congenital Virilizing Adrenocorticism Compared. *Psychoneuroendocrinology* 7 (4): 339–346.

Money, J., and D. Mathews. 1982. Prenatal Exposure to Virilizing Progestins: An Adult Follow-Up Study of Twelve Women. *Archives of Sexual Behavior* 11 (1): 73–83.

Money, J., and C. Ogunro. 1974. Behavioral Sexology: Ten Cases of Genetic Male Intersexuality with Impaired Prenatal and Pubertal Androgenization. *Archives of Sexual Behavior* 3 (3): 181–205.

Money, J., and M. Schwartz. 1977. Dating, Romantic and Nonromantic Friendships, and Sexuality in 17 Early-Treated Andrenogenital Females, Aged 16–25. In *Congenital Adrenal Hyperplasia,* ed. P. A. Lee, L. P. Plotnick, A. A. Kowarski, and C. J. Migcon. Baltimore: University Park Press.

Money, J., M. Schwartz, and V. G. Lewis. 1984. Adult Erotosexual Status and Fetal Hormonal Masculinization and Demasculinization: 46,XX Congenital Virilizing Adrenal Hyperplasia and 46,XY Androgen-Insensitivity Syndrome Compared. *Psychoneuroendocrinology* 9 (4): 405–414.

Moore, T., C. Quinter, and L. M. Freeman. 2005. Lack of Correlation between 2D:4D Ratio and Assertiveness in College Age Women. *Personality and Individual Differences* 39 (1): 115–121.

Morgan, J. F., H. Murphy, J. H. Lacey, and G. Conway. 2005. Long Term Psychological Outcome for Women with Congenital Adrenal Hyperplasia: Cross Sectional Survey. *British Medical Journal* 330 (7487): 340–341.

Morris, J. A., K. L. Gobrogge, C. L. Jordan, and S. M. Breedlove. 2004. Brain Aromatase: Dyed-in-the-Wool Homosexuality. *Endocrinology* 145 (2): 475–477.

Morrissette, D., and L. S. Myers. 2002. In Memoriam: Julian M. Davidson. *Archives of Sexual Behavior* 31 (4): 311–318.

Mosse, G. L. 1985. *Nationalism and Sexuality: Middle Class Morality and Sexual Norms in Modern Europe.* Madison: University of Wisconsin Press.

Mulaikal, R. M., C. J. Migeon, and J. A. Rock. 1987. Fertility Rates in Female Patients with Congenital Adrenal Hyperplasia due to 21-Hydroxylase Deficiency. *New England Journal of Medicine* 316:178–182.

Murnen, S. K., and M. Stockton. 1997. Gender and Self-Reported Sexual Arousal in Response to Sexual Stimuli: A Meta-Analytic Review. *Sex Roles* 37 (3–4): 135–153.

Mustanski, B. S., J. M. Bailey, and S. Kaspar. 2002. Dermatoglyphics, Handedness, Sex, and Sexual Orientation. *Archives of Sexual Behavior* 31 (1): 113–122.

Mustanski, B. S., M. L. Chivers, and J. M. Bailey. 2002. A Critical Review of Recent Biological Research on Human Sexual Orientation. *Annual Review of Sex Research* 13:89–140.

National Academy of Sciences. 2006. *Beyond Bias and Barriers: Fulfilling the Potential of Women in Academic Science and Engineering.* Washington, DC: National Academy of Sciences.

National Center for Education Statistics. 2007. *Digest of Education Statistics.* U.S. Department of Education, Institute of Education Sciences. nces.ed.gov/programs/digest/d07/tables/dt07_009.asp.

———. 2009. *Fast Facts: What Is the Percentage of Degrees Conferred by Sex and Race?* U.S. Department of Education, Institute of Education Sciences. nces.ed.gov/fastfacts/display.asp?id=72.

Neave, N., M. Menaged, and D. R. Weightman. 1999. Sex Differences in Cognition: The Role of Testosterone and Sexual Orientation. *Brain and Cognition* 41 (3): 245–262.

Nehm, R. H., and R. M. Young. 2008. "Sex Hormones" in Secondary School Biology Textbooks. *Science and Education* 17 (10): 1175–90.

Nelkin, D. 1992. *Controversy: Politics of Technical Decisions.* Newbury Park, CA: Sage.

Neustadt, R., and A. Myerson. 1940. Quantitative Sex Hormone Studies in Homosexuality, Childhood, and Various Neuropsychiatric Disturbances. *American Journal of Psychiatry* 97 (3): 524–551.

New, M. I. 2006. Extensive Clinical Experience: Nonclassical 21-Hydroxylase Deficiency. *Journal of Clinical Endocrinology and Metabolism* 91 (11): 4205–14.

Newbold, R. R. 1993. Gender-Related Behavior in Women Exposed Prenatally to Diethylstilbestrol. *Environmental Health Perspectives* 101 (3): 208–213.

Noble, R. G. 1979. Male Hamsters Display Female Sexual Responses. *Hormones and Behavior* 12 (3): 293–298.

NOVA. *See* "Sex Unknown."

Nunez-Farfan, J., and C. D. Schlichting. 2001. Evolution in Changing Environments: The "Synthetic" Work of Clausen, Keck, and Hiesey. *Quarterly Review of Biology* 76 (4): 433–457.

Okten, A., M. Kalyoncu, and N. Yaris. 2002. The Ratio of Second- and Fourth-Digit Lengths and Congenital Adrenal Hyperplasia due to 21-Hydroxylase Deficiency. *Early Human Development* 70 (1–2): 47–54.

Oliver, M. B., and J. S. Hyde. 1993. Gender Differences in Sexuality: A Meta-Analysis. *Psychological Bulletin* 114 (1): 29–51.

Ong, K. K., M. L. Ahmed, and D. B. Dunger. 2006. Lessons from Large Population Studies on Timing and Tempo of Puberty (Secular Trends and Relation to Body Size): The European Trend. *Molecular and Cellular Endocrinology* 254:8–12.

Oosterhuis, H. 1997. Richard Von Krafft-Ebing's "Step-Children of Nature": Psychiatry and the Making of Homosexual Identity. In *Science and Homosexualities*, ed. V. A. Rosario. New York: Routledge.

Osterman, K., K. Bjorkqvist, and K. M. J. Lagerspetz. 1998. Cross-Cultural Evidence of Female Indirect Aggression. *Aggressive Behavior* 24 (1): 1–8.

Oudshoorn, N. 1994. *Beyond the Natural Body: An Archeology of Sex Hormones.* London: Routledge.

Paredes, R. G., and M. J. Baum. 1995. Altered Sexual Partner Preference in Male Ferrets Given Excitotoxic Lesions of the Preoptic Area Anterior Hypothalamus. *Journal of Neuroscience* 15 (10): 6619–6630.

Parish, A., and F. B. M. de Waal. 2000. The Other "Closest Living Relative": How Bonobos *(Pan paniscus)* Challenge Traditional Assumptions about Females, Dominance, Intra- and Intersexual Interactions and Hominid Evolution. In *Evolutionary Perspectives on Human Reproductive Behavior*, ed. D. Lecroy and P. Moller. New York.

Pasterski, V. L., M. E. Geffner, C. Brain, P. Hindmarsh, C. Brook, and M. Hines. 2005. Prenatal Hormones and Postnatal Socialization by Parents as Determinants of Male-Typical Toy Play in Girls with Congenital Adrenal Hyperplasia. *Child Development* 76 (1): 264–278.

Pasterski, V., P. Hindmarsh, M. Geffner, C. Brook, C. Brain, and M. Hines. 2007. Increased Aggression and Activity Level in 3- to 11-Year-Old Girls with Congenital Adrenal Hyperplasia (CAH). *Hormones and Behavior* 52 (3): 368–374.

Payne, B. K. 2001. Prejudice and Perception: The Role of Automatic and Controlled Processes in Misperceiving a Weapon. *Journal of Personality and Social Psychology* 81 (2): 181–192.

Perkins, A., J. A. Fitzgerald, and G. E. Moss. 1995. A Comparison of LH Secretion and Brain Estradiol Receptors in Heterosexual and Homosexual Rams and Female Sheep. *Hormones and Behavior* 29 (1): 31–41.

Perkins, M. W. 1981. Female Homosexuality and Body Build. *Archives of Sexual Behavior* 10 (4): 337–345.

Perlman, S. M. 1973. Cognitive Abilities of Children with Hormone Abnormalties: Screening by Psychoeducational Tests. *Journal of Learning Disabilities* 6 (1): 21–29.

Petersen, A. C., J. E. Schulenberg, R. H. Abramowitz, D. Offer, and H. D. Jarcho. 1984. A Self-Image Questionnaire for Young Adolescents (SIQYA): Reliability and Validity Studies. *Journal of Youth and Adolescence* 13 (2): 93–111.

Phoenix, C. H., R. W. Goy, A. A. Gerall, and W. C. Young. 1959. Organizing Action of Prenatally Administered Testosterone Propionate on the Tissues Mediating Mating Behavior in the Female Guinea Pig. *Endocrinology* 65 (3): 369–382.

Pinckard, K. L., J. Stellflug, and F. Stormshak. 2000. Influence of Castration and Estrogen Replacement on Sexual Behavior of Female-Oriented, Male-Oriented, and Asexual Rams. *Journal of Animal Science* 78 (7): 1947–53.

Pinker, S. 2002. *The Blank Slate: The Modern Denial of Human Nature.* New York: Viking.

———. 2005. Sex Ed: The Science of Difference. *New Republic,* February 14.

———. 2008. *The Sexual Paradox: Men, Women, and the Real Gender Gap.* New York: Scribner.

Pinto-Correia, C. 1997. *The Ovary of Eve: Egg and Sperm and Preformation.* Chicago: University of Chicago Press.

Plante, E., C. Boliek, A. Binkiewicz, and W. K. Erly. 1996. Elevated Androgen, Brain Development and Language Learning Disabilities in Children with Congenital Adrenal Hyperplasia. *Developmental Medicine and Child Neurology* 38 (5): 423–437.

Postma, A., R. Izendoorn, and E. H. F. De Haan. 1998. Sex Differences in Object Location Memory. *Brain and Cognition* 36 (3): 334–345.

Prentice, D. A., and E. Carranza. 2002. What Women and Men Should Be, Shouldn't Be, Are Allowed to Be, and Don't Have to Be: The Contents of Prescriptive Gender Stereotypes. *Psychology of Women Quarterly* 26 (4): 269–281.

Proctor, R. 2008. Agnotology: A Missing Term to Describe the Cultural Production of Ignorance (and Its Study). In *Agnotology: The Making and Unmaking of Ignorance,* ed. R. N. Proctor and L. Schiebinger. Stanford: Stanford University Press.

Putz, D. A., S. J. C. Gaulin, R. J. Sporter, and D. H. McBurney. 2004. Sex Hormones and Finger Length: What Does 2D:4D Indicate? *Evolution and Human Behavior* 25 (3): 182–199.

Puts, D. A., M. A. McDaniel, C. L. Jordan, and S. M. Breedlove. 2008. Spatial Ability and Prenatal Androgens: Meta-analyses of Congenital Adrenal Hyperplasia and Digit Ratio (2D:4D) Studies. *Archives of Sexual Behavior* 37 (1): 100–111.

Quadagno, D. M., R. Briscoe, and J. S. Quadagno. 1977. Effect of Perinatal Gonadal Hormones on Selected Nonsexual Behavior Patterns: A Critical Assessment of the Nonhuman and Human Literature. *Pyschological Bulletin* 84 (1): 62–80.

Rahman, Q. 2005. Fluctuating Asymmetry, Second to Fourth Finger Length Ratios and Human Sexual Orientation. *Psychoneuroendocrinology* 30 (4): 382–391.

Rahman, Q., S. Abrahams, and G. D. Wilson. 2003. Sexual-Orientation-Related Differences in Verbal Fluency. *Neuropsychology* 17 (2): 240–246.

Rahman, Q., D. Andersson, and E. Govier. 2005. A Specific Sexual Orientation-Related Difference in Navigation Strategy. *Behavioral Neuroscience* 119 (1): 311–316.

Rahman, Q., M. Korhonen, and A. Aslam. 2005. Sexually Dimorphic 2D:4D Ratio, Height, Weight, and Their Relation to Number of Sexual Partners. *Personality and Individual Differences* 39 (1): 83–92.

Rahman, Q., and G. D. Wilson. 2003. Sexual Orientation and the 2nd to 4th Finger Length Ratio: Evidence for Organising Effects of Sex Hormones or Developmental Instability? *Psychoneuroendocrinology* 28 (3): 288–303.

Rahman, Q., G. D. Wilson, and S. Abrahams. 2003. Sexual Orientation Related Differences in Spatial Memory. *Journal of the International Neuropsychological Society* 9 (3): 376–383.

———. 2004. Biosocial factors, sexual orientation and neurocognitive functioning. *Psychoneuroendocrinology* 29 (7): 867–881.

Rammsayer, T. H., and S. J. Troche. 2007. Sexual Dimorphism in Second-to-Fourth Digit Ratio and Its Relation to Gender-Role Orientation in Males and Females. *Personality and Individual Differences* 42 (6): 911–920.

Redick, A. 2004. *American History XY: The Medical Treatment of Intersex, 1916–1955.* Ph.D. diss., American Studies, New York University, New York.

Reiner, W. G. 2004. Psychosexual Development in Genetic Males Assigned Female: The Cloacal Exstrophy Experience. *Child and Adolescent Psychiatric Clinics of North America* 13 (3): 657–674.

———. 2005. Gender Identity and Sex-of-Rearing in Children with Disorders of Sexual Differentiation. *Journal of Pediatric Endocrinology and Metabolism* 18 (6): 549–553.

Reiner, W. G., and B. P. Kropp. 2004. A 7-Year Experience of Genetic Males with Severe Phallic Inadequacy Assigned Female. *Journal of Urology* 172 (6): 2395–2398.

Reinisch, J. M. 1977. Prenatal Exposure of Human Foetuses to Synthetic Progestin and Oestrogen Affects Personality. *Nature* 266:561–562.

———. 1981. Prenatal Exposure to Synthetic Progestins Increases Potential for Aggression in Humans. *Science* 211 (4487): 1171–73.

Reinisch, J. M., and W. G. Karow. 1977. Prenatal Exposure to Synthetic Progestins and Estrogens: Effects on Human Development. *Archives of Sexual Behavior* 6 (4): 257–288.

Reinisch, J. M., M. Ziemba-Davis, and S. Sanders. 1991. Hormonal Contributions to Sexually Dimorphic Behavioral Development in Humans. *Psychoneuro-endocrinology* 16 (1–3): 213–278.

Reite, M., J. Sheeder, D. Richardson, and P. Teale. 1995. Cerebral Laterality in Homosexual Males: Preliminary Communication Using Magnetoencephalography. *Archives of Sexual Behavior* 24 (6): 585–593.

Resko, J. A., A. Perkins, C. E. Roselli, J. A. Fitzgerald, J. V. A. Choate, and F. Stormshak. 1996. Endocrine Correlates of Partner Preference Behavior in Rams. *Biology of Reproduction* 55 (1): 120–126.

Resnick, S. M., I. I. Gottesman, S. A. Berenbaum, and T. J. Bouchard. 1986. Early Hormonal Influences on Cognitive Functioning in Congenital Adrenal-Hyperplasia. *Developmental Psychology* 22 (2): 191–198.

Resnick, S. M., I. Gottesman, and M. McGue. 1993. Sensation Seeking in Opposite-Sex Twins: An Effect of Prenatal Hormones. *Behavior Genetics* 23: 323–329.

Rieger, G., M. L. Chivers, and J. M. Bailey. 2005. Sexual Arousal Patterns of Bisexual Men. *Pyschological Science* 16 (8): 579–584.

Rieger, G., J. A. W. Linsenmeier, L. Gygax, and J. M. Bailey. 2008. Sexual Orientation and Childhood Gender Nonconformity: Evidence from Home Videos. *Developmental Psychology* 44 (1): 46–58.

Robinson, S. J., and J. T. Manning. 2000. The Ratio of 2nd to 4th Digit Length and Male Homosexuality. *Evolution and Human Behavior* 21 (5): 333–345.

Rodgers, C. S., B. I. Fagot, and A. Winebarger. 1998. Gender-Typed Toy Play in Dizygotic Twin Pairs: A Test of Hormone Transfer Theory. *Sex Roles* 39 (3–4): 173–184.

Rose, R. J., J. Kaprio, T. Winter, D. M. Dick, R. J. Viken, L. Pulkkinen, and M. Koskenvuo. 2002. Femininity and Fertility in Sisters with Twin Brothers: Prenatal Androgenization? Cross-Sex Socialization? *Psychological Science* 13 (3): 263–267.

Roselli, C. E., K. Larkin, J. A. Resko, J. N. Stellflug, and F. Stormshak. 2004. The Volume of a Sexually Dimorphic Nucleus in the Ovine Medial Preoptic Area/ Anterior Hypothalamus Varies with Sexual Partner Preference. *Endocrinology* 145 (2): 478–483.

Roselli, C. E., K. Larkin, J. M. Schrunk, and F. Stormshak. 2004. Sexual Partner Preference, Hypothalamic Morphology and Aromatase in Rams. *Physiology and Behavior* 83 (2): 233–245.

Roselli, C. E., J. A. Resko, and F. Stormshak. 2002. Hormonal Influences on Sexual Partner Preference in Rams. *Archives of Sexual Behavior* 31 (1): 43–49.

Rosenstein, L. D., and E. D. Bigler. 1987. No Relationship between Handedness and Sexual Preference. *Psychological Reports* 60:704–706.

Rothman, B. K. 1986. *The Tentative Pregnancy: Prenatal Diagnosis and the Future of Motherhood*. New York: Viking.

Rothman, E. K. 1984. *Hands and Hearts: A History of Courtship in America*. New York: Basic Books. Quoted in D'Emilio and Freedman 1988, 261.

Roughgarden, J. 2004. *Evolution's Rainbow: Diversity, Gender and Sexuality in Nature and People*. Berkeley: University of California Press.

Rouse, J. 2004. Barad's Feminist Naturalism. *Hypatia* 19 (1): 142–161.

Rubin, G. 1975. The Traffic in Women: Notes on the "Political Economy" of Sex. In *Toward an Anthropology of Women,* ed. R. R. Reiter. New York: Monthly Review Press.

———. 1992. Thinking Sex: Notes for a Radical Theory of the Politics of Sexuality. In *Pleasure and Danger: Exploring Female Sexuality,* ed. C. S. Vance. London: Pandora/HarperCollins.

Rudman, L. A., and J. E. Phelan. 2008. Backlash Effects for Disconfirming Gender Stereotypes in Organizations. *Research in Organizational Behavior* 28:61–79.

Russett, C. E. 1989. *Sexual Science: The Victorian Construction of Womanhood.* Cambridge, MA: Harvard University Press.

Sandberg, D. E., H. F. L. Meyer-Bahlburg, T. J. Yager, T. W. Hensle, S. B. Levitt, S. J. Kogan, and E. F. Reda. 1995. Gender Development in Boys Born with Hypospadias. *Psychoneuroendocrinology* 20 (7): 693–709.

Sanders, G., and L. Ross-Field. 1986. Sexual Orientation and Visuo-Spatial Ability. *Brain and Cognition* 5:280–290.

Sanders, G., and M. Wright. 1997. Sexual Orientation Differences in Cerebral Asymmetry and in the Performance of Sexually Dimorphic Cognitive and Motor Tasks. *Archives of Sexual Behavior* 26 (5): 463–480.

Sanders, S. A., and J. M. Reinisch. 1999. Would You Say You "Had Sex" If . . . ? *Journal of the American Medical Association* 281 (3): 275–277.

Satz, P., E. N. Miller, O. Selnes, W. Van Gorp, L. F. D'Elia, and B. Visscher. 1991. Hand Preference in Homosexual Men. *Cortex* 27:295–306.

Sax, L. 2005. *Why Gender Matters.* New York: Doubleday.

Sayers, D. L. 1971. *Are Women Human?* Grand Rapids, Mich.: Eerdmans.

Schachter, S. C. 1994. Handedness in Women with Intrauterine Exposure to Diethylstilbestrol. *Neuropsychologia* 32 (5): 619–623.

Schardein, J. L. 1980. Congenital Abnormalities and Hormones during Pregnancy: A Clinical Review. *Teratology* 22 (3): 251–270.

Scheirs, J. G. M., and A. Vingerhoets. 1995. Handedness and Other Laterality Indexes in Women Prenatally Exposed to DES. *Journal of Clinical and Experimental Neuropsychology* 17 (5): 725–730.

Schmidt, G., and U. Clement. 1990. Does Peace Prevent Homosexuality? *Archives of Sexual Behavior* 19 (2): 183–187.

Schneider, H. J., J. Pickel, and G. K. Stalla. 2006. Typical Female 2nd–4th Finger Length (2D:4D) Ratios in Male-to-Female Transsexuals: Possible Implications for Prenatal Androgen Exposure. *Psychoneuroendocrinology* 31 (2): 265–269.

Schwartz, J. 2007. Of Gay Sheep, Modern Science, and Bad Publicity. *New York Times,* January 25.

Sell, R. L. 1997. Defining and Measuring Sexual Orientation: A Review. *Archives of Sexual Behavior* 26 (6): 643–658.

Sell, R. L., and J. B. Becker. 2001. Sexual Orientation Data Collection and Progress toward Healthy People 2010. *American Journal of Public Health* 91 (6): 876–882.

Sell, R. L., J. A. Wells, and D. Wypij. 1995. The Prevalence of Homosexual Behavior and Attraction in the United States, the United Kingdom and France: Results of National Population-Based Samples. *Archives of Sexual Behavior* 24 (3): 235–248.

Sengoopta, C. 1998. Glandular Politics: Experimental Biology, Clinical Medicine, and Homosexual Emancipation in Fin-de-Siècle Central Europe. *Isis* 89:445–473.

———. 2006. *The Most Secret Quintessence of Life: Sex, Glands, and Hormones, 1850–1950*. Chicago: University of Chicago Press.

Servin, A., A. Nordenstrom, A. Larsson, and G. Bohlin. 2003. Prenatal Androgens and Gender-Typed Behavior: A Study of Girls with Mild and Severe Forms of Congenital Adrenal Hyperplasia. *Developmental Psychology* 39 (3): 440–450.

"Sex: Unknown." 2001. Ed. A. Ritsko. *NOVA*. PBS.

Seyler, L. E., E. Canalis, S. Spare, and S. Reichlin. 1978. Abnormal Gonadotropin Secretory Responses to LRH in Transsexual Women after Diethylstilbestrol Priming. *Journal of Clinical Endocrinology and Metabolism* 47:176–183.

Shapiro, B. H., A. S. Goldman, A. M. Bongiovanni, and J. M. Marino. 1976. Neonatal Progesterone and Feminine Sexual Development. *Nature* 264 (5588): 795–796.

Shifren, K., A. Furnham, and R. L. Bauserman. 2003. Emerging Adulthood in American and British Samples: Individuals' Personality and Health Risk Behaviors. *Journal of Adult Development* 10 (2): 75–88.

Sidorowicz, L. S., and G. S. Lunney. 1980. Baby X Revisited. *Sex Roles* 6 (1): 67–73.

Simon, N. G., and R. Gandelman. 1978. Estrogenic Arousal of Aggressive Behavior in Female Mice. *Hormones and Behavior* 10 (2): 118–127.

Simos, P. G., J. M. Fletcher, E. Bergman, J. I. Breier, B. R. Foorman, E. M. Castillo, R. N. Davis, M. Fitzgerald, and A. C. Papanicolaou. 2002. Dyslexia-Specific Brain Activation Profile Becomes Normal following Successful Remedial Training. *Neurology* 58 (8): 1203–13.

Sinforiani, E., C. Livieri, M. Mauri, P. Bisio, L. Sibilla, L. Chiesa, and A. Martelli. 1994. Cognitive and Neuroradiological Findings in Congenital Adrenal-Hyperplasia. *Psychoneuroendocrinology* 19 (1): 55–64.

Slijper, F. M. E. 1984. Androgens and Gender Role Behavior in Girls with Congenital Adrenal Hyperplasia (CAH). *Progress in Brain Research* 61:417–422.

Slijper, F. M. E., S. L. S. Drop, J. C. Molenaar, and S. M. P. F. D. Keizer-Schrama. 1998. Long-Term Psychological Evaluation of Intersex Children. *Archives of Sexual Behavior* 27 (2): 125–144.

Smallwood, S. 2001. Report Questions MIT's Study on Treatment of Female Professors. *Chronicle of Higher Education*, February 16, 17.

Smith, B. C. 1989. *Men and Women: A History of Costume, Gender, and Power.* Washington, D.C.: Smithsonian Institution.

Smith, L. L., and M. Hines. 2000. Language Lateralization and Handedness in Women Prenatally Exposed to Diethylstilbestrol (DES). *Psychoneuroendocrinology* 25 (5): 497–512.

Smith, T. M. 1995. *The Educational Progress of Women: Findings from "The Con-*

dition of Education 1995." Washington, D.C.: National Center for Education Statistics.

Snowden, R. J., J. Wichter, and N. S. Gray. 2008. Implicit and Explicit Measurements of Sexual Preference in Gay and Heterosexual Men: A Comparison of Priming Techniques and the Implicit Association Task. *Archives of Sexual Behavior* 37 (4): 558–565.

Solarz, A. L., ed. 1999. *Lesbian Health: Current Assessment and Directions for the Future.* Washington DC: National Academy Press.

Sommer, I. E., A. Aleman, M. Somers, M. P. Boks, and R. S. Kahn. 2008. Sex Differences in Handedness, Asymmetry of the Planum Temporale and Functional Language Lateralization. *Brain Research* 1206:76–88.

Song, S., and S. A. Burgard. 2008. Does Son Preference Influence Children's Growth in Height? A Comparative Study of Chinese and Filipino Children. *Population Studies* 62 (3): 305–320.

Spelke, E. S. 2005. Sex Differences in Intrinsic Aptitude for Mathematics and Science? A Critical Review. *American Psychologist* 60 (9): 950–958.

Stake, J. E. 2000. When Situations Call for Instrumentality and Expressiveness: Resource Appraisal, Coping Strategy Choice, and Adjustment. *Sex Roles* 42 (9–10): 865–885.

Star, S. L. 1989. *Regions of the Mind: Brain Research and the Quest for Scientific Certainty.* Stanford: Stanford University Press.

Stein, R. 2009. Born to Be a Trader? Fingers Point to Yes; Study Cites Early Testosterone Exposure. *Washington Post,* January 13.

Stellflug, J. N., N. E. Cockett, and G. S. Lewis. 2006. Relationship between Sexual Behavior Classifications of Rams and Lambs Sired in a Competitive Breeding Environment. *Journal of Animal Science* 84 (2): 463–468.

Stikkelbroeck, N. M. M. L., C. C. M. Beerendonk, W. N. P. Willemsen, C. A. Schreuders-Bais, W. F. J. Feitz, P. N. M. A. Rieu, A. R. M. M. Hermus, and B. J. Otten. 2003. The Long Term Outcome of Feminizing Genital Surgery for Congenital Adrenal Hyperplasia: Anatomical, Functional and Cosmetic Outcomes, Psychosexual Development, and Satisfaction in Adult Female Patients. *Journal of Pediatric and Adolescent Gynecology* 16 (5): 289–296.

Stikkelbroeck, N. M. M. L., A. R. M. M. Hermus, D. D. M. Braat, and B. J. Otten. 2003. Fertility in Women with Congenital Adrenal Hyperplasia Due to 21-Hydroxylase Deficiency. *Obstetrical and Gynecological Survey* 58 (4): 275–284.

Stoler, A. L. 2002. *Carnal Knowledge and Imperial Power: Race and the Intimate in Colonial Rule.* Berkeley: University of California Press.

Story, L. 2005. Many Women at Elite Colleges Set Career Path to Motherhood. *New York Times,* September 20.

Strand, S., I. J. Deary, and P. Smith. 2006. Sex Differences in Cognitive Abilities Test Scores: A UK National Picture. *British Journal of Educational Psychology* 76:463–480.

Su, R., J. Rounds, and P. I. Armstrong. 2009. Men and Things, Women and People: A Meta-Analysis of Sex Differences in Interests. *Psychological Bulletin* 135 (6): 859–884.

Suen, H. K. 1990. *Principles of Test Theories*. Hillsdale, NJ: L. Erlbaum Associates.

Summers, L. H. 2005. Remarks at NBER Conference on Diversifying the Science and Engineering Workforce. In *National Bureau of Economic Research Conference on Diversifying the Science and Engineering Workforce: Women, Underrepresented Minorities, and Their S&E Careers*. Cambridge, MA.

Swaab, D. F. 2004. Sexual Differentiation in the Human Brain: Relevance for Gender Identity, Transsexualism, and Sexual Orientation. *Gynecological Endocrinology* 19:301–312.

———. 2007. Sexual Differentiation of the Brain and Behavior. *Best Practice and Research Clinical Endocrinology and Metabolism* 21 (3): 431–444.

Swaab, D. F., and E. Fliers. 1985. A Sexually Dimorphic Nucleus in the Human Brain. *Science* 228 (4703): 1112–15.

Swaab, D. F., and M. A. Hofman. 1990. An Enlarged Suprachiasmatic Nucleus in Homosexual Men. *Brain Research* 537:141–148.

Taha, W., D. Chin, A. I. Silverberg, L. Lashiker, N. Khateeb, and H. Anhalt. 2001. Reduced Spinal Bone Mineral Density in Adolescents of an Ultra-Orthodox Jewish Community in Brooklyn. *Pediatrics* 107 (5): e79.

Tellegen, A. 1982. Brief Manual for the Multidimensional Personality Questionnaire. Unpublished manuscript, University of Minnesota.

Tenhula, W. N., and J. M. Bailey. 1998. Female Sexual Orientation and Pubertal Onset. *Developmental Neuropsychology* 14:369–383.

Terry, J. 1997. The Seductive Power of Science in the Making of Deviant Subjectivity. In *Science and Homosexualities*, ed. V. A. Rosario. New York: Routledge.

———. 1999. *An American Obsession: Science, Medicine, and Homosexuality in Modern Society*. Chicago: University of Chicago Press.

Therrell, B. L., S. A. Berenbaum, V. Manter-Kapanke, J. Simmank, K. Korman, L. Prentice, J. Gonzalez, and S. Gunn. 1998. Results of Screening 1.9 Million Texas Newborns for 21-Hydroxylase-Deficient Congenital Adrenal Hyperplasia. *Pediatrics* 101 (4): 583–590.

Thomas, W. I., and D. S. Thomas. 1928. *The Child in America: Behavior Problems and Programs*. New York: Johnson Reprint Corp. Original edition, 1928.

Tiefer, L. 2004. *Sex Is Not a Natural Act and Other Essays*. Boulder, CO: Westview Press.

Tierney, J. 2006. Academy of P.C. Sciences. *New York Times*, Sept. 26, 2006, 23.

Titus-Ernstoff, L., K. Perez, E. E. Hatch, R. Troisi, J. R. Palmer, P. Hartge, M. Hyer, R. Kaufman, E. Adam, W. Strohsnitter, K. Noller, K. E. Pickett, and R. Hoover. 2003. Psychosexual Characteristics of Men and Women Exposed Prenatally to Diethylstilbestrol. *Epidemiology* 14 (2): 155–160.

Tuana, N. 2004. Coming to Understand: Orgasm and the Epistemology of Ignorance. *Hypatia* 19 (1): 194–232.

Tuttle, G. E., and R. C. Pillard. 1991. Sexual Orientation and Cognitive Abilities. *Archives of Sexual Behavior* 20 (3): 307–318.

Tyre, P. 2005. Boy Brains, Girl Brains: Are Separate Classrooms the Best Way to Teach Kids? *Newsweek*, Sept. 19, 2005, 59.

Udry, R. 2000. Biological Limits of Gender Construction. *American Sociological Review* 65 (3): 443–457.

Udry, J. R., and K. Chantala. 2006. Masculinity-Femininity Predicts Sexual Orientation in Men but not in Women. *Journal of Biosocial Science* 38 (6): 797–809.

Udry, R., N. M. Morris, and J. Kovenock. 1995. Androgen Effects on Women's Gendered Behavior. *Journal of Biosocial Science* 27 (3): 359–368.

U.S. House of Representatives, Committee on Government Reform, Minority Staff. 2004. The Content of Federally Funded Abstinence-Only Education Programs. Washington, DC: United States Congress.

Valentine, D. 2007. *Imagining Transgender: An Ethnography of a Category*. Durham, N.C.: Duke University Press.

Valian, V. 1999. *Why So Slow? The Advancement of Women*. Cambridge, MA: MIT Press.

van den Wijngaard, M. 1997. *Reinventing the Sexes: The Biomedical Construction of Femininity and Masculinity* Bloomington: Indiana University Press.

van der Meer, T. 1997. Sodom's Seed in The Netherlands: The Emergence of Homosexuality in the Early Modern Period. *Journal of Homosexuality* 34 (1): 1–16.

Vance, C. S. 1989. Social Construction Theory: Problems in the History of Sexuality. In *Homosexuality, Which Homosexuality?* Edited by D. Altman and C. S. Vance. Holland: An Dekker.

———. 1991. Anthropology Rediscovers Sexuality: A Theoretical Comment. *Social Science in Medicine* 33 (8): 875–884.

Vogel, J. L., C. A. Bowers, and D. S. Vogel. 2003. Cerebral Lateralization of Spatial Abilities: A Meta-Analysis. *Brain and Cognition* 52 (2): 197–204.

Volkl, T. M. K , D. Simm, C. Beier, and H. G. Dorr. 2006. Obesity among Children and Adolescents with Classic Congenital Adrenal Hyperplasia due to 21-Hydroxylase Deficiency. *Pediatrics* 117 (1): E98–E105.

Voracek, M., and S. G. Dressler. 2006. Lack of Correlation between Digit Ratio (2D:4D) and Baron-Cohen's "Reading the Mind in the Eyes" Test, Empathy, Systemising, and Autism-Spectrum Quotients in a General Population Sample. *Personality and Individual Differences* 41(8): 1481–91.

Voracek, M., and L. M. Loibl. 2009. Scientometric Analysis and Bibliography of Digit Ratio (2d:4d) Research, 1998–2008. *Psychological Reports* 104 (3): 922–956.

Voracek, M., J. T. Manning, and I. Ponocny. 2005. Digit Ratio (2D:4D) in Homosexual and Heterosexual Men from Austria. *Archives of Sexual Behavior* 34 (3): 335–340.

Voracek, M., and S. Stieger. 2009. Replicated Nil Associations of Digit Ratio (2D:4D) and Absolute Finger Lengths with Implicit and Explicit Measures of Aggression. *Psicothema* 21 (3): 382–389.

Voyer, D., A. Postma, B. Brake, and J. Imperato-McGinley. 2007. Gender Differences in Object Location Memory: A Meta-analysis. *Psychonomic Bulletin and Review* 14 (1): 23–38.

Voyer, D., S. Voyer, and M. P. Bryden. 1995. Magnitude of Sex Differences in Spatial Abilities: A Meta-analysis and Consideration of Critical Variables. *Psychological Bulletin* 117 (2): 250–270.

Vreugdenhil, H. J. I., F. M. E. Slijper, P. G. H. Mulder, and N. Weisglas-Kuperus. 2002. Effects of Perinatal Exposure to PCBs and Dioxins on Play Behavior in Dutch Children at School Age. *Environmental Health Perspectives* 110 (10): A593–A598.

Wade, Nicholas. 2007. Pas de Deux of Sexuality Is Written in the Genes. *New York Times,* April 10.

Wahabi, H. A., N. F. A. Althagafi, and M. Elawad. 2007. Progestogen for Treating Threatened Miscarriage. *Cochrane Database of Systematic Reviews.*

Wakshlak, A., and M. Weinstock. 1990. Neonatal Handling Reverses Behavioral Abnormalities Induced in Rats by Prenatal Stress. *Physiology and Behavior* 48 (2): 289–292.

Wallen, K. 2005. Hormonal Influences on Sexually Differentiated Behavior in Nonhuman Primates. *Frontiers in Neuroendocrinology* 26 (1): 7–26.

Ward, I. L. 1972. Prenatal Stress Feminizes and Demasculinizes the Behavior of Males. *Science* 175:82–84.

———. 1984. The Prenatal Stress Syndrome: Current Status. *Psychoneuroendocrinology* 9:3–11.

Ward, M. 2003. Census Bureau Report Oversimplifies Discrepancy in Wages. *The Battalion Online,* media.www.thebatt.com/media/storage/paper657/news/2003 /04/01/Opinion/Breaking.The.Glass.Ceiling-514315.shtml.

Watabe, T., and A. Endo. 1994. Sexual Orientation of Male-Mouse Offspring Prenatally Exposed to Ethanol. *Neurotoxicology and Teratology* 16 (1): 25–29.

Weeks, J. 1986. *Sexuality.* Ed. P. Hamilton. 6th ed. New York: Routledge.

Wegesin, D. J. 1998. A Neuropsychologic Profile of Homosexual and Heterosexual Men and Women. *Archives of Sexual Behavior* 27 (1): 91–108.

Weil, E. 2008. Teaching Boys and Girls Separately. *New York Times Magazine,* March 2, MM38.

Weis, S. E., A. Firker, and J. Hennig. 2007. Associations between the Second to Fourth Digit Ratio and Career Interests. *Personality and Individual Differences* 43 (3): 485–493.

Weise, E. 2006. Maybe We Are Different: New Book Argues Female Brain Wired to Nurture. *USA Today,* August 21.

Weisen, M., and W. Futterweit. 1983. Normal Plasma Gonadotropin Response to Gonadotropin-Releasing Hormone after Diethylstilbestrol Priming in Transsexual Women. *Journal of Clinical Endocrinology and Metabolism* 57 (1): 197–199.

Whalen, R. E. 1964. Hormone-Induced Changes in the Organization of Sexual Behavior of the Male Rat. *Journal of Comparative and Physiological Psychology* 57:175–182.

Whalen, R. E., and D. A. Edwards. 1967. Hormonal Determinants of Development of Masculine and Feminine Behavior in Male and Female Rats. *Anatomical Record* 157 (2): 173.

Whitam, F. L., and R. M. Mathy. 1991. Childhood Cross-Gender Behavior of Homosexual Females in Brazil, Peru, the Philippines, and the United States. *Archives of Sexual Behavior* 20 (2): 151–170.

White, P. C., and P. W. Speiser. 2000. Congenital Adrenal Hyperplasia due to 21-Hydroxylase Deficiency. *Endocrine Reviews* 21 (3): 245–291.

Wieringa, S. E., E. Blackwood, and A. Bhaiya, eds. 2007. *Women's Sexualities and Masculinities in a Globalizing Asia*. New York: Palgrave MacMillan.

Wilkins, L. 1960. Masculinization of Female Fetus due to Use of Orally Given Progestins. *Journal of the American Medical Association* 172:1028–32.

Williams, J. 2000. *Unbending Gender: Why Family and Work Conflict and What to Do about It*. New York: Oxford University Press.

Williams, T. J., M. E. Pepitone, S. E. Christensen, B. M. Cooke, A. D. Huberman, N. J. Breedlove, T. J. Breedlove, C. L. Jordan, and S. M. Breedlove. 2000. Finger-Length Ratios and Sexual Orientation. *Nature* 404 (6777): 455–456.

Willmott, M., and H. Brierley. 1984. Cognitive Characteristics and Homosexuality. *Archives of Sexual Behavior* 13 (4): 311–319.

Wilson, B. E. 2006. 5-Alpha-Reductase Deficiency: Treatment and Medication. *eMedicine by WebMD*. Omaha, NE: Medscape.

Wilson, G., and Q. Rahman. 2005. *Born Gay: The Psychobiology of Sexual Orientation*. London: Peter Owen.

Wilson, B. E., and W. G. Reiner. 1998. Management of Intersex: A Shifting Paradigm. *Journal of Clinical Ethics* 9 (4): 360–369.

Wisniewski, A. B., and C. J. Migeon. 2002. Long Term Perspectives for 46,XY Patients Affected by Complete Androgen Insensitivity Syndrome or Congenital Micropenis. *Seminars in Reproductive Medicine* 20 (3): 297–304.

Wisniewski, A. B., C. J. Migeon, H. F. L. Meyer-Bahlburg, J. P. Gearhart, G. D. Berkovitz, T. R. Brown, and J. Money. 2000. Complete Androgen Insensitivity Syndrome: Long-Term Medical, Surgical, and Psychosexual Outcome. *Journal of Clinical Endocrinology and Metabolism* 85 (8): 2664–69.

Wisniewski, A. B., C. J. Migeon, M. A. Malouf, and J. P. Gearhart. 2004. Pyschosexual Outcome in Women Affected by Congenital Adrenal Hyperplasia due to 21-Hydroxylase Deficiency. *Journal of Urology* 171 (June 2004): 2497–2501.

Witelson, S. F. 1991. Neural Sexual Mosaicism: Sexual Differentiation of the Human Temporo-Parietal Region for Functional Asymmetry. *Psychoneuroendocrinology* 16 (1–3): 131–153.

Witelson, S. F., D. L. Kigar, A. Scamvougeras, D. M. Kideckel, B. Buck, P. L. Stanchev, M. L. Bronskill, and S. Black. 2007. Corpus Callosum Anatomy in Right-Handed Homosexual and Heterosexual Men. *Archives of Sexual Behavior* 37 (6): 857–863.

Wootton, C. W., and B. E. Kemmerer. 2000. The Changing Genderization of Bookkeeping in the United States, 1930–1990. *Accounting, Business and Financial History* 10 (2): 169–190.

Yalom, I. D., R. Green, and N. Fisk. 1973. Prenatal Exposure to Female Hormones: Effect on Psychosexual Development in Boys. *Archives of General Psychiatry* 28:554–561.

Yao, H. H. C. 2005. The Pathway to Femaleness: Current Knowledge on Embryonic Development of the Ovary. *Molecular and Cellular Endocrinology* 230 (1–2): 87–93.

Yasuhara, F., W. G. Kempinas, and O. C. M. Pereira. 2005. Reproductive and Sexual Behavior Changes in Male Rats Exposed Perinatally to Picrotoxin. *Reproductive Toxicology* 19 (4): 541–546.

Young, C. 2006. Women, Science Deserve Better. *Seattle Post-Intelligencer,* October 2.

Young, R. M. 2000. Sexing the Brain: Measurement and Meaning in Biological Research on Human Sexuality. Ph.D. diss., Division of Sociomedical Sciences, Columbia University, New York.

———. 2004. Measuring Sexual Orientation in a Highly-Marginalized Population. Paper Read at Proceedings of the American Statistical Association, Section on Government Statistics.

Young, R. M., and E. Balaban. 2003. Aggression, Biology and Context: Dejá-Vù All Over Again? in *Neurobiology of Aggression,* ed. M. P. Mattson. Totowa, NJ: Humana Press.

Young, R. M., S. R. Friedman, P. Case, M. W. Asencio, and M. Clatts. 2000. Women Injection Drug Users Who Have Sex with Women Exhibit Increased HIV Infection and Risk Behaviors. *Journal of Drug Issues* 30 (3): 499–523.

Young, R. M., and I. H. Meyer. 2005. The Trouble with "MSM" and "WSW": Erasure of the Sexual-Minority Person in Public Health Discourse. *American Journal of Public Health* 95 (7): 1144–49.

Young, W. C., ed. 1961. *Sex and Internal Secretions.* Baltimore: Williams and Wilkins.

Young, W. C., R. W. Goy, and C. H. Phoenix. 1965. Hormones and Sexual Behavior. In *Sex Research: New Developments,* ed. J. Money. New York: Holt, Rinehart and Winston.

Zeavey, C., P. A. Katz, and S. R. Zalk. 1975. Baby X. *Sex Roles* 1 (2): 103–109.

Zhou, J. N., M. A. Hofman, L. J. G. Gooren, and D. F. Swaab. 1995. A Sex Difference in the Human Brain and Its Relation to Transsexuality. *Nature* 378 (November 2): 68–70.

Zondek, B. 1934. Mass Excretion of Oestrogenic Hormone in the Urine of the Stallion. *Nature* 133:209–210.

Zucker, K. J. 2002. Evaluation of Sex- and Gender-Assignment Decisions in Patients with Physical Intersex Conditions: A Methodological and Statistical Note. *Journal of Sex and Marital Therapy* 28 (3): 269–274.

Zucker, K. J., N. Beaulieu, S. J. Bradley, G. M. Grimshaw, and A. Wilcox. 2001. Handedness in Boys with Gender Identity Disorder. *Journal of Child Psychology and Psychiatry and Allied Disciplines* 42 (6): 767–776.

Zucker, K. J., S. J. Bradley, G. Oliver, J. Blake, S. Fleming, and J. Hood. 1996. Psychosexual Development of Women with Congenital Adrenal Hyperplasia. *Hormones and Behavior* 30 (4): 300–318.

Zucker, K. J., S. J. Bradley, G. Oliver, J. Blake, S. Fleming, and J. Hood. 2004. Self-Reported Sexual Arousability in Women with Congenital Adrenal Hyperplasia. *Journal of Sex and Marital Therapy* 30 (5): 343–355.

Acknowledgments

I am the ninth of ten children, reared in a loud, loving, and argumentative family in Missouri. Early on I learned the power of a good story. My father was a charismatic, hilarious storyteller who could spin a tale that would suck in all but the most dedicated skeptic. My siblings and I had a favorite game of wits that involved packaging improbable propositions in elaborate stories and sticking to them as long as it took to wear down the wary. As one of the youngest and initially more gullible, I learned to demand evidence. In the midst of all these tall tales, there was my mother with her exacting disposition. Amused but also exasperated by my father's unwillingness to let facts get in the way of a good story, she would drive him nuts by correcting details that would then make his whole beautiful narrative collapse. I learned the value of precision as a defensive move against compelling but unreliable stories. So my birth family—Leo and Pinkey, Susan, Stan, John, Mark, Tom, Saphronia, Nancy, and also my oldest brothers Ron and Dave—get some credit for my approach to this material, if not the content. The cousins, too—especially "Little" Bill, Donna, David, and Lynn—because they are a delightfully argumentative bunch.

It is a pleasure to also acknowledge the enormous help and support I received while writing this book. I give special thanks to my teachers and mentors Carole S. Vance and Evan Balaban, who have been inspiring, challenging, and extraordinarily generous with time and ideas. I hope this book reflects some of their light. For critical and enthusiastic reading of the entire manuscript, I thank Sally Cooper, David Lurie, Geertje Mak, and Afsaneh Najmabadi; Evan Balaban, Raffaella Rumiati, Ross Nehm, and Jennifer Terry gave indispensable feedback on certain sections. My editor at Harvard University Press, Elizabeth Knoll, has brought wit and fierce support to this project. Sheila Goloborotko produced beautiful graphics that absolutely fit my ideas. For additional discussions that have helped me focus and sharpen my argument, I thank Deborah Cameron, Anke Ehrhardt, Steven Epstein, John Gagnon, Amber Hollibaugh, Janice Irvine, Katrina Karkazis, Judith Levine, Bruce Link, Jonathan Ned Katz, Mark Liberman, Emily Martin, Sherry Martin, Joanne Meyerowitz, Annemarie Mol, the late Dorothy Nelkin, Qazi Rahman, Gayle Rubin, Raffaella Rumiati, Svati Shah, Ann Snitow, Jeanne Stellman, Leonore Tiefer, Theo van der Meer, Amy Villarejo, and Alison Wylie. Two brilliant friends, Adina Back and Allan Berubé, did not live to see this book in print, but nurtured it in especially critical and practical ways. I miss them.

Friends and colleagues on various AIDS projects in Washington, D.C., New York, and Providence, Rhode Island, contributed to my ability to ask critical questions about gender and sexuality, especially Donna Marie Alexander, Marty Camacho, Patricia Case, Dwight Clark, Jeanne Flavin, Eddie Flowers, Sam Friedman, Myron Johnson, A. Billy S. Jones, Beverly (Candy) Jones, Marianne Krayer, and Troy Werley. I thank the hundreds of people who were willing to tell the intimate details of their lives in hopes that it would somehow help in the AIDS epidemic, especially Affreekka, Dee, and Gigi.

My colleagues and students at Barnard College and Columbia University have been a sweet dream. Each of my department chairs—Janet Jakobsen, Natalie Kampen, Laura Kay, Neferti Tadiar, and Alison Wylie—has been a model for teaching and scholarship. Each has also, in Neferti's words, acted as my champion rather than my boss. A series of wonderful research and teaching assistants have tracked down obscure references, wrestled with the EndNote database, and—often without pay—helped me in numerous other ways to get my head around the gigantic set of studies I've worked with: Howard Hsueh-Hao Chiang, E. Grace Glenny, Rebecca de Guzman, Jonah Hassenfeld, Sahar Sadjadi, Rayna Sobieski, Lucy Trainor, and Natalie Wittlin.

Money, time, and the opportunity to present my work in progress were provided by several fellowships and visiting scholar positions. The Sexuality Research Fellowship Program of the Social Science Research Council, expertly directed by Diane di Mauro, brought me into an amazing network of fellows and mentors who have left an indelible mark on my thinking and teaching. Grants from the National Institute on Drug Abuse and a Health Disparities Scholar award from the National Center on Minority Health & Health Disparities (NCMHD) supported my research on HIV. As a visiting scholar at the Cognitive Neurosciences Department of the International School for Advanced Studies (SISSA) in Trieste, Italy, I had the chance to refine my analysis, especially the section related to sex-typed interests. Dr. Raffaella Rumiati generously hosted me during that visit, and her friendship and collegiality have been vital to me ever since.

I have also learned a great deal at conferences, lectures, and workshops where I presented sections of this work. I especially wish to thank audiences at the Barnard Center for Research on Women; the Sexuality Network of the European Social Science History Conference; the Center for Lesbian and Gay Studies at the Graduate Center of the City University of New York; the Center for the Study of Gender and Sexuality at New York University; the Columbia University Seminar on Sexuality, Gender, Health, and Human Rights; the Institute for Social Sciences and the Department of Feminist, Gender & Sexuality Studies at Cornell University; the Lesbian, Gay, Bisexual and Transgender Community Center of New York; the FEST Trieste (Italy) International Science Media Fair; the Shelter Rock congregation of Unitarian Universalists; the Gender Studies Program at Whitman College; and the Departments of Medical History, Philosophy, and Women's Studies, and the Programs in Critical Medical Humanities and the History and Philosophy of Science, at the University of Washington, Seattle.

To enumerate all the other specific debts I owe is impossible, but the following list includes some of the people who have supported me with friendship, encouragement, intellectual stimulation, productive arguments, healing, food, money, and a quiet place to write: Marysol Asencio, Lila Abu-Lughod, Eleanor Bell, Elizabeth Bernstein, Dr. Carol Brown, the Broholm-Vail family, the Brown-Leonard family, Janet Cooper, Rachel Efron, Linda Gaal and the entire Back-Gaal family, Bob and Mindy Fullilove, Katherine Franke, the Garrison-Bedell family, Peter Flom, Marianne Fox, Claudia Ginanni, Gayann Hall, Dave Jordan, Amanda Joseph, Kerwin Kaye, Irene Lambrou, James Learned, Lauren Liss, Geertje Mak, Hans Meij, Susan Messina, Ilan Meyer, Carolyn Patierno, Penny Saunders, Monica Schoch-Spana, Lesley Sharp, Team Teverow, Miriam Ticktin and Patrick Dodd, David Valentine, and Theo van der Meer. A few of my family have, in addition to generally being supportive, read, nagged, encouraged, sent money or endured my inattention when they deserved better: Sally Cooper, Katrina Santana, Cuba Jimenez, Saphronia Young, Stanley Jordan Young, and Verlie J. (Pinkey) Young. Sal, above all, has been unswervingly convinced of the importance of the book, even when it has gotten in the way of so many other things. She has kept me going.

Index